The Quest for Relevant Air Power

Air University Series on
Airpower and National Security

The Quest for Relevant Air Power

Continental European Responses to the Air Power Challenges of the Post–Cold War Era

Christian F. Anrig, PhD

Air University Press
Air Force Research Institute
Maxwell Air Force Base, Alabama

August 2011

Library of Congress Cataloging-in-Publication Data

Anrig, Christian F.
The quest for relevant air power : continental European responses to the air power challenges of the post–Cold War era / Christian F. Anrig.
    p. cm.
Includes bibliographical references and index.
ISBN 978-1-58566-216-6
1. Air power—Europe. 2. Air forces—Europe. I. Title.
UG635.E85A57 2011
358.4'03094—dc23

2011024833

Cover photo courtesy of Gripen International; photo by FMV (Swedish Defence Materiel Administration)

Disclaimer

Air University Press
Air Force Research Institute
155 North Twining Street
Maxwell AFB, AL 36112-6026
http://aupress.au.af.mil

Dedicated to my parents, Olga and Christian,
and
to the memory of my beloved sister, Judith

Contents

Illustrations

Photo *Page*

Foreword

During the Cold War, the comparatively minor contributions of European air forces to the North Atlantic Treaty Organization (NATO) were not visible to the outside world because of their incorporation within the alliance structure and strategy. The first Gulf War starkly revealed the disparity between the air power of the United States and that of any other country. Subsequent operations in the Balkans, Iraq, and Afghanistan have demanded even more timely and accurate intelligence, with swifter response and greater precision in attack.

The US accretion of all-weather precision munitions; stealth technology; netted real-time command, control, communications, and intelligence; unmanned aerial vehicles; and satellite systems has widened the gap with European air forces still further. The evolution and contribution of continental European air forces to recent operations remain largely unexplored, partly because of their limitations and partly because of Anglo-Saxon intellectual domination of air power analysis and concepts.

Christian Anrig examines the responses of four countries to the challenges of air power in the last two decades. He has selected four very different air forces: the French Air Force (FAF), German Air Force (GAF), Royal Netherlands Air Force (RNLAF), and Swedish Air Force (SwAF). All four were influenced by the Cold War period. The GAF and RNLAF were embedded in Cold War NATO, the SwAF maintained a well-armed neutrality, and the FAF reflected the semi-independent strategic stance of post–de Gaulle France.

The author addresses four questions: how have these air forces responded to post–Cold War political uncertainties, how have they operated, how have they responded to new air power thinking, and how have they adapted to the challenges of costs and technologies? He convincingly argues that budgetary provision has not been the most important factor in generating effective air power.

He traces the different interactions of political will, procurement, and operational effectiveness among the four air forces. All are highly professional, but only France has sought to sustain a capability across all air power roles.

Christian Anrig's examination is both instructive and disheartening. The United States operator and student will be enlightened by awareness of the quality, value, and *proportional* scale of recent and current European air operations. Anyone who is detailed to work alongside these air forces will benefit considerably from understanding how and why they do what they do. He or she will recognise the circumstances which have precluded them from the doctrinal path of John Warden. Sadly, the author has only too clearly identified the national features which, with one or two exceptions, are likely to inhibit the creation of European air power in the foreseeable future.

The author brings deep scholarship to his study, reinforced by his national objectivity. It is a unique and indispensable contribution to international awareness of twenty-first-century air power.

Air Vice-Marshal Tony Mason
Professor, R A Mason CB CBE MA DSc FRAeS DL
University of Birmingham, England
1 March 2011

About the Author

Christian F. Anrig holds a PhD from King's College London, University of London. He is deputy director, doctrine research and education, Swiss Air Force. From early 2007 until September 2009, he was a lecturer in air power studies in the Defence Studies Department of King's College London while based at the Royal Air Force (RAF) College. He was one of two lead academics who created the new distance-learning master's degree, air power in the modern world—a course especially, but not exclusively, designed for RAF officers. In November 2009, he became a member of the RAF Centre for Air Power Studies academic advisory panel. Dr. Anrig began his professional career in defence studies as a researcher at the Center for Security Studies, Swiss Federal Institute of Technology (ETH Zurich), in January 2004.

He has published articles and book chapters, primarily in German, covering topics from the European Union battle groups and military transformation to modern air power and its ramifications for small nations. Whilst working in the United Kingdom, he was on the editorial board of the *Royal Air Force Air Power Review*.

Dr. Anrig spent the first half of his military career in the artillery, Swiss Army. Currently, he is a reserve captain assigned to the Air Staff, Swiss Air Force. He is a dual-national Swiss and Liechtensteiner.

Acknowledgments

The completion of this book, based on the doctoral thesis "Continental European Responses to the Air Power Challenges of the Post-Cold War Era" submitted at King's College London (KCL), would not have been possible without the support of many people and institutions. This PhD research project represents a successful joint effort between the Department of War Studies, KCL, University of London, and the Center for Security Studies, ETH Zurich (or Swiss Federal Institute of Technology).

I am greatly beholden to my supervisors and mentors, Prof. Philip Sabin (KCL) and Prof. Andreas Wenger (ETH). Prof. Philip Sabin encouraged me to ask important questions and to produce the best answers I could both by the example he set and the counsel he gave at every stage. He was endlessly generous in helping my thesis approach rigorous standards. Professor Wenger proved relentless in his demand for analytical clarity and was supremely supportive. My appreciation also extends to my external examiners, Air Vice-Marshal Prof. R. A. Mason, University of Birmingham, and Prof. Ron Smith, Birkbeck College, University of London, who helped to fine-tune the thesis in its final stages.

The thesis focused on four major Continental European air forces in the post–Cold War era—those of France, Germany, the Netherlands, and Sweden. Given its contemporary focus, the project would not have been possible without the dedicated help of some outstanding individuals. The main contributors are listed below in a primarily alphabetical order. Since my interviews, many of those may have risen in rank or moved on to other positions.

For my research on the French Air Force, I am especially grateful to Capt Stéphane Bréart, Sirpa Air (Service d'information et de relations publiques de l'Armée de l'Air); Col Régis Chamagne, retired; Prof. Patrick Facon, director of the Air Historical Branch, Château de Vincennes, who proved invaluable in providing a historical outlook on French air power thought; Lt Col Jean-Luc Fourdrinier, Air Staff, Armée de l'Air; Col Denis Gayno, deputy director, Centre d'enseignement supérieur aérien (CESA [Centre of Strategic Aerospace Studies]); Bastien Irondelle, senior researcher, Centre d'études et de recherches internationales (CERI [Centre for International Studies and Research]), for his view on the latest air power doctrine development in France; Cdt Jean-Patric Le Saint, Cellule études et stratégie aérienne militaire (CESAM [air power strategy and studies cell]); Col Gilles Lemoine, Collège interarmées de défense (CID [Joint Forces Defence College]); Col Denis Mercier, Air Staff, Armée de l'Air, for his insights into procurement; Col Jean-Christophe Noël, director, CESAM; Commandant Ollier for organising the various interviews within the Air Staff, Armée de l'Air; Prof. Pierre Pascallon; Gen Jean Rannou, retired, chief of the Air Staff, 1995–2000; and Col Marc Weber, Air Staff, Armée de l'Air. Without these individuals, I would not have been able to explore the latest developments in French air power thinking and education as well as to investigate aspects relating to operations and procurement in the required depth.

In the case of the German Air Force, I owe particular thanks to Lt Col (General Staff [GS]) Jens Asmussen, Joint Air Power Competence Centre (JAPCC); Lt Col Manfred Brose, FüAk (Führungsakademie), who proved invaluable in providing me with information on German air power education; Lt Gen Hans-Werner Jarosch, retired, Lt Gen Walter Jertz, retired, and Lt Gen Axel Kleppien, retired, for their specific insights into the German reunification's impact upon the Luftwaffe and other aspects; Lt Col (GS) Arnt Kuebart, Luftmachtzentrum (LMZ [Air Power Centre]), Col (GS) Thomas Lorber, FüAk (Führungsakademie); Lt Col Dr. Wolfgang Schmidt, Military-Historical Research Office, Potsdam; Lt Gen Hans-Joachim Schubert, executive director, JAPCC; and Col (GS) Reinhard Vogt, LMZ. Finally, my sincere thanks also go to the chief of the Air Staff, Lt Gen Klaus-Peter Stieglitz, and to his staff members in Bonn for helping to develop this work—Lt Col (GS) Armin Havenith, Colonel (GS) Hewera, Lt Col (GS) Stefan Klenz, Col (GS) Jörg Lebert and his staff, Lt Col (GS) Guido Leiwig, Col (GS) Burkhard Potozky, Lt Col (GS) Nicolas Radke, Lt Col (GS) Dr. Michael Wolfgang Romba, Col (GS) Hans-Dieter Schön and his staff, Col (GS) Lothar Schmidt, and Lt Col (GS) Michael Trautermann.

In the Netherlands, I am especially indebted to Maj Edwin Altena, Headquarters Royal Netherlands Air Force (RNLAF), for going to great lengths to provide me with policy documents; Col Jan F. W. van Angeren, retired, who gave me a particularly valuable insight into the early development stages of modern air power education in the Netherlands; Col Henk Bank, former head, policy integration branch, RNLAF; Air Cdre J. T. Broedersz, retired, former deputy director, European Air Group, who was particularly helpful as a generous gate-opener; Air Cdre J. L. H. Eikelboom, Headquarters RNLAF; Col Lex Kraft van Ermel, former director of studies, Netherlands Defence College; Col Robert J. Geerdes, F-16 Replacement Office; Lt Gen P. J. M. Godderij, military representative of the Netherlands to the North Atlantic Treaty Organization; Lt Col Dr. Marcel de Haas, Netherlands Institute of International Relations Clingendael; Maj Gen Kees Homan, retired, Clingendael; Lt Col W. M. Klumper, Netherlands Defence College; Lt Col Eric A. de Landmeter, director, joint doctrine branch; Joris Janssen Lok, *Jane's International Defence Review*; Erwin van Loo, Air Force History Unit, for sparing no effort in dealing with my numerous requests; Lt Gen H. J. W. Manderfeld, retired, commander in chief, RNLAF, 1992–95; Col Frans Osinga, then senior researcher at the Netherlands Institute of International Relations Clingendael, for his invaluable insights into Dutch air power education; Dr. Sebastian Reyn, senior policy advisor, Ministry of Defence; Prof. Joseph Soeters, Royal Netherlands Military Academy and War College; Lt Col Peter Tankink, commander 323 TACTESS (Tactical Training, Evaluation, and Standardisation Squadron), for sharing his view on specific operational aspects; Lt Col Bob Verkroost, JAPCC; Col P. Wijninga, director, Policy Integration Branch; and Rolf de Winter, Air Force history unit.

With regard to doctrine development and air power education in the Swedish Air Force (SwAF), I owe a great debt of gratitude to Maj Tobias Harryson, former member of the doctrine development cell, and to Prof. Nils Marius Rekkedal and Lt Col Jan Reuterdahl, Swedish National Defence College (SNDC). The director of the Swedish Defence Commission, Michael Mohr, proved particularly helpful in shedding light on the complexities of Scandinavian threat and risk perceptions, and Lt Col Torbjörn A. Olsson provided a unique insight into the SwAF's support of UN operations in Africa. Lt Col Anders Persson, armed forces headquarters, spared no effort in updating me with the latest developments in SwAF planning. My appreciations extend to Brig Gen Göran Tode, retired, and Maj Gen Bert Stenfeldt, retired, as well as to the Swedish Air Transport Unit commanders Lars-Eric Blad, Lt Col Pepe Brolén, Lt Col Bertil Höglund, and, in particular, Brig Gen Åke Svedén, retired. My special thanks go to Col Bertil Wennerholm, retired, who helped me to overcome various linguistic challenges and provided me with a distinctive insight into the SwAF's current and past force structure. I am also grateful to Bengt Andersson and Bengt-Göran Bergstrand, senior analysts, Swedish Defence Research Agency; Per Klingvall, Joint Forces Command; Capt Åsa Schön, Skaraborg Wing; and Peter Liander, editor, *Joint Armed Forces Magazine*. I am also indebted to a host of others in Sweden who generously supported me. These include researchers at the Swedish Defence Research Agency, civil servants and scholars at the Ministry of Defence and SNDC, officers across the three services, and industrialists and engineers of the Swedish aerospace industry.

Particular thanks and consideration also go to both air power practitioners and theorists in the United States, the United Kingdom, and Norway—Col (Prof.) Dennis M. Drew, USAF, retired, former associate dean of the School of Advanced Air and Space Studies; Gp Capt Chris Finn, RAF, retired; Dr. Grant T. Hammond, USAF, Air University; Col John Olsen, Royal Norwegian Air Force; Gp Capt Neville Parton, director, defence studies, RAF; and Air Vice-Marshal Andrew Vallance, RAF. I would also like to thank my friends Arti and Dr. Andrew Conway, lecturer in air power studies at the Royal Air Force College, whose endless support and patience smoothed the edges of my English. Moreover, I owe a great debt of gratitude to Dr. Daniel Mortensen, dean of the Air Force Research Institute at Maxwell AFB, Alabama, and the Air University Press team—Jeanne Shamburger, Carolyn Burns, Daniel Armstrong, Susan Fair, and Ann Bailey—for their efforts in turning my dissertation into a book. Jeanne excelled in her endless patience with me, and her work added greatly to the quality of this book. I am also very grateful to my former boss, Dr. Joel Hayward, dean of the Royal Air Force College, who enlightened my views on air power.

To all those who supported me on this journey, I would like to express my sincerest gratitude. But for the facts narrated and the conclusions drawn in this work, I bear sole responsibility.

Finally, I would like to thank my mother, Olga, and my father, Christian. Their unwavering love and infinite support over the past years have been invaluable.

Chapter 1

Introduction

Continental Europe was at the vanguard of air power in the early days of military aviation. The Italian invasion of Libya in 1911 saw the first use of aircraft in combat, and France soon seized the lead in the development of aircraft engines. By the dawn of World War I, Germany and France wielded the most potent air arms. After the cancellation of the Versailles treaty, German air power was again on the rise, with the Luftwaffe playing a crucial role in enabling the sweeping blitzkrieg campaigns.

Yet in the course of World War II, the United States and the United Kingdom emerged as the leading nations in the application of air power—particularly in strategic bombing. The dominance of American air power continued throughout the Cold War and became particularly apparent again in Desert Storm and later conflicts. Against this setting, a prominent British air power scholar and practitioner coined the term "differential air power," referring to the transatlantic air power capability gap.[1] The United Kingdom and France were the only European military actors to maintain anything approaching a full spectrum of air power capabilities for outside interventions, while the majority of Western European states were primarily concerned with territorial defence throughout the Cold War. As regards the doctrinal and intellectual mastery of air power, the United States, along with Great Britain and to a lesser extent Australia, has dominated the debate in the post–Cold War era.

Despite this transatlantic air power capability gap, Continental European air forces remain significant. Some of them have been regularly deployed to out-of-area operations in the Balkans, in Afghanistan, and elsewhere. In the shadow of the Anglo-Saxon air power debate, however, Continental European approaches to air power have largely been ignored in the academic literature. In contrast, Russian and, to a lesser degree, Chinese air power have received attention. *The Quest for Relevant Air Power* attempts to redress this imbalance by contributing to a more inclusive picture of Western air power. To do so, it analyses the following key question:

Courtesy Gripen International; photo by FMV (Swedish Defence Materiel Administration)

FAF C-135FR tanker refuelling a Gripen C and D in midair

how have Continental European air forces responded to the air power challenges of the post–Cold War era?

Air Power Challenges

This book identifies four major challenges that Continental European air forces have confronted in the post–Cold War era:

1. *How have Continental European air forces adapted to the uncertainties created by shifting defence and alliance policies?*

2. *How have these air forces responded to the challenges of real operations?*

3. *How have they responded to the new intellectualism in air power thinking and doctrine?*

4. *How have defence planners attempted to maintain a relevant air force in light of escalating costs and advanced technologies?*

Shifting Defence and Alliance Policies

Bipolar tension in the Cold War provided a clear framework for a nation's defence policy. In the post–Cold War era, however, most Continental European states have struggled to reorientate their national defence policies as the threat and risk spectrum has become more diffuse. In sum, the strategic orientation of a particular country and its defence policy became more complex. This has had far-reaching implications for the application of air power. Continental European air forces no longer know where, when, with whom, and under what circumstances they are going to fight.

Devoid of its major opponent, the North Atlantic Treaty Organization's (NATO) main purpose shifted from deterrence to crisis management. NATO engaged in its first out-of-area combat missions in the Balkan civil wars of the 1990s. In parallel it began its ongoing enlargement process in the second half of the 1990s. Simultaneous to these NATO developments, the European Union (EU) has emerged as a global strategic player and has built its own defence structure. Although the European Security and Defence Policy (ESDP) is intended to be a complementary arrangement to NATO, tensions have arisen (and are likely to continue to arise): some European nations put a premium on a strong transatlantic partnership with the United States, and others on a more autonomous European defence architecture. Failures in alliance policy have led to ad hoc coalitions such as the US-led "coalition of the willing" against Iraq in 2003. Neutral and non-aligned countries have found it more and more difficult to legitimise their passive stance, particularly against the backdrop of an increase in deployed operations for humanitarian purposes.

Challenge of Real Operations

With a few exceptions—such as airlift in humanitarian and peacekeeping operations and the participation of the French Air Force (FAF) in peripheral conflicts—Continental European air power was relegated to deterrence postures in the Cold War era. Since the end of the Cold War, however, a shift away from a deterrence posture and towards real operations has occurred. In an environment of highly unpredictable and diverse conflicts, air power was called upon to carry out missions across the spectrum of force. The post–Cold War era has seen the employment of air power in high-intensity conflicts, in "air policing" operations, and in peace support and humanitarian operations.

In the post–Cold War era, the demand for airlift in humanitarian and peace support operations has increased. In the early 1990s, Bosnia in particular called for military airlift. Under the most adverse conditions, Western air forces kept Sarajevo alive by mounting an air bridge. Airlift was also used for interventions and humanitarian relief operations in Africa and elsewhere.

New for Continental European air forces was the employment of air power in conventional high-intensity conflicts. In the 1990–91 Gulf War, the United Kingdom, France, and Italy contributed their combat air power. Their assistance, however, was dwarfed by the massive use of American air power. In the remainder of the 1990s, Western air power was primarily employed in intrastate conflicts. Western military interventions in the Balkans led to three major air campaigns with the aim of stopping ethnic cleansing—Operation Deny Flight over Bosnia from 1993 to 1995, Operation Deliberate Force in August and September 1995, and Operation Allied Force over Kosovo and Serbia in 1999. In each air campaign, Continental European air forces participated. Yet their employment was dependent upon American air power.

In the wake of the 1990–91 Gulf War, air policing operations were also conducted over northern and southern Iraq to check the Iraqi dictator Saddam Hussein, who proved reluctant to comply with United Nations (UN) resolutions. France was the only Continental European combat aircraft contributor. After the terrorist attacks on 11 September 2001, Continental European air forces went into air policing action over Afghanistan. Only a few of them, however, such as the Royal Netherlands Air Force (RNLAF), contributed to stabilisation operations in Iraq.

This shift from Cold War deterrence postures to real operations across the spectrum of force has represented a major challenge for most Continental European air forces, as deployed operations require a very different organisational structure and mindset.

New Intellectualism

Whereas in the Cold War a strict distinction existed between tactical/conventional and strategic/nuclear air power, the Gulf War revealed that conventional military air power can do more than execute a supportive role for land warfare. Consequently, a new intellectualism in air power thinking and doctrine arose, particularly in the Anglo-Saxon countries.

Today's conflicts are generally complex and multilayered. Armed forces have to prepare not only for interstate conflict but also for a broad range of conflicts across the spectrum of force.

Moreover, no clear line can be drawn between the different kinds of conflict; they often merge into each other. On top of that, the unchallenged superiority of Western military technology, particularly American air power, has created a shift from symmetrical to asymmetrical warfare.

In light of these new kinds of operations, Anglo-Saxon air forces recognised early that sound doctrine and education are of utmost importance. Moreover, many officers and scholars alike started to publish their views on air power. A central question that emerged from Operation Desert Storm is whether air power is a panacea instrument in the new strategic environment. Other widely discussed topics have been the concepts of parallel warfare and effects-based operations. In contrast to serial warfare, parallel warfare—as applied in the air campaign of Desert Storm—aims to paralyse the opponent by simultaneously striking at so-called tactical, operational, and strategic targets. In the second half of the 1990s, the need to limit physical damage and to achieve limited goals led to the discussion of effects-based operations.

The intellectual mastery of air power is essential in tackling the complexities of today's conflicts and is indispensable for effective interoperability with allied air forces. This discussion therefore includes how Continental European air forces cope with a new intellectualism in air power thinking and doctrine and to what extent they have been influenced by their Anglo-Saxon counterparts.

Procurement

With its clearly defined threats, the Cold War provided governments and militaries alike with a more or less reliable framework for planning, force structuring, and procurement. This certainty has ceased to exist, and European politicians have proved reluctant to keep defence spending on a Cold War level. Cashing in the peace dividend has seemed to be more popular. In parallel, costs for advanced air power technology have risen almost exponentially in the information age. Political, financial, and technological constraints have prevented most European nations from acquiring air power capabilities comparable to those of the United States.[2] Consequently, the transatlantic air power capability gap continued to widen in the 1990s, as Operation Allied Force clearly demonstrated.

Cooperative ventures can help to reduce costs. Multinational aircraft programmes had already been undertaken in the Cold War, such as the development of the Franco-British Jaguar or the trinational Panavia Tornado combat aircraft. Pressure for industrial cooperation has increased since then. Role specialisation offers another way to reduce costs.

To add to the complexity of procurement, the provenance and technical standard of air power assets can have far-reaching repercussions on options for participation in combined operations. Nowadays, interoperability is dependent upon complex and costly data links, sensors, armaments, and weapons platforms. In particular, interoperability with the United States Air Force (USAF) is a major challenge for Continental European air forces. Thus, defence planners and militaries alike have to think hard about how to maintain a relevant air force. Procurement in the face of declining defence budgets and soaring costs for air power hardware has become a major challenge for Continental European air forces.

Scope of Discussion

The 1999 air publication *British Air Power Doctrine* defines *air power* as "the ability to project force in air or space by or from a platform or missile operating above the surface of the earth." It further defines *air platforms* as "any aircraft, helicopter or unmanned air vehicle."[3] This discussion also includes platforms such as ground based air defence (GBAD) systems. The study as a whole focuses on air power as exercised by air forces. Though air power assets are often not confined to one single service, air forces remain the chief practitioners of air power among the armed services. They assume major responsibility for specifically air power-related issues, ranging from the protection of the national airspace to air power doctrine development. Aspects of interservice rivalry are examined as exogenous factors.

How Air Forces Were Selected

The choice of the air forces required a balance between depth and breadth. The EU has 27 member states, most of which have an air force. Several non-EU member states also possess air power assets. Since it is not feasible to cover all Continental European air forces in the required depth in this study, criteria for the choice of air forces were devised. Foremost, it was decided to look at the major Continental European players, of which there were several at the outset of the post–Cold War era. In contrast to smaller European air forces—primarily relegated to defensive counterair, air policing, and airlift missions—the forces chosen offer a larger air power spectrum for examination and are more comparable to the major Anglo-Saxon players. Diversity in defence and alliance policies, real operations, and air power hardware was another selection criterion.

The air forces that best fit these criteria are the Armée de l'Air (French Air Force), Luftwaffe (German Air Force), Koninklijke Luchtmacht (Royal Netherlands Air Force), and Flygvapnet (Swedish Air Force). The study of these four air forces highlights the impact of profoundly different defence policies upon air power and compares NATO air forces with a non-aligned air force. Within the NATO air forces, this choice also provides a broad range. At one end of the scale is the Netherlands with its transatlantic orientation and its strong emphasis upon NATO as its most important security pillar. At the other end is France, which only recently reentered NATO's integrated military structure. France emphasises an independent European security and defence policy, emancipated from American leadership. The air forces selected also highlight the impact of professional and conscript armed forces on air power. While France and the Netherlands professionalized their armed forces in the 1990s, Germany has thus far retained conscript armed forces with a professional core component. In July 2011, however, Germany plans to implement "its biggest military reform in more than 50 years" and go to an all-volunteer force. The German chancellor also agreed to reduce troops from 250,000 to 185,000.[4] Sweden suspended conscription only in mid-2010, turning its armed forces into all-volunteer forces. Regarding the challenge of real operations, this discussion analyses air forces that have been regularly deployed to missions abroad as well as those that have been relegated to a rather defensive stance. At one end of the scale is the FAF. During the Cold War, France was already mounting air campaigns in peripheral conflicts in Africa and elsewhere. At the other end is Sweden with its legacy of neutrality. Additionally, this study encompasses the main types of modern combat aircraft available in

Continental Europe and, more importantly, the main models of procurement. The RNLAF is equipped with American F-16 fighter aircraft bought off the shelf, the French and Swedish Air Forces have pursued indigenous fighter programmes, and the German Air Force (GAF) has participated in multinational development programmes. Moreover, the GAF opens a window to Eastern European experience as it had to integrate the East-German Air Force in the early 1990s, and the RNLAF covers many aspects of the European F-16 operators. Though France, Germany, the Netherlands, and Sweden operated some of the most capable European air forces at the outset of the post–Cold War era, their national economies were of very different sizes. This variation allows for comparing the relationship of a country's economy to the generation of air power in the post–Cold War era. Analysing these diverse Continental European air forces allows one to view the four subquestions from various angles.

Why the Post–Cold War Era Was Selected

Because the end of the Cold War brought about far-reaching changes in the global strategic setting, the research is chronologi-

Courtesy Netherlands Institute for Military History; photo by R. Frigge

Against the backdrop of Operation Allied Force, a Dutch pilot inspects his low-altitude navigation and targeting infrared for night (LANTIRN) laser-designator pod prior to takeoff.

cally confined to the post–Cold War era. Militarily, this period constitutes a transitional phase from a potential symmetric clash between East and West through intrastate conflicts to asymmetric attacks against the West and military operations in the context of the so-called war on terror. This era saw Western military forces engaged in a greater variety of operations than at any time during the Cold War.[5] For the reader to understand current Continental European air power, the post–Cold War period as a whole is discussed. Many of the decisions that led to the current force structures have their roots in the 1990s and sometimes beyond. Also, given the rise of deployed operations, this period forms an entity.

Overview

The next chapter, "Post–Cold War Challenges," lays down the background for examining the selected air forces. Subsequent chapters examine the French, German, Royal Netherlands, and Swedish Air Forces.

French Air Force

Undoubtedly, the FAF is the most experienced Continental European air force when it comes to power projection beyond the borders of Europe. Apart from Operations Northern Watch, Desert Fox, and Iraqi Freedom, it participated in all major Western air campaigns during the post–Cold War era. To be an instrument of French foreign policy, the FAF is geared towards rapid intervention missions. It is capable of independently mounting medium-sized operations in Africa and elsewhere. To fulfil its missions and to prevent infringements upon the sovereign political decision making process, the FAF has strived to maintain a balanced force structure and has mainly relied upon indigenous air assets and equipment.

German Air Force

Due to its historical legacy, the GAF was almost exclusively bound to the defence of allied territory in NATO's central region throughout the Cold War. The question as to whether Germany should leave its emphasis upon territorial defence and shift to a more expeditionary stance has been a major bone of contention in German politics throughout the post–Cold War era. The first steps towards a more active defence policy were taken in 1995 and 1999 in the milieu of Operations Deliberate Force and Allied Force, respectively. Significant reunification costs also led to severely tight defence budgets. Nevertheless, the GAF remains one of the most important European air forces, and NATO alliance partners welcome a more active involvement of the GAF in alliance operations.

Royal Netherlands Air Force

The RNLAF has participated in air operations over the Balkans since 1993, making substantial contributions out of proportion to its relatively small size. Among the leading air forces in European air power cooperation, the RNLAF has launched several initiatives for improving interoperability. Its most prominent initiative is the establishment of a European expeditionary F-16 wing. European cooperation, distinct doctrinal features, a strong transatlantic partnership, and the political will to engage in operations across the spectrum of military force have made the RNLAF one of the most effective European air forces of the post–Cold War era.

Swedish Air Force

Though Sweden's economy and population are relatively small, the Swedish have pursued ambitious air power programmes. Given the country's neutral stance throughout the Cold War era, its goal was to be as independent as possible in the development and production of military assets. Therefore, Sweden is unique among the smaller European countries in the sense that its air force employs indigenously designed and manufactured aircraft. At the end of the Cold War, the Swedish Air Force (SwAF) was one of the largest Western European air forces. A RAND study published in 1991 regarded it as one of the best European air forces, particularly in the area of air defence.[6] Almost exclusively geared up for autonomous territorial defence, the SwAF has made significant steps towards power projection and interoperability in the post–Cold War era.

The concluding chapter assesses European air power across the four guiding challenges of shifting defence and alliance policies, real operations, new intellectualism in air power thinking and doctrine, and procurement. While each air force has responded to the air power challenges of the post–Cold War era according to its context, the conclusion nevertheless attempts to identify overall characteristics of Continental European air power.

Notes

1. Tony Mason, *Air Power: A Centennial Appraisal*, rev. ed. (London: Brassey's, 2002), 236–38.

2. David Gates, "Air Power: The Instrument of Choice?," in *Air Power 21: Challenges for the New Century*, ed. Peter W. Gray (London: The Stationery Office, 2000), 26.

3. Air Publication 3000, *British Air Power Doctrine*, 3d ed. (London: HMSO, 1999), 1.2.1.

4. Associated Press, "Germany to Scrap Conscription in Mid-2011," *Washington Post*, 15 December 2010.

5. Timothy Garden, "European Air Power," in Gray, *Air Power 21*, 9.

6. Richard A. Bitzinger, *Facing the Future: The Swedish Air Force, 1990–2005* (Santa Monica, CA: RAND, 1991), 13, 27.

Chapter 2

Post–Cold War Challenges

Our discussion now turns to the uncertainties created by shifting defence and alliance policies and to the challenge of real operations, thereby providing the background for the air forces selected for examination. These issues closely relate as they show how air power has both been shaped by the political level and been used as a means to pursue political goals. We then consider how Anglo-Saxon (US, UK, and Australian) air forces have responded to the new intellectualism in air power thinking and doctrine and how defence planners have attempted to maintain relevant air forces in light of escalating costs. These questions are tied in that the first deals with conceptual and doctrinal aspects of air power and the second with its physical components.

Changing Strategic Context and Alliances

With the end of the Cold War, threats and risks have become complex and multilayered, making it difficult for European states to adapt their defence policies. Simultaneously, alliance structures forged during the Cold War have evolved to respond to the changing strategic context.

Changing Strategic Context

The end of the Cold War marked a profound change in the international system as the antagonism between two opposing social and value systems ceased to be the defining criterion. Consequently, the 1990s became a period of instability, continuous change, and a quest for new structures.[1]

Although the 1990–91 Gulf War was a clear set-piece conflict, this type of clash was not the rule throughout the 1990s. Instead, Western states were confronted with an increasing number of intrastate conflicts in the Balkans, Africa, and elsewhere. Western crisis management responses, however, often proved unsatisfactory. Military means, developed to deter and contain an all-out war between two major blocs, were not adequate for dealing with intrastate conflicts.[2] Philip Sabin points out that concurrently, conflicts have become more asymmetric due to factors including the demise of bipolarity in international affairs; Western military superiority, especially in air power; and differing levels of commitment and ruthlessness.[3]

Ethnic clashes and crisis interventions have not been the only concerns for the international community. By the end of the 1990s, the nuclear non-proliferation regime had obviously suffered serious setbacks.[4] Moreover, at the dawn of the new century, the 11 September attacks brought international terrorism prominently onto the international agenda. These events also had far-reaching consequences on the non-proliferation front, as the United States extended the war against global terrorism to so-called rogue regimes, trying to draw a connection between weapons of mass destruction (WMD) terrorism and Iraq, Iran, and North Korea.[5]

To sum up, developments during the post–Cold War era have led to increased uncertainty and instability in many parts of the world. Due to globalisation, regional tensions and conflicts generate ripple effects that influence apparently secure Western states. Traditional concepts of national security have proven inadequate to deal with today's contingencies. Territorial defence is no longer deemed an effective strategy in an interconnected world.

Changing Paradigms for Defence Planning. Whereas the Cold War was dominated by nuclear deterrence and preparations for a symmetric large-scale war in Continental Europe, the post–Cold War era has required a different approach. Latent or future threats—so-called risks—superseded present and direct threats as key determinants of defence policy in the 1990s. With an attendant loss of predictability, a conceptual realignment of the paradigms for defence planning became necessary.[6]

At the end of the Cold War, the Soviet Union was still viewed as a continuing security threat chiefly because of instability within the Soviet Union itself.[7] Therefore, the possibility of a renewed confrontation with Soviet or Russian military power could not be entirely discarded. Very early, however, it was recognised that future security challenges were more likely to stem from regional conflicts in the Balkans, the Middle East, or elsewhere.[8] The altered security context thus led to force reductions among NATO states in the early 1990s as well as an increasing requirement for strategic mobility of Western armed forces in the context of out-of-area operations.[9] As a result, France and the Netherlands, for instance, abandoned conscription in the mid-1990s with the aim of creating fully professional forces.[10]

In an era of uncertainty and tightened defence budgets, European defence planners faced difficult choices such as readiness versus reconstitution, independence of their forces versus full integration into an alliance, mobility versus punch, and quality versus quantity.[11] Though Europe as a whole has attempted to transform its military by strengthening the deployability and professionalism of its forces, critical capability gaps remain in such vital areas as strategic lift.

Impact upon Air Power. In the aftermath of Desert Storm, the inherent characteristics of air power such as speed, range, flexibility, precision, and lethality had a certain appeal to Western defence planners and experts, particularly in the United States. Air power seemed to many a versatile military instrument in an era of uncertainty.[12] Indeed, air power was used in many instances throughout the 1990s as the weapon of choice largely because it is the military instrument of least commitment. Since it can reach into a conflict zone from outside, its use limits exposing ground troops. These characteristics are particularly important in operations where national stakes are not too high and commitment is optional, as has been the case with a number of Western interventions throughout the 1990s.[13]

The United States' defence spending reflects its emphasis upon air power. From 1990 to 2004, the Army received on average 30 per cent of the services' budget share, the Navy 36 per cent, and the Air Force 34 per cent.[14] If expenses for Army and naval air assets are added to the balance, it can be assumed that the United States spends up to 50 per cent on air power. In 1990 US Army aviation consisted of some 700 fixed-wing and 8,500 rotary-wing aircraft, plus an impressive number of surface-to-air missile (SAM) systems. The Navy operated 14 aircraft carriers together with 13 air wings, with an average size of 86 aircraft each. In addition, US Marine Corps (USMC) aviation contained 503 combat aircraft and 96 armed helicopters.[15] In 2003 Army aviation strength tapered to some 300 fixed-wing and 4,600 rotary-wing aircraft,

Courtesy US Navy

USNS *Alan Shepard* and aircraft carrier USS *George Washington* are under way together during a replenishment at sea during ANNUALEX 21G in 2009.

while the number of Navy aircraft carriers was trimmed to 12.[16] Despite these reductions, the numbers are still impressive.

In Continental Europe, on the contrary, air power has never enjoyed such a prominent role as in the United States, leading to a major transatlantic capability gap. This gap became particularly visible in the post–Cold War era, when air forces were actually employed in major air campaigns. As late as 2003, a Dutch air power expert complained that Europe was still not able to mount an operation similar in complexity to Allied Force on its own.[17]

Alliances

In the wake of the Cold War, NATO underwent a major transformational process, culminating in an enlargement process. Simultaneously, the European Union started to build its own defence architecture. The ESDP is commonly perceived to be a complementary arrangement to NATO. Both institutions have influenced the development of European air power.

European Security and Defence Policy. The end of the Cold War prompted the European Community member states to consider extending cooperation to the sphere of security and defence policy. With the Treaty on the European Union officially signed in Maastricht, the Netherlands, in February 1992, the development of a Common Foreign and Security Policy (CFSP) was agreed upon. At the time, the prospect of a future common defence was held out and an explicit link to the Western European Union (WEU) as an integral part of EU security and defence matters was established.[18]

Almost concurrently with these developments, WEU member states declared the necessity for a genuine European security and defence identity.[19] For this purpose, the WEU adopted in June 1992 the Petersberg Tasks Declaration, encompassing humanitarian and rescue tasks, peacekeeping, and peace enforcement. Moreover, WEU member states declared their intention to make available military units for the accomplishment of these tasks in the context of WEU, NATO, or EU operations.[20]

With the ratification of the Amsterdam Treaty in June 1997, the European Council incorporated the WEU Petersberg Tasks into the EU's CFSP.[21] Despite the declaration of broad political intentions, however, no firm and concrete action plan for common defence matters was decided upon. Moreover, Europe's difficulties in dealing with the crisis in the Balkans during the 1990s seriously put into question the effectiveness of the CFSP. It was essential to improve European military capabilities if the EU wanted to take on strategic responsibilities. This realisation led Great Britain and France—the two critical European military actors—to reconcile opposing views. For France, a more effective use of military force meant a more pragmatic approach towards NATO. Great Britain, for its part, was apprehensive that Europe's military powerlessness might imperil the very foundation of the Atlantic partnership. The rapprochement between France and Great Britain led to the bilateral Franco-British St. Malo Declaration in late 1998.[22] France and the UK jointly declared that the CFSP had to be backed by credible military forces, decision-making bodies, and sources of intelligence and analysis.[23]

In the ensuing years, a rapid Europeanisation of the St. Malo Declaration took place. The institutional changes for an effective European security and defence policy as an integral part of the EU's CFSP—decided at the summit in Cologne, Germany, elaborated on in Helsinki, Finland, and finalised at Santa Maria de Feira, Portugal—were agreed upon by the member states at the

EU summit in Nice in December 2000.[24] Particularly, the apparent European shortfalls during the Kosovo air campaign were a catalyst for making swift progress in common European defence matters.[25] These developments resulted in a transfer of almost all WEU functions to the EU, but they do not imply the end of the WEU treaty as such.[26]

Despite the European divide over the Iraq crisis as of 2003, a further important step was taken in the same year. The document *A Secure Europe in a Better World*, written under the direction of the High Representative for the CFSP, Javier Solana, for the first time framed a common European security strategy.[27] The document basically recognises that traditional territorial defence is a thing of the past and that crises have to be tackled at their roots; they require a multilateral approach and preventive engagement.[28]

To create the required military capabilities for the ESDP, the EU agreed on two European headline goals. The Helsinki Headline Goal (HHG), established at the European Council in Helsinki in December 1999, foresaw the capability of being able, by 2003, to deploy within 60 days and to sustain for at least one year up to a corps-size force.[29] Though the HHG goals were formally declared to be met in 2003, limitations in the ESDP's military capabilities were acknowledged.[30]

With the adoption of a common European security strategy, a new headline goal was issued in 2004, reflecting the evolution of the strategic environment. Headline Goal 2010 (HG 2010) builds on and complements the previous HHG. It foresees the creation of national and international battle groups for rapid response operations—the so-called EU battle groups—and the enhancement of European lift capacities, including the development of a European airlift command.[31] The EU battle groups represent a key element of HG 2010. They are the minimum military effective and coherent force packages capable of stand-alone operations in rapid response scenarios.[32] In contrast to the HHG with its quantitative approach, the adoption of HG 2010 led to a concerted attention to quality.[33]

To avoid unnecessary duplications and to make NATO assets available for EU operations, a close relationship between the ESDP and NATO has been established. In particular, finalisation of the "Berlin Plus" agreement in December 2002 paved the way for a strategic partnership between the EU and NATO.[34] It allows the EU to "borrow" common military assets and capabilities from NATO.[35] HG 2010 also emphasises military interoperability between NATO and ESDP concepts and procedures, mainly because only a single set of forces is available for both organisations.[36] Hence, the EU aims at arrangements for full consultation, cooperation, and transparency with NATO.[37]

NATO after the Cold War. NATO responded to the end of the Cold War by developing a new strategic concept completed in November 1991. It stressed cooperation with former adversaries, as working towards improved and expanded security for Europe as a whole was considered a new goal.[38] The concept also provided for major changes in NATO's defence posture, such as reductions in the size of military forces on the one hand and improvements in their mobility, flexibility, and adaptability to different contingencies on the other.[39] Nevertheless, it was written in the context of a still-existing Soviet Union and, hence, made an explicit reference to its large conventional and nuclear arsenals.[40]

Out-of-area operations had a major impact upon the evolution of NATO in the following years. Whereas deterrence and collective defence had provided the fulcrum during the Cold War era and have continued to be a central alliance issue, the emphasis de facto shifted towards peace support and crisis management operations. At the ministerial meeting of the North Atlantic Council in December 1992, member states confirmed their preparedness to support, on a case-by-case basis, peace-keeping operations under the authority of the UN Security Council and, hence, to provide forces for out-of-area contingencies.[41] Throughout the 1990s, NATO shifted from a relatively limited role in supporting UN peacekeeping efforts to assuming full control of complex peace support operations, culminating in Operation Allied Force in 1999.[42]

The emergence of a new out-of-area role for NATO also led to institutional changes and innovations, such as the adoption of the combined joint task force (CJTF) concept. Endorsed by NATO member states at the January 1994 summit in Brussels, the CJTF aimed at improving NATO's ability to deploy multinational forces into out-of-area operations on short notice.[43] On the same occasion, NATO announced its support for the European Security and

Defence Identity (ESDI), agreed on two years earlier by the WEU. Emerging as the defence component of the EU, the WEU was intended to strengthen the European pillar of the alliance. Collective alliance assets were to be made available for WEU operations undertaken by European allies in the context of the CFSP. Support for the ESDI was believed to enable European allies to take greater responsibility for their common security and defence and hence to reinforce the transatlantic link.[44] Yet it did not imply NATO's unconditional assistance for autonomous European operations. The June 1996 Berlin summit foresaw that a CJTF could be WEU-led but not WEU-authorised, and European assets were to be "separable but not separate" from NATO assets.[45] Moreover, to streamline the alliance's command structure and to adapt it to the changed strategic setting, the top echelon was reduced from three to two major NATO commands by June 1994 by disbanding Allied Command Channel and restructuring NATO under the remaining Allied Command Europe (ACE) and Allied Command Atlantic (ACLANT).[46]

At the January 1994 Brussels summit, the Partnership for Peace (PfP) programme was also launched. Operating under the authority of the North Atlantic Council, it was supposed to forge new security relationships with non-NATO states.[47] Membership was offered to states participating in the Conference for Security and Cooperation in Europe.[48] On a strategic level, the aim of PfP has been to enhance stability and security throughout Europe. On an operational level, its purpose has amongst others been to enhance the partner countries' ability to operate with alliance forces.[49] Originally relegated to the lower spectrum of military force, the PfP operation-spectrum was extended in 1997 to include peace enforcement operations.[50]

In the second half of the 1990s, cooperation with former adversaries went beyond the PfP programme. The extent of the alliance's commitment to external transformation was particularly demonstrated at the 1997 NATO summit in Madrid, when accession invitations were issued to the Czech Republic, Hungary, and Poland. In early 1999, these countries became formal members of the alliance.[51] Two further enlargement rounds have occurred since then. In 2004 seven further Eastern European states formally entered the alliance—Slovakia, Slovenia, Romania, Bulgaria,

Estonia, Latvia, and Lithuania. In 2009 Croatia and Albania became members of NATO.

In concert with NATO's enlargement, power projection became an increasingly important focus for the alliance. The strategic concept, adopted at the 1999 Washington summit, underlined NATO's willingness to support operations under UN authority.[52] To respond to the demands of deployed operations, particularly in the wake of 11 September, the alliance launched the NATO Response Force (NRF) project at the 2002 NATO summit in Prague. The NRF is a quickly deployable and technologically advanced joint force numbering up to 25,000 troops.[53] It is designed for worldwide employment across the spectrum of military operations, from evacuation operations and deployment as a mere demonstrative force to Article 5 (collective defence) operations. The response force is inherently joint and contains land, air, sea, and special operations forces components, including a brigade-size land component, a naval task force with a carrier battle group and an amphibious task group, and an air component capable of 200 combat sorties a day.[54] It reached initial operational capability on 15 October 2003, and full operational capability was declared at the November 2006 Riga summit in Latvia.[55] The process of setting up the NRF was considered to be a major driving factor behind the alliance's military transformation towards expeditionary forces.[56]

Requirements of expeditionary warfare, especially in the context of operations in Afghanistan, led to a further adaptation of NATO's military command structure. Operational responsibilities, formerly shared by ACE and ACLANT, were merged into Allied Command Operations (ACO), based in Mons, Belgium, under the responsibility of the supreme allied commander, Europe (SACEUR). To deal with the transformational challenges, the new Allied Command Transformation (ACT) was established in the United States. ACO and ACT are NATO's two strategic commands.[57] As such, geographical boundaries delineating operational command authorities had been removed, reflecting the nature of deployed operations that might occur anywhere. Secondly, ACT advances NATO's continuing transformation against the backdrop of a highly fluid threat and risk spectrum.

Alliance Policy and Its Impact upon European Air Power

European alliance frameworks have either directly or indirectly influenced cooperative ventures in the arena of air power. Cooperative ventures have taken place in the context of the ESDP, NATO, and bilateral or multilateral organisational relationships.

ESDP Context. In the domain of air power, the HHG conceived the ability to deploy up to 400 combat aircraft for EU operations.[58] As mentioned, the HHG turned out to be too ambitious at the time, and an approach of incremental improvements had to be adopted. Analogous to the EU battle group concept, basically a land-centric approach, consideration was given to developing a rapidly deployable EU air component within the framework of HG 2010.[59] Originally, the lead nation of each EU battle group was responsible for providing a non-specified air component. Germany and France took the view that a rapidly deployable common air component would be necessary to strengthen the EU's rapid response capability. Thus, representatives of both countries' air forces drafted a food-for-thought paper, establishing the EU Rapid Response Air Initiative (EU RRAI). At the Franco-German summit on 14 March 2006, the defence ministers formalised this proposal, which was subsequently submitted to the European Union Military Committee.[60]

The initiative came into effect as the EU Air Rapid Response Concept (EU Air RRC), which the European Union Military Committee agreed to on 21 December 2007. The concept has an air database at its core, supporting the force generation process by indicating potentially available air assets and capabilities for EU operations. The database was established for the first time in October 2008.[61] Since the main challenge is not the availability of combat aircraft but rather the scarce European logistics for deployed operations, the deployable air base is at the core of the concept. In particular, it is a matter of identifying modules that national air forces can contribute to a multinational air base.[62]

NATO Context. The NATO Airborne Early Warning and Control Force (NAEW&CF), established in the early 1980s, represents one of the most prominent and successful cooperative ventures in the arena of European air power. It now consists of two operational elements—a multinational NATO fleet and the UK's airborne early warning component. The multinational NATO fleet is based in Germany and operates 17 Boeing NATO E-3A Airborne Warning and Control System (AWACS) aircraft. The aircraft are manned by international crews from 15 NATO nations (Belgium, Canada, Denmark, Germany, Greece, Hungary, Italy, the Netherlands, Norway, Poland, Portugal, Romania, Spain, Turkey, and the United States).[63]

Courtesy US Air Force

NATO E-3A AWACS aircraft. A GAF NATO communications technician stands at the entrance to the E-3A AWACS, Cold Lake, Alberta. The aircraft and its crew, stationed at NATO Air Base Geilenkirchen, Germany, were participating in Maple Flag 37, a multinational air combat exercise.

More recently, the NRF has been a major catalyst for air force cooperation. The NRF's air component is multinational and contains approximately 6,000 personnel, 84 combat aircraft, 80 support aircraft, and three deployable air bases.[64] So far, combined joint force air component (CJFAC) commands have rotated between NATO's regional air component command headquarters at Ramstein AB, Germany, and at Izmir, Turkey, as well as between the RAF and the French Air Force.[65]

Within a NATO context, cooperation has recently also been fostered on a conceptual level through the setting up of the Joint Air Power Competence Centre (JAPCC). The German Air Force was a key player in setting up the JAPCC, which became operational in Kalkar, Germany, in early 2005.[66] It had been argued that air and space power was not adequately represented as an integral part within NATO's command structure and that a specialised centre of excellence would provide the required competence as an added value. To that end, the JAPCC provides innovative advice and subject matter expertise for the transformation of alliance air forces.[67]

Bilateral and Multilateral Organisational Relationships. In the post–Cold War era, instituting the Franco-British European Air Group (FBEAG) was a major achievement in European air power cooperation. In light of real operations in the Balkans and elsewhere, the RAF and FAF realised a need for improving interoperability between the two air forces and established the FBEAG at High Wycombe, UK, in 1995.[68] One of the FBEAG's primary tasks was facilitating combined air operations to support Petersberg missions, thereby reflecting the emerging European security and defence policy.[69] The title of the organisation was formally changed to the European Air Group (EAG) in January 1998. The same year, Italy reached full membership status, and on 12 July 1999, Belgium, Germany, the Netherlands, and Spain joined the EAG.[70]

The official goal of the EAG is "to improve the operational capabilities of the Parties' Air Forces to carry out operations in pursuit

Courtesy FAF, Sirpa Air

Franco-British cooperation. FAF Mirage 2000D and RAF Tornado GR4 flying in formation.

of shared interests, primarily through mechanisms which enhance interoperability."[71] To implement this objective, the EAG produces various outputs on the tactical/technical level such as technical arrangements, tactical planning guides, operating procedures, interoperability handbooks, or coordination of training activities.[72] Current EAG activities specifically underline its role in making European air forces fit for deployed operations. Its endeavors include force protection, joint personnel recovery, and deployable multinational air wings.[73] The deployable multinational air wing (DMAW) project, for instance, aims at providing support tools and framework arrangements to facilitate combined and deployed operations. A DMAW consists of up to three squadrons and can be tailored for a specific task. To avoid duplications, the DMAW project is closely coordinated with corresponding NATO and ESDP efforts.[74]

The EAG has been crucial in improving and coordinating existing European airlift capacities. Some European countries had already begun to exchange military airlift capacities on the basis of bilateral memoranda of understanding (MOU) in the 1980s. In an era of primarily multilateral action, however, these bilateral agreements became increasingly inflexible. To remedy this, the seven EAG member states established the Air Transport and Air-to-Air Refuelling Exchange of Services (ATARES) programme to allow for cash-free trading of services.[75]

Moreover, in the wake of the Franco-German summit of 30 November 1999, the EAG was tasked with a European airlift study to identify better use of the available European airlift means. France and particularly Germany considered this a first step towards a common European airlift command.[76] The European airlift study the EAG conducted actually built upon already existing bilateral agreements, and it finally resulted in the establishment of the European Airlift Coordination Cell (EACC) at Eindhoven, Holland, in September 2001. Though the EACC had been successful in coordinating air transport and air-to-air refuelling (AAR) activities among EAG air forces, in terms of both operational and financial aspects, the need to further develop the EACC was soon recognized. Subsequently, the EACC was transferred into the European Airlift Centre (EAC) on 1 July 2004. In contrast to the EACC, the EAC's broader scope of planning authority and responsibilities includes conceptual development and training.[77] Along with the EAG member countries, Norway became an associated EAC member.[78] Subsequently, the EAC was merged with the Sealift Coordination Centre into the Movement Coordination Centre Europe (MCCE) in Eindhoven on 1 July 2007.[79] This step allows for drawing upon synergies in the joint domain.

Though neither the EACC nor the EAC was set up at the direct request of the ESDP, European HG 2010 explicitly refers to the further development of the EACC into the EAC and welcomed the development of a European airlift command by 2010.[80] As is discussed later, France and Germany have been the lead nations in promoting the establishment of a European air transport command (EATC).

On 23 March 2006, the chartering of Ukrainian wide-body, long-range strategic transport aircraft became institutionalised through the Strategic Airlift Interim Solution (SALIS). A contract with a private company provides for two Antonov An-124-100s permanently ready in Germany, with the option of chartering another four aircraft within nine days. The SALIS initiative was launched at the Prague NATO summit in November 2002 and contributes to bridging the European capability gap in strategic airlift until the commissioning of the European A400M military transport aircraft.[81] SALIS is available to both NATO and the ESDP operations, underlining the strategic partnership between the two alliances. Among the 17 participating nations, 14 are both EU and NATO member states, with Canada, Norway, and Sweden being the exceptions.[82] As an interim solution, SALIS ensures timely availability of a capability to deploy outsized cargo, benefitting European rapid reaction forces (RRF) such as the NRF or EU battle groups.[83]

Cooperation on an Operational Level. NATO exchange programmes have contributed significantly to the interoperability of alliance air forces on both a tactical and personal level. As is described in the chapters on the air forces, major integrated air exercises such as Red Flag in the United States or Maple Flag in Canada have increasingly attracted Continental European air forces against the backdrop of real multinational operations. Moreover, Continental Europeans have begun to organise their own integrated air exercises.

Employment of Western Air Power in Real Operations

Since the early 1990s, Western air power has been applied across the spectrum of force. The utilization of Western air power in the geographical hotspots of the Middle East and the Balkans is examined next, followed by its operations in other areas of the world.

Western Air Operations in the Middle East

The Iraqi invasion of Kuwait in August 1990 triggered the beginning of a series of air operations in the Middle East. Besides high-intensity air campaigns as conducted in the course of Operations Desert Storm and Iraqi Freedom, Western air power was used in a constabulary role for implementing the no-fly zones (NFZ) over northern and southern Iraq.

Desert Storm. On 2 August 1990, Iraq invaded Kuwait with the aim of annexing it. In the wake of this aggressive act, the United States embarked upon Operation Desert Shield. Its purpose was to deter further aggression against Saudi Arabia and to protect an allied force buildup in the region. Due to its agility, air power proved to be the most essential element in the early phase of Operation Desert Shield. Swiftly, a force consisting of fighters, fighter-bombers, AWACS aircraft, and a number of combat support aircraft was assembled. By 23 August, approximately 500 allied

Courtesy US Air Force

USAF A-10 during Operation Desert Shield. An air-to-air view of an A-10A Thunderbolt II attack aircraft from the 354th Tactical Fighter Wing, Myrtle Beach AFB, SC.

aircraft had already been dispatched to the crisis area, forestalling any potential Iraqi thrust into Saudi Arabia.[84] Unlike ground power, air power enabled the allies to bring to bear their numerical and technological superiority at an early stage.

The European combat aircraft contributors to Operations Desert Shield and Desert Storm were the United Kingdom, France, and Italy. In variety and scale, the UK contribution was second only to that of the United States.[85] The French deployed the third largest air force contingent to the Gulf area. While France was willing to commit combat aircraft to Operation Desert Shield at an early stage, it was more hesitant to commit forces to Desert Storm, the offensive campaign. The French government only decided to do so just prior to the attack. The Italian Air Force contributed by deploying eight Tornado GR1 bombers. Belgium, Germany, and the Netherlands augmented the protection of Turkey by dispatching combat aircraft as well as GBAD batteries. Yet, these aircraft were not involved in combat action during the war.[86] In the course of the air campaign, European losses amounted to eight aircraft out of a total of 38 lost allied aircraft.[87]

Despite these various European contributions, a transatlantic gap in the air war effort was obvious. In total, fixed-wing aircraft flew 118,661 sorties in Desert Storm. Of these, US services logged 85.4 per cent. The UK, France, and Italy flew 6.7 per cent, with the UK contributing more than two-thirds of the European share.[88]

Early on 17 January 1991, after the diplomatic negotiations had failed, the allies embarked upon their concentrated air campaign against Iraq and the Iraqi armed forces in Kuwait. In a delicately orchestrated effort, the coalition hit virtually every target set in the initial strikes, from strategic leadership targets in Baghdad to Iraqi ground forces in the Kuwaiti theatre of operation.[89] Attacking so-called strategic, operational, and tactical target sets simultaneously was referred to later as parallel warfare.[90] Two main enablers of Operation Desert Storm were precision-guided munitions (PGM), which offset the need for mass attacks to achieve a high probability of success, and stealth aircraft, which provided access into high-threat environments.[91] Whereas the number of aircraft equipped to laser-designate targets was relatively small, the larger number of aircraft not equipped to do so dropped unguided munitions in large quantities to reduce Iraqi combat strength in the

Kuwaiti theatre of operation.[92] Early in February, aircraft equipped for precision engagement were shifted to contribute to this attrition effort and proved to be very effective.[93]

Particularly, the smashing of the Iraqi ground offensive against the Saudi coastal town of Al Khafji by coalition air power in the early stages of the air campaign revealed a new relationship between air and land power.[94] Though air power could not hold ground, it denied it to Iraqi ground forces.[95] Desert Storm turned out to be a catalyst for the evolution of European air power in the post–Cold War era irrespective of direct air force involvement in the campaign (discussed in more detail in the next chapters).

Courtesy US Marine Corps

US Marine Corps artillery during Operation Desert Storm. Marine artillerymen from the 2d Marine Expeditionary Force fire their M-198 155 mm howitzer in support of the opening of the ground offensive to free Kuwait during Operation Desert Storm.

No-Fly Zones over Northern and Southern Iraq. To protect the Iraqi Kurds from air strikes in the wake of Desert Storm, the USAF, RAF, and FAF kept a wing-sized force of aircraft in Turkey.[96] The operation was superseded by Operation Northern Watch in January 1997. The United States, Great Britain, and Turkey provided approximately 45 combat and support aircraft for this new operation.[97]

Analogous to the operations in the north, Operation Southern Watch was mounted over southern Iraq by American, British, and French forces from August 1992. The area south of the 32d parallel was prohibited to Iraqi aircraft.[98] Various Iraqi violations of the NFZ occurred.[99] In August 1996, for instance, the Iraqi Air Force launched air strikes to intervene in the fighting between Kurdish factions. The United States responded by launching cruise missiles and by extending the Southern Watch NFZ up to the 33d parallel, just south of Baghdad.[100] The French refused to back the extension of the southern NFZ.[101]

In 1997 and 1998, Iraq continued to engage in a repeated pattern of obstructing and deceiving United Nations Special Command (UNSCOM) weapons inspections. This situation finally culminated in Operation Desert Fox, a limited punitive air campaign. The Anglo-American operation entailed 650 aircraft sorties and 415 cruise missile launches. Ninety-seven targets in Iraq were attacked, mainly related to leadership and, to a lesser degree, to alleged WMD facilities.[102]

In late 2001 and 2002, US and British forces stepped up their efforts and embarked upon an intensive destruction of enemy air defences campaign.[103] From July 2002 to 19 March 2003, coalition forces flew no fewer than 8,600 sorties to reduce the Iraqi air defence capability and C2 networks.[104] In fact, Operations Northern and Southern Watch turned into a preliminary campaign for an actual invasion of Iraq.

Iraqi Freedom. After Afghanistan, the "war on terror" turned to Iraq. Operation Iraqi Freedom commenced on 19 March 2003 with a failed decapitation strike against Saddam and his sons by means of Tomahawk land-attack cruise missiles and F-117s. To achieve surprise and to secure the Rumilyah oil fields, however, the Anglo-American ground incursion preceded the actual air campaign, with the former starting early in the morning and the latter in the evening of 21 March.[105]

The air war that followed was twofold. A strategic air campaign struck against selected targets in Baghdad. An operational air war relentlessly hammered Iraqi ground forces.[106] The vast majority of air strikes, approximately 80 per cent, were allocated to the latter. Close air support (CAS) decisively shaped ground engagements, and, at an operational level, aerial firepower spearheaded ground manoeuvres.[107] When ground forces were pinned down by sand-storms, American air power was able to strike decisive blows against the Republican Guard. By means of air-to-ground sur-veillance radars and Global Positioning System (GPS) guided bombs, air power could hit point targets under most unfavourable weather conditions.[108] It is estimated that all but 19 of the Repub-lican Guard's 850 tanks were destroyed or abandoned during the air strikes.[109]

A particularly high level of jointness was displayed in the course of Operation Iraqi Freedom during March and April 2003.[110] Con-fronted with a swift and massive ground assault, the Iraqi military had to move and concentrate its divisions, which were then anni-hilated by coalition air power. Air power also covered the flanks and the rear of the advancing coalition forces.[111] Air Chief Mar-shal Sir Glenn Torpy points out that while Operation Desert Storm involved a basically distinct air campaign followed by a discrete land campaign, the high-intensity phase of Iraqi Freedom was a truly integrated operation.[112] On top of that, the number of preci-sion munitions employed rose from a small percentage in Desert Storm to a majority in Operation Iraqi Freedom.[113] Yet in contrast to Desert Storm, the air coalition was much smaller. Besides American and British aerial forces, the Royal Australian Air Force (RAAF) participated in the air campaign.

After these major combat operations had ceased, Western air power continued to play an important role in Iraq by providing air surveillance, reconnaissance, precision air strikes, and airlift. Air power was particularly applied in so-called urban joint warfare, as the battle for Fallujah in November 2004 demonstrated.[114]

Western Air Operations over the Balkans

Immediately after Desert Storm, the Balkans emerged as the second major trouble spot for Western governments in the post–Cold War era. Unlike the case of the Gulf War in 1991, it was not easy to reach political consensus on how to tackle the situation in the Balkans throughout the 1990s. Thus, the execution of air opera-tions was influenced by political, diplomatic, and alliance friction.

Deny Flight. Operation Deny Flight started on 12 April 1993. It was first supposed to enforce a declared NFZ over Bosnia by means of round-the-clock combat air patrols (CAP).[115] In the months that immediately followed, the mission spectrum was extended to include air-to-ground strikes.[116] Among NATO countries, providing aircraft for combat missions over Bosnia from the beginning of the campaign were the United States, France, Great Britain, the Netherlands, and Turkey.[117] In late November 1994, Spain deployed eight F-18 Hornets to take part in Operation Deny Flight.[118]

Due to the narrow rules of engagement (ROE), the civil war parties could execute airspace violations with near impunity.[119] Also, with regard to CAS missions, the ROEs were tight. The CAS request system worked on a dual key command chain, with both the UN and NATO having to authorise air strikes. While NATO responded quickly to CAS requests, the UN approval process worked slowly and most often too slowly.[120] Consequently, a total of only four authorized CAS attacks were conducted over two years.[121]

It was only in 1995 that consensus for more vigorous air strikes began to build. Yet, NATO bombing of Bosnian Serb ammunition depots in May 1995 resulted in hundreds of UN peacekeepers being taken hostage. A further humiliation for the UN followed when in July 1995, Dutch peacekeepers failed to protect Srebrenica, Bosnia-Herzegovina. NATO jets were ready to provide CAS within minutes, but the UN refused to approve any air strikes for two entire days.[122] In response to these developments, the dual key command chain was abandoned in July 1995.[123]

Deliberate Force. The immediate event that triggered a more robust air campaign—Operation Deliberate Force—NATO's first major combat mission ever, was the shelling of a marketplace in Sarajevo, Bosnia-Herzegovina, on 28 August 1995. Early on 30 August 1995, NATO aircraft took off to strike targets in Bosnia. The campaign itself was halted for negotiations. After these faltered, the bombing was resumed. On 14 September, the Bosnian Serbs agreed to UN terms, and offensive operations were suspended; the campaign was declared closed on 20 September.[124] Political sensitivity had an overarching impact upon the conduct of Operation Deliberate Force. The American air component commander considered the avoidance of collateral damage to be of pivotal strategic importance.[125]

Alliance air forces flew approximately 3,500 sorties, including 750 strike sorties.[126] A total of about 300 aircraft were assigned for Operation Deliberate Force, including approximately 20 air-to-air refuelling aircraft.[127] In terms of sorties, the US services accomplished by far the most (66 per cent), followed by the UK (10 per cent) and France (8 per cent), with Dutch, German, Italian, Spanish, Turkish, and common NATO aircraft flying the remainder.[128] Deliberate Force was the first air campaign to see the predominant use of PGMs—69 per cent out of 1,026 weapons released.[129] Albeit on a significantly lower level, this stood in stark contrast to the Desert Storm air campaign, where 7.6 per cent out of approximately 227,000 weapons expended were precision-guided.[130] International cooperation was tight; a typical strike mission was normally conducted by a multinational fighter package.[131] German combat aircraft were in the air during the operation, though not officially integrated.

The air campaign was part of a larger package, finally producing the November 1995 Dayton Accords. The international sanctions began to have an effect, and the Bosnian Muslims and Croats

Courtesy US Army

US Army units entering Bosnia-Herzegovina in late 1995. In the wake of Operation Deliberate Force and the Dayton Accords, a US Army Bradley fighting vehicle leads a column of Humvees across a pontoon bridge into Bosnia-Herzegovina.

launched a ground offensive alongside the air campaign.[132] Moreover, Deliberate Force included a ground component. On a French initiative, a multilateral RRF was created and inserted into Bosnia. The combat element of the RRF was a French, British, and Dutch multinational brigade. According to the French general commanding the brigade, allied artillery fire paralysed military movements around Sarajevo and produced synergies with air power.[133]

Air Bridge to Sarajevo and Eastern Bosnia. Simultaneously with Operations Deny Flight and Deliberate Force, Western air power was used in a major humanitarian relief effort. As the convoy routes to Sarajevo were often obstructed by the warring parties, a substantial part of aid and supply had to be flown in. On 3 July 1992, the efforts of the UN high commissioner for refugees to airlift supplies to the Sarajevo airport officially started. At the beginning, 14 nations contributed to the relief effort by providing military transport aircraft. Additionally, the UN chartered airlifters. European contributions came from Belgium, Great Britain, Denmark, France, Germany, Greece, Italy, Norway, Spain, and Sweden. Two months after the official start of the operation, on 3 September, an Italian aircraft—lacking any self-protection system—was shot down by a SAM. The UN reacted by shutting down the airlift effort for over one month. As a consequence of this fatal incident, most nations withdrew their transport aircraft.[134]

The air bridge was reactivated in October 1992 and lasted until January 1996. It was mainly operated by military aircraft from the United States, Germany, France, the UK, and Canada. From time to time, other aid organisations or air forces, such as the Swedish Air Force, supported the airlift effort. Aircraft were often shot at, though without any serious consequences. When the air bridge was finished in 1996, it was the longest running in history. Often, it provided the only way to get aid and supply into Sarajevo.[135]

Aid and supply were also air-dropped over the Muslim enclaves in Eastern Bosnia and the Muslim-held eastern sector of Mostar city, Bosnia-Herzegovina. In February 1993, a USAF airlift wing was given the task of conducting the airdrop missions. When the operation gathered momentum one month later, the American C-130s were augmented by three German and two French C-160 Transall aircraft. All missions were flown at night due to the ground threat. The airdrop operation lasted until August 1994. Simultaneously, British and French transport helicopters secured the evacuation of wounded from the enclaves.[136]

Allied Force. Between 24 March and 9 June 1999, NATO embarked upon the last major air campaign over the Balkans with the goal of stopping the suppression of the Albanian majority in Kosovo.[137] It was widely assumed that Operation Deliberate Force would serve as a pattern for Kosovo. Hence, NATO planned to bomb only a specified number of selected targets. A short bombing campaign was expected to make Slobodan Milosevic concede to Western demands. However, there were two crucial differences between Deliberate Force and Allied Force. First, in contrast to Bosnia, Kosovo possessed special historical meaning for the Serbs.[138] Second, a Croat and Muslim offensive was putting pressure on the Bosnian Serbs in 1995, and there was a Western combat component on the ground—both were lacking in 1999.

From the very beginning, an overwhelming force–type operation was excluded.[139] Instead, NATO settled for a gradual approach to coerce Milosevic.[140] Only in the second half of Operation Allied Force did NATO strike with determination.[141] Undoubtedly, most European allies would have been reluctant to approve a massive use of force in advance. Yet, according to Ivo Daalder and Michael O'Hanlon, the American charge that the air campaign was primarily hampered by political interference from European NATO member states does not carry substance. The decisions to adopt a strategy of gradual escalation and to rule out the deployment of ground forces were made in Washington, as it was believed—particularly by Pres. Bill Clinton and Secretary of State Madeleine Albright—that Milosevic would give in quickly.[142] Since the United States and NATO had ruled out a ground option from the outset, the Yugoslav Army was able to survive the air attacks by spreading out and concealing its equipment.[143] Moreover, alliance friction was created by the United States using a parallel but separate mechanism for mission planning and air tasking regarding sensitive systems such as the B-2, F-117, and cruise missile. While all other alliance air assets were tasked by NATO, these systems were tasked by the US European Command.[144]

Why Milosevic gave in is still a controversial issue. A bundle of factors was identified, with the air campaign at the core and underpinning all the remaining factors—declining support from Russia,

NATO's cohesion as an alliance, diplomatic interventions, and the increasing threat of a NATO ground intervention.[145]

The United States was shouldering by far the largest effort. While some European allies were able to make valuable suppression of enemy air defences (SEAD) contributions or to deliver PGMs, many European allies lacked the capabilities to operate effectively with the US services, which contributed 59 per cent of all allied aircraft involved in the air campaign and released over 80 per cent of the expended munitions.[146] Approximately 23,000 bombs and missiles were used, of which 35 per cent were precision-guided, including 329 cruise missiles.[147]

Courtesy US Air Force

USAF B-2 Spirit. The B-2 Spirit made its combat debut during Operation Allied Force.

Examining NATO's air campaign as a model for possible future European military operations, a British defence expert argued in 2000 that the approximately 500 all-weather bombers that the UK, France, Germany, and Italy could field at the time needed to be increased by about 50 per cent.[148] Furthermore, the transatlantic capability gap became visible in the fields of AAR and airborne standoff jamming. With regard to the latter, NATO's air campaign hinged entirely upon US capabilities. Regarding AAR, American aircraft flew about 90 per cent of the sorties. Yet, the published sources on European tanker contributions are inconsistent. For instance, while a RAND report concluded that France deployed only three dedicated tanker aircraft, the French Ministry of Defence (MOD) indicated that France was able to deploy 10 C-135FR tankers.[149]

Besides the United States, Canada, and Turkey, European nations that participated in Operation Allied Force were Belgium, Denmark, France, Germany, Italy, the Netherlands, Norway,

Portugal, Spain, and the United Kingdom. Of the European allies, the French provided the most deployed aircraft and sorties flown, followed by the British. The Royal Netherlands Air Force as well as the German and Italian Air Forces made other important contributions, the latter two particularly in the domain of SEAD.[150]

Western Air Operations in the Rest of the World

The application of Western air power in areas outside the Middle East and the Balkans has been manifold—though more in a supportive role and therefore less prominent—with the exception of Operation Enduring Freedom over Afghanistan in the aftermath of 11 September.

Air Operations against "Terrorism." The air campaign against the Taliban regime and al-Qaeda in late 2001 had a precursor in 1998. After the US embassies in Kenya and Tanzania were bombed on 7 August 1998, the Clinton administration embarked upon Operation Infinite Reach. Approximately 70 Tomahawk cruise missiles were launched from naval platforms against allegedly al-Qaeda-related targets in Sudan and Afghanistan.[151]

After the attacks of 11 September, Operation Enduring Freedom aimed at chasing down al-Qaeda by depriving it of its safe haven in Afghanistan and toppling the Taliban regime. The first strikes were conducted by US forces on 7 October 2001, with British submarines striking at targets by means of Tomahawk cruise missiles.[152]

Courtesy US Air Force

F-15E Strike Eagle aircraft from the 335th Expeditionary Fighter Squadron approach a mission objective in eastern Afghanistan in November 2009. The 335th deployed to Bagram Air Field, Afghanistan, from Seymour Johnson AFB, NC.

The campaign turned out to be a major struggle between the Taliban and the northern alliance forces, with precision air power—directed by special forces—affecting the outcome decisively.[153] The delivery of firepower reached a new level in Enduring Freedom. During Desert Storm, there was an average of 10 aircraft sorties per target. In Enduring Freedom, on the other hand, a combat aircraft attacked on average two targets per sortie.[154] Furthermore, the sensor-to-shooter cycle was considerably enhanced. In slightly more than two years, the Tomahawk targeting process had been reduced from 101 minutes in Operation Allied Force to 19 minutes in Enduring Freedom. Despite these new technological possibilities, 10 opportunities to attack senior Taliban leadership did not receive clearance in time.[155]

Among the Continental European air forces that provided combat aircraft to later phases of Operation Enduring Freedom were France and a combined Danish, Dutch, and Norwegian detachment. Moreover, during the second half of December 2001, combat aircraft from the French carrier *Charles de Gaulle* and the Italian carrier *Garibaldi* joined the air campaign.[156]

Third-Party Intervention in Failed States. With the end of the Cold War, states started to disintegrate not only in Europe but also elsewhere, particularly in Africa. As a consequence, Western powers intervened several times attempting to stabilise the situation. In these conflicts, air power executed an enabling and supporting role, such as power projection, intratheatre airlift, reconnaissance, and CAS.

In 1993 a multinational force attempted to restore order in Somalia. This operation came to an abrupt halt when US transport helicopters were shot down by a warlord's faction on 3 October 1993, resulting in 18 American Soldiers killed. The Clinton administration subsequently ordered a phased withdrawal.[157] Soon afterwards, genocide unfolded in Rwanda. In response, the French embarked upon Operation Turquoise, aimed at stabilising the situation in western Rwanda.[158] This and other French operations in Africa are explored later. Sharing a similar colonial legacy, the UK also became involved in conflicts in Africa. In 2000, for instance, Britain dispatched a battle-group-sized force with corresponding air and maritime support to civil-war-torn Sierra Leone.[159] Essential for Operation Palliser was rapid deployment, which could be achieved only through airlift.[160] In recent years, the European Union has taken on an increasing responsibility for the conduct of joint military interventions on behalf of the UN.

Outside Africa, it was primarily Haiti and East Timor that saw Western third-party intervention in the wake of internal tensions and clashes.[161] In both cases, air power proved crucial for timely power projection.

Air Policing and Constabulary Tasks. The events of 11 September also put an unprecedented emphasis upon aerial constabulary tasks and reemphasised the importance of national airspace protection. In the aftermath of the terrorist attacks against the United States, Western air forces undertook special efforts to protect national airspace. Furthermore, though air power alone cannot fight terrorism, it is regarded as an important enabler for the other services in combating terrorism, particularly through the gathering of intelligence.

Humanitarian Relief Missions. Throughout the post–Cold War era, military airlift has played a major role in disaster relief missions. These disasters have been caused by civil wars, famine, or natural catastrophes, and they have required responsive delivery of medical supplies, food, shelter, and mobile communications. As is shown in the ensuing chapters, Continental European air forces have been heavily engaged in humanitarian relief missions by providing airlift. Yet, disaster relief missions are not relegated to airlift alone. In 1996, for instance, an RAF aerial reconnaissance detachment was dispatched to former Zaire in support of Operation Purposeful. An assessment of aerial imagery revealed the refugees' requirements.[162]

Air Power Doctrine and Thinking

In British parlance, *military doctrine* is defined as a collection of fundamental principles that serve as a guideline for military operations but should not be adhered to dogmatically.[163] The second half of the Cold War era has often been characterised as a period inimical to doctrinal thought, with the primary concern of airmen relegated to tactical and technical issues. In contrast, the end of the Cold War together with Desert Storm provoked a major surge in conventional air power thinking both within and outside the air forces of the Anglo-Saxon world. An early indication of this rising air power debate was John Warden's book *The Air Campaign: Planning for Combat*, published in 1988, on the precipice of the end of the Cold War. However controversial Warden's ideas have become, his book contributed considerably to the reemergence of a doctrinal debate embracing air power in its entirety instead of focusing upon single air power roles.[164]

We next review the evolution of air power doctrine in the USAF, RAF, and RAAF. This provides a yardstick to examine the doctrine development process in Continental European air forces and also helps to illuminate to what extent Anglo-Saxon air power thinking has influenced Continental European thinking. The USAF was chosen for its leading role in air power and the RAF for its prominent air power history. The RAAF is an example of a relatively small- to medium-sized air force that—despite its size constraints—has made significant contributions to air power thinking. The discussion concentrates upon basic air power doctrine—the so-called strategic doctrine—as it is the link between national defence policy and the operational level. We also consider the evolution of air power teaching as well as the dissemination of air power thinking within these air forces.

The resurgence of intellectual air power thinking within these air forces has also been accompanied by scores of books and articles written by scholars and airmen alike. In the aftermath of Desert Storm, the literature chiefly reflected the intense debate over the effectiveness of air power. The air campaign seemed to prove the views of air power zealots, who believed in the distinct effectiveness of air power and who voiced their standpoints in a number of articles. From their point of view, the distinct employment of air power against leadership-related target sets is synonymous with the term *strategic*.[165] Yet, soon these views began to be contested by critics such as Robert A. Pape, who argued that the true value of precision air power lies in integrated air-land operations and not in distinct strategic air campaigns.[166] Besides these air power controversies, Anglo-Saxon authors also presented more balanced and encompassing views on air power evolution.[167] Primary references for this research were limited to official and unofficial writings within the air forces.

Specific topics have emerged in the post–Cold War air power debate. In the aftermath of Desert Storm, the concept of parallel warfare came to the fore, and in the second half of the 1990s, the term *effects-based operations* (EBO) began to dominate the doctrinal debate. Both concepts imply that achieving strategic goals is within the sphere of not only nuclear forces but also conventional air power. PGMs proved so successful in Desert Storm that they offset the need for mass attacks. Many different target sets could be attacked simultaneously versus sequentially, hence, the term *parallel warfare.* This approach has been thought to create paralysis within the enemy system.[168]

In the post–Cold War era, air campaign planners have also been confronted with various peace support and coercive operations, where pure military victory has not been the primary strategic criterion. This situation has in turn led to the EBO debate in air power doctrine circles. While during the Cold War, strategic effect was relegated to the nuclear strike role, the term *strategic* had to be extended in the 1990s to make it fit the new environment.[169] It no longer connotes only the kinetic effect of air power but instead refers to all of its roles.[170] According to current NATO air power doctrine, it is the effect on the strategic objective that determines the strategic nature of such operations (air operations for strategic effect) and not the range, type of platform, or weapon used.[171] It was further suggested that it was no longer useful to strictly distinguish between tactical and strategic target sets: strategic effect can be achieved by hitting targets such as fielded forces and their logistics tail.[172]

Air Power Doctrine Development

The USAF released its first post–Cold War basic aerospace doctrine, Air Force Manual (AFM) 1-1, in March 1992. It superseded AFM 1-1 from 5 January 1984 and served to usher the way into a new air power era by formalizing the Gulf War air campaign lessons. As such, it highlighted aerospace power's inherent capability to rapidly or simultaneously engage objectives at any level from strategic through theatre to tactical, thereby embracing the concept of parallel warfare, though not explicitly mentioning it.[173] It was particularly Maj Gen David A. Deptula who conceptually established the concept of parallel warfare in the ensuing years.[174] *Strategic attack* was defined as "achiev[ing] maximum destruction of the enemy's ability to wage war."[175] This definition implied a shift from a concept based on target types to a concept based on combat out-comes and thus anticipated the debate over EBO.[176] Moreover, parallel and independent aerospace campaigns centrally con-trolled by an airman were considered to be most effective.[177] Do-ing so, AFM 1-1 particularly highlighted the success of the joint force air component commander (JFACC) concept as employed in Desert Storm.

To institutionalise the doctrine development process, the Air Force Doctrine Center (AFDC), with headquarters at Maxwell AFB, Alabama, was established in February 1997. Ten years later, the AFDC was merged with the College of Aerospace Doctrine, Research, and Education into the LeMay Center for Doctrine De-velopment and Education. The AFDC was responsible for re-searching, developing, and producing basic and operational doc-trine and for reviewing doctrine education. Moreover, the centre participated in developing and investigating future aerospace concepts and technologies. In September 1997, the AFDC re-leased a revised edition of *Air Force Basic Doctrine*.[178] The so-called Air Force Doctrine Document 1 (AFDD 1) was approved by Gen Michael E. Ryan, the newly appointed USAF chief of staff, who had been commander, Allied Air Forces Southern Europe (NATO), between September 1994 and April 1996 and who was the combined joint force air component commander (CJFACC) during Operation Deliberate Force. So it was not surprising that the document took into account the employment of air power in peace support operations, referred to as "military operations other than war."[179] The 1997 edition also introduced the term *strategic effect*. In accordance with this concept, strategic attack is not aimed at particular target sets but at the "adversary's capability to continue the conflict."[180] Two years earlier, Deliberate Force was perceived to produce strategic outcomes by concentrating strikes at the mobility and command infrastructure of the Bosnian Serb army, without having to go beyond the immediate area of the ad-versary's ground operations.[181] Given that American involvement in the Balkans primarily hinged upon air power, the 1997 edition not surprisingly identified Air Force operations as the potentially most decisive force in demonstrating the nation's will to counter an adversary's aggression, thereby underlining the distinctiveness of the service with regard to the other services.[182]

The next edition of AFDD 1 in November 2003 was designed to provide guidance for the application of air power in the war on terror. According to Gen John P. Jumper, key themes included contingency operations, network-centric warfare, and American air power's contribution to the joint battle.[183] Regarding the latter, AFDD 1 argues that effectively integrating the four service branches' capabilities remained pivotal to successful joint war fighting.[184] This stands in stark contrast to the 1992 AFM 1-1, which emphasises the effectiveness of air power in relatively independent aerospace campaigns. Operational experience in Afghanistan and Iraq in 2003—as opposed to Desert Storm with a distinct air and ground campaign—certainly caused this shift in doctrine.

Devoid of any national strategic air power doctrine, the RAF began to realise towards the end of the Cold War the importance of establishing a coherent and systematic doctrine to think about air power in overall and integrated campaign terms. This differed from the stand-alone roles characterising the RAF's presentation of air power since the 1957 defence review and the handover of principal responsibilities for the nuclear deterrent to the Navy.[185] Against the backdrop of inherent scepticism within the RAF towards doctrine and concern amongst senior RAF officers that a national doctrine might conflict with NATO manuals, an attempt at producing a new RAF doctrine manual was undertaken in 1988. With the approval of the assistant chief of the Air Staff, the first edition of the RAF air power doctrine, Air Publication (AP) 3000, was produced with a limited print run.[186]

The doctrine was finally published in 1991. In his foreword, Air Chief Marshal Sir Peter Harding, chief of the Air Staff, identifies two fundamental reasons for establishing a national air power doctrine. First, quantum leaps in air power technology had made a conceptual grasp of the new capabilities and limitations for informed doctrinal discussion indispensable. Secondly, according to Harding, the new doctrine was supposed to amalgamate past experience with forethought to prepare airmen to meet the challenges of the day.[187] A reference to the turmoil in the international arena was not yet made, showing that the doctrine was still embedded into a Cold War type of setting. Typical Cold War features of AP 3000 were its emphasis upon offensive counterair missions using the JP233 airfield attack system, particularly designed for attacks against Warsaw Pact airfields; layered air defence; and nuclear strike operations. Also, regarding the spectrum of conflict, AP 3000 bore the hallmarks of a Cold War paradigm. While it identified nuclear, non-nuclear, and insurgent warfare, no explicit reference was made to peace support operations.[188] Besides these Cold War vestiges, however, the quantum leap, brought to light by the Gulf air campaign, was cautiously being grasped by referring to the outstanding effectiveness of PGMs.[189]

Regarding AP 3000, Air Chief Marshal Harding particularly identified three specific aims: "To foster a more cohesive approach to air power education within the Service; second, to be the foundation of our contribution to the formulation of joint-service doctrine, and alliance doctrine with NATO or other allies; and, finally, to enhance the understanding of air power within our sister-Services, the Civil Service, Parliament and the general public."[190]

Though conceived during the last months of the Cold War, AP 3000 nevertheless provided a sound starting point for post–Cold War doctrine development. The fundamental changes in the international arena as well as Desert Storm made a revision of AP 3000 necessary. In the eyes of the then-director of defence studies RAF, the Gulf War air campaign opened eyes that had previously been closed and as such was a statement of progress.[191] Consequently, within two years the RAF published AP 3000's second edition. In its foreword, Air Chief Marshal Sir Michael Graydon, chief of the Air Staff, particularly highlighted the importance of recent operational experience for the doctrine development process: "Since AP 3000's first edition was published, the RAF's air power doctrine has been reviewed and refined in the light of operational experience in the Gulf and elsewhere."[192] While the framework in terms of chapter structure remained, themes such as the role of air power in managing international crises or the effects of strategic air campaigns were newly introduced or redefined, underlining the perceived effectiveness of simultaneous strikes against a number of target sets.[193] Despite these amendments, vestiges of a NATO Cold War setting can be found, such as the concept of the layered air defence of the UK or of a strategic nuclear campaign.[194]

AP 3000's third edition, published in 1999, was an effort to harmonise British air power doctrine with the expeditionary strategy, as laid down in the *Strategic Defence Review* of 1998, and it anticipated the development of joint doctrine and defence structures.[195] Accordingly, Air Chief Marshal Sir Richard Johns, chief of the Air Staff, argues in the foreword that jointness and multinational operations must be a central theme to the generation of air power.[196] In this vein, three entire chapters in the 1999 edition were dedicated to joint force employment. Furthermore, its subtitle *British Air Power Doctrine*, as opposed to the previous subtitles *Royal Air Force Air Power Doctrine*, underlined a single environmental focus: establishing the RAF's role as the prime custodian of British air power thinking.[197] Besides this thrust towards jointness, AP 3000's third edition introduced the concept of "air operations for strategic effect," thereby reflecting the debate on effects-based operations.[198] Since this edition predated much of contemporary UK joint doctrine, it also contained large segments that are generic across defence studies, such as the principles of war.[199]

In comparison to its predecessor, which had not been firmly embedded into a doctrine hierarchy, AP 3000's fourth edition aimed at distilling the underlying philosophy of air power into a concise format by complementing joint doctrine rather than replicating it.[200] One of the chief reasons to produce a revised edition of AP 3000 was the need to reflect contemporary and enduring operations doctrinally.[201] The revised AP 3000's subtitle of *British Air and Space Power Doctrine* particularly underlines the importance of space to modern Western military operations.[202] In this domain, the doctrine is relatively explicit about the UK's dependency upon US assets.[203]

The RAAF for its part embarked upon an innovative air power doctrine development process towards the end of the Cold War. Being aware of the importance of sound air power theory and doctrine, the chief of the Air Staff ordered the formation of the RAAF Air Power Studies Centre in August 1989 at RAAF Fairbairn. The centre was renamed Aerospace Centre in 2000 and then Air Power Development Centre in 2004, but the primary purpose remained the development of strategic air power doctrine and the promotion of air power thinking. The centre developed the RAAF's first air power manual in 1990, which at that time was the second of its type in the world.[204] Describing air power as the dominant component of combat power in modern warfare, Air Marshal R. G. Funnell, chief of the Air Staff, declared in the foreword that "the Royal Australian Air Force cannot fully discharge its responsibility to the nation until it can, through rigorous analysis, explain the best use of air power. The first essential step in this responsibility is the establishment of Australian air power doctrine."[205] Though still embedded in a Cold War context, the air power doctrine manual actually anticipated the coming of age of air power as experienced in the Gulf War air campaign. It also argued that conventional air power has developed a capability far beyond the mere extension of surface forces.[206] The RAAF produced several revised editions in the ensuing years.[207]

The NATO alliance has published air power doctrine documents since the days of the Cold War. These doctrine documents have been jointly agreed upon by member states, and they have left, as one RAND analyst argued in the mid-1980s, considerable room for national interpretation.[208] NATO alliance air power doctrine cannot be examined in depth, as not all standardisation documents are publicly available.[209] Generally, it can be concluded that in the 1980s, NATO tactical air doctrine was almost exclusively designed to deter and counter an attack by Warsaw Pact forces.[210] In the late 1990s, NATO allied air power doctrine in particular began to elaborate on non-Article 5 crisis response operations.[211] Moreover, it embraced the concept of EBO.[212] Institutionally, NATO's doctrine and concept development was strengthened by the foundation of the JAPCC in 2005, as previously mentioned.

Teaching of Air Power Thought

In the United States, the USAF's leading school in air power theory and history is the School of Advanced Air and Space Studies (SAASS). SAASS runs a master's degree programme specifically structured to prepare a select group of officers to become the military's future air and space power strategists.[213] A catalyst for setting up high-quality teaching institutions within the US armed forces and particularly within the USAF was the frustrations of the Vietnam War. During the late 1970s and early 1980s, courses and curricula at the USAF's Air University (AU) were thoroughly reexamined and changed. As a consequence, the number of civilian faculty positions was increased and greater emphasis was placed on research and publication. SAASS originated from this revision process and was officially established in 1988.[214] Its curriculum encompasses classical military, air power, and space power theory and history. During this intensive 48-week programme, weekly reading loads average 1,200–1,500 pages.[215]

Moreover, SAASS has actively published, thereby establishing a linkage between air power education and the dissemination of air power thought to a broader audience. Of particular interest was the 1997 publication *The Paths of Heaven*, edited by the SAASS commandant and with individual chapters written by members of the SAASS faculty. It is considered to be the first definitive history of the evolution of air power theory.[216] Furthermore, individual SAASS faculty members have published books and have made major contributions to the professional literature in various journals on international security, strategy, and defence issues.[217]

In the UK, a major reform in the education of selected midcareer officers developed in autumn 1997. Marshalling this change was, first, a thrust towards jointness by collocating the three services' advanced command and staff courses at the RAF Staff College in Bracknell and, second, the integration of academics from the newly founded Department of Defence Studies, King's College London, into the delivery of air power teaching. Six weeks of the Advanced Command and Staff Course were exclusively devoted to an air power module encompassing air power history, theory, and campaign planning. The module itself was divided into two parts. While the academics focused upon air power history and theory, the military teaching staff dealt with air power doctrine and campaign planning—a fruitful military-academic symbiosis. In September 2000, new teaching facilities were provided at the newly established Joint Services Command and Staff College (JSCSC) in Shrivenham, ushering in an ever greater thrust towards a joint curriculum. Since late 2005, King's College London has also been providing academic education to initial officer training at the RAF College, and in 2009 staff based at the college developed a distance-learning master's degree, air power in the modern world.[218]

The latest edition of AP 3000, *British Air and Space Power Doctrine*, predominantly targets a military audience at the staff training establishments. A link between doctrine and the teaching of air power theory and concepts is clearly reflected.[219]

With the setting up of the Air Power Studies Centre in August 1989, the RAAF embarked not only on an innovative approach to doctrine development but also on teaching air power doctrine, theory, and history. Presently, the RAAF Air Power Development Centre runs the Advanced Air Power Course, designed to give a greater understanding of air power theory and doctrine. The course covers air power topics from World War I to the present day. Its quality is strengthened by the provision of academic supervision, ensuring a productive link between the military and academia.[220] Moreover, the chief of Air Force fellowships has fostered proactive air power thinking. The fellowship programme was established in 1990 to develop greater awareness of Australian air power. It offers the opportunity to spend considerable time preparing a piece of substantial work relevant to Australian air power. The programme is also open to representatives from regional air forces and to those of Australia's major allies, thereby fostering mutual understanding and a combined approach to air power theory, doctrine, and history.[221]

Dissemination of Air Power Thought

The USAF has been at the leading edge of furthering the discussion of air power concepts. One vehicle for this dialog is the *Air and Space Power Journal*, first published in 1947. Through this journal, the Air Force Research Institute at Maxwell Air Force Base, Alabama, endeavours to promote worldwide air and space power thought. The journal is now published not only in English but also in five other languages, thereby reaching out to a vast audience. The English version is intended to foster a professional and intellectual dialogue within the USAF.[222] Another important institution for the dissemination of American air power thought is the Air Force Historical Foundation, founded in 1953. It is dedicated to the preservation and publication of the history of American military aviation. For this purpose, the foundation's activities include a book programme, an awards programme, and symposia.[223] It also publishes the journal *Air Power History*.[224]

By combining doctrine development with semiofficial publications, the RAF embarked upon an innovative path in the 1990s. Concurrently with the development of AP 3000's first edition, for instance, the director of defence studies RAF published a collection of essays in early 1990 on the doctrine and air power of various nations.[225] In 1994 the chief of the Air Staff established the air power working group, initially bringing together air force officers and senior academics but later expanded to include representatives from the British Army, Royal Navy, USAF, and RAAF. The goal was not only to promote air power thinking in the wake of Desert Storm but also to advocate the British view on air power within the UK as well as with the USAF and the RAAF. The first workshop resulted in the publication *The Dynamics of Air Power*.[226] It examined evolving air power theory and the role of air power in peace support operations, a highly topical area given the operations over the Balkans.[227] In the ensuing years, a succession of directors of defence studies RAF compiled and edited a growing body of topical doctrinal literature evolving from the chief of the Air Staff's air power workshops, which had emerged from the air power working group.[228] Besides the activities of the air power working group and the workshops, air power conferences have been organised not only to promote air power thinking or to inform initial thinking on recent campaigns but also to influence British decision makers.[229]

Another significant step in broadening the doctrinal debate was the creation of the *Royal Air Force Air Power Review* in 1998. The first issue's opening pages point out the RAF's historical role as being in the vanguard of air power thinking as well as the journal's intent to foster that leading posture through robust and informed debate. The journal has been distributed free of charge across and beyond the service.[230] This platform was intended to provide an open forum for study, stimulating discussion and thinking on air power in its broadest context.[231]

One of the UK's most recently established vehicles for promoting air power thinking is the Royal Air Force Centre for Air Power Studies. Established in August 2007, it synergistically draws upon the competencies of the Air Power Studies Division of King's College London, the Directorate of Defence Studies RAF, and the Air Historical Branch RAF. The centre aims to strengthen the relationship between academia and the RAF.[232]

The post–Cold War era has also seen the reemergence of major analytical surveys analogous to World War II's *United States Strategic Bombing Survey*. The *Gulf War Air Power Survey* endeavoured to compile the conduct and the lessons of the Gulf War air campaign in five volumes and a summary volume. Sponsored by the USAF, the study was headed by an academic and harnessed people from both academia and the military.[233] Another USAF-sponsored survey in the ensuing years includes an examination of the campaigns over Bosnia, published in January 2000.[234]

Force Structuring and Procurement

Western air campaigns of the post–Cold War era reveal a large discrepancy between American and European air power. Inadequate equipment largely accounts for this variance.

Approaches to Procurement

While Europe as a whole is not far behind the United States in defence spending, its military capabilities nevertheless lag significantly behind. Duplications, lack of integration, and retaining Cold War structures within Europe have all contributed to this gap.[235] In several European countries, defence planners continued throughout the 1990s to focus upon worst-case scenarios—including a major conventional military onslaught on Western Europe—instead of shifting their primary focus to the most probable tasks.[236] The issues of retaining Cold War structures and inefficient spending, two related subjects, have been aggravated by a price escalation for sophisticated weapons and shrinking defence budgets. Price escalation has particularly related to combat aircraft, as each aircraft generation has been considerably more expensive in real terms than the one it replaced.[237] This surge in costs has been coupled in the post–Cold War era with extreme delays in aircraft programmes. Financial, technical, bureaucratic, and political problems have created routine deferments of up to 10 years.[238] According to distinguished British defence analysts, while defence planners are not unfamiliar with affordability issues, they can no longer be disguised or circumvented.[239] Further obstacles to military innovation in Continental Europe have been the limited resources devoted to research and development, with the exceptions of the UK, France, and Sweden.[240]

With the end of the Cold War, a major shift from threat-based to capability-based force structuring has occurred. To retain freedom of action, the UK and France have aspired to cover as full a spectrum of capabilities as possible.[241] Yet, owing to technological and financial constraints, no other single state can cover an air power panoply comparable to that of the United States.[242] In 2000, after significant post–Cold War cuts, the United States could still muster over 4,500 combat aircraft as well as over 1,500 transport and air-to-air refuelling aircraft.[243] To preserve at least certain relevant capabilities, it has been argued that some NATO member states might consider pursuing a path of role specialisation despite the inherent drawbacks of this policy such as a diminution in their political autonomy and in their scope for exercising discretion.[244]

National Autarchy versus International Cooperation. Simultaneously with these developments, industrial autonomy has become elusive. Maintaining an industry that designs and produces only small numbers of sophisticated weapons would yield products that are cost prohibitive. Even Sweden, which placed a premium upon industrial autarchy, has undergone significant defence economic changes in the post–Cold War era. Regarding the latest Swedish combat aircraft design, the Gripen, essential components had to be imported from overseas, and one of the Swedish defence industry's key roles has been to act as a system coordinator and integrator.[245] This led one commentator to conclude that second-tier states appear to be more dependent than ever on the first-tier producers for critical technologies, components, and capital.[246] France has been the only Western European state in the post–Cold War era that has pursued the development and production of its latest combat aircraft, the Rafale, as an exclusively national venture.[247] But when it comes to specialised assets such as airborne early warning aircraft, France has had to rely upon American supply as well.

The FAF operates four E-3 AWACS aircraft.

Besides national ventures, international cooperation has always been an important consideration in European aircraft development. For instance, Britain, Germany, and Italy joined together to develop and produce the most important European all-weather strike aircraft of the Cold War, the Tornado. This project has been followed by the currently most important European combat aircraft programme, the Eurofighter Typhoon. Planning started in the UK as early as 1979. By the mid-1980s, France, Germany, Italy, and Spain had joined the programme. The early stages of the programme were plagued by disagreements over operational requirements, and France finally ceased its participation in the multinational programme in August 1985.[248] In the ensuing years, the Eurofighter programme has been exposed to severe delays "caused by disagreements over specification, work-sharing, the placing of contracts internationally, and by political considerations in more than one country."[249] Despite the inherent drawbacks of international cooperation, strong voices in Europe have argued that the future successor of the Rafale, Eurofighter, and Gripen has to be a single aircraft programme, as Europe will no longer be able to afford a fragmentation of its aerospace industry.[250]

In the domain of airlift, European nations have been pursuing a cooperative path in the A400M programme. Multilateral efforts to develop a successor for the C-130 Hercules and the C-160 Transall began in 1982. This joint venture underwent several changes not only in the managerial structures and number of participating nations but also in the project and aircraft names. In the second half of the 1990s, Airbus Industries was officially handed the main responsibility for the programme. The programme was also affected by several political problems. In 1997, for instance, the German defence minister suddenly pushed ahead with the option of developing a common European transport aircraft together with Russia and the Ukraine on the basis of the Antonov An-70.[251] Later on, Germany proved reluctant to commit itself to a definite order and thereby delayed the whole programme, whereas France and the UK were pressing for early production.[252] In 2003, when a multinational order was submitted, first deliveries were expected before the end of October 2009. Yet due to serious delays in the programme, as of 2009, the earliest possible deliveries were expected not to take place before late 2012.[253] Despite this setback,

Courtesy Airbus Military; photo by S. Ramadier, Airbus Operations

A400M on its first flight. The A400M will significantly enhance European intertheatre airlift capacities.

the European defence ministers involved in the A400M programme commonly declared in July 2009 their intention to continue the project.[254]

Off the Shelf. An alternative to European aircraft programmes is buying American or Russian. This option, however, has serious industrial and political consequences, such as potential US market domination, which are likely to widen the transatlantic air power capability gap. Buying Russian is often incongruent with common European defence interests, and supply of spare parts might not meet Western reliability standards.[255] Nevertheless, buying off the shelf has proved to be a cost-effective solution, particularly for smaller European states. Moreover, instead of being involved in long-lasting development programmes, these states have been able to quickly react to urgent operational demands by buying on the international market. Surveys of the 1980s indicated that European co-development of aircraft might cost from 15 to 35 per cent more than hypothetical direct purchases from US firms.[256] In the absence of a European aerospace industry, however, US firms might have had little incentive to charge less than monopoly prices.[257] Regarding the latest American combat aircraft design, the F-35 Joint Strike Fighter (JSF), a new approach has been embarked upon concerning foreign sales. The US government offered international participation in the programme, allowing foreign partners to participate in an advanced combat aircraft programme at relatively low costs but as an unequal partner. The leadership is clearly with the United States.[258]

Leasing, Chartering, and Life Extension of Airframes. In the face of severe delays in aircraft programmes, chartering or leasing provides interim solutions. In 2001 the RAF leased four C-17 Globemasters from the Boeing Company with the option to buy or extend at the end of a seven-year period.[259] Lacking adequate airlift capacities, the German armed forces chartered Ukrainian and Russian An-124 transport aircraft for operations in Afghanistan.[260]

Moreover, given the severe delays and cost escalation in current aircraft programmes, many air forces have been obliged to extend the life of their current inventories. Upgrading and modernization can offer either improvement in an existing role or an interim capability.[261] In the post–Cold War era, upgrade programmes have presented both a necessity and a dilemma. While they have served as necessary gap fillers, they have simultaneously contributed to further delays of the major aerospace programmes by diverting already scarce resources.

Shared Ownership. For certain excessively expensive key assets, shared ownership has long been pursued as a necessary option in the context of NATO. The prime example is the NATO Airborne Early Warning and Control Force. Through the NAEW&CF, NATO members were able to acquire an airborne early warning capability at much lower day-to-day operating costs than would have been the case if operated on an individual national basis.[262] As regards wide area air-to-ground surveillance, NATO has launched the Alliance Ground Surveillance (AGS) Core project. Similar to the NAEW&CF, AGS Core is intended to be procured and operated by NATO but financed by alliance members. The system originally was supposed to integrate synthetic aperture radar on the basis of the Airbus A321 and the high-altitude, long-endurance remotely piloted aircraft (HALE RPA) Global Hawk.[263] In the meantime, however, the projected air segment of AGS Core had been reduced to its HALE RPA component. On 25 September 2009, 15 NATO nations signed an MOU, and AGS Core is anticipated to be operational as of 2012.[264]

Another NATO initiative is the Strategic Airlift Capability programme. A letter of intent to commence contract negotiations to multilaterally acquire C-17 Globemasters was signed in September 2006. Apart from 10 NATO member countries (Bulgaria, Estonia, Hungary, Lithuania, the Netherlands, Norway, Poland, Romania, Slovenia, and the United States), participants include Finland and Sweden, both Partnership for Peace nations.

Courtesy US Air Force

NATO C-17 Globemaster III from the multinational Heavy Airlift Wing, Papa AB, Hungary, makes the new unit's first landing at Baghdad International Airport, Iraq. The airlift into Iraq was a first by the wing and facilitated the deployment for members of the NATO Training Mission–Iraq.

The first of three C-17 Globemasters arrived at its home air base in Papa, Hungary, on 27 July 2009.[265] Concurrently, the European Defence Agency (EDA) began to explore the possibilities of pooling acquisition, maintenance, and training in case of additional A400M purchases.[266] In the long term, these efforts might result in a multinational A400M unit.[267]

NATO and EU Initiatives. NATO has attempted to improve military effectiveness not only through shared ownership but also through strategic initiatives. At the Washington summit meeting in April 1999, NATO's Defence Capabilities Initiative (DCI) was launched. It was primarily designed to improve alliance defence capabilities in combined and deployed operations.[268] Though a biannual NATO Force Goal Process had already attempted to address apparent shortfalls, increased political attention could be gained through the DCI.[269] In the domain of air power, the DCI particularly identified capability gaps in areas such as strategic airlift, combat search and rescue (CSAR), AAR, airborne standoff jamming, and air-to-ground surveillance systems.[270]

Three and a half years later, at the November 2002 summit in Prague, NATO members adopted a new capabilities initiative, the Prague Capabilities Commitment. The new initiative differed from its predecessor in that individual allies made firm political commitments for improvements. Air power requirements were basically reiterated, with an added emphasis upon improvements in combat effectiveness, including PGMs and SEAD.[271]

In parallel, the EU has undertaken Helsinki Headline Goal and HG 2010 initiatives aimed at improving its military crisis management capabilities.[272] Identified areas of air power shortfalls were generally identical to those that the NATO initiatives had pinpointed—strategic air mobility, AAR, PGMs, theatre ballistic missile defence (TBMD), and RPAs.[273] HG 2010 laid the foundation for the EDA, established to coordinate development, research, and acquisition.[274] On 10 November 2008, the EDA facilitated a meeting of European defence ministers, who declared their intention of establishing a future European Air Transport Fleet. Its objective is to provide a framework for facilitating European cooperation through such existing structures as the European Air Transport Command.[275]

The Pillars of Air Power

The physical components of air power are essentially made up of four pillars—combat aircraft; air mobility; command, control, communications, computers, information/intelligence, surveillance, target acquisition, and reconnaissance (C4ISTAR); and ground based air defence. While the first pillar has always been regarded as the backbone of air forces, the second and third pillars have gained in importance. This has to do with the increasing thrust towards deployed operations and with technological developments in the information age. The fourth pillar, ground based air defence, has also experienced an evolution, with its tasks having been extended from primarily defending against combat aircraft to defending against a panoply of airborne threats, especially theatre ballistic missiles.

Combat Aircraft. Though Western fighter designs of the 1960s and the 1970s were built to perform secondary ground-attack roles, this supposedly multirole capability did not stretch much beyond qualifying a once pure fighter to deliver air-to-ground munitions with mediocre accuracy. Complex air-to-ground missions still required dedicated all-weather strike aircraft such as the A-6, the F-111, or the Tornado. Only since the later stages of the Cold War era have true multirole aircraft designs become a viable option due to a number of remarkable technological achievements. This step was further facilitated through the introduction of PGMs because they allow even relatively small aircraft to achieve very good weapons-delivery accuracy.[276]

This thrust towards multirole aircraft is also reflected in the latest European combat aircraft programmes. The French Rafale and the Swedish Gripen were designed from the outset as true multirole aircraft. The Eurofighter Typhoon was originally conceived as an air superiority fighter with only secondary ground-attack capability. A reason for this lay in the fact that three out of the four partner nations had already been operating the very capable Tornado in the attack role. However, the practical experiences of the post–Cold War era as well as considerations for potential export customers have made the development of a truly multirole fighter-bomber necessary.[277] Alongside these European programmes, the F-35 JSF is likely to become the future backbone of a number of European NATO air forces.[278]

Courtesy FAF, Sirpa Air

Rafale flying over Afghanistan armed with *armement air-sol modulaire* (AASM), or modular air-to-surface armament, for precision air strikes

Nowadays, many European air forces, such as the RAF, see multirole fighters as a necessity both for financial and operational reasons. Only multirole aircraft can meet the increasing demands of flexibility that have resulted from changes in the overall security environment.[279] Regarding costs, multirole aircraft have the potential to significantly reduce the logistical tail as compared to a fleet of two types of specialised aircraft. To exploit the full potential of multirole aircraft, however, they have to be combined with multirole weapons and crews. Multirole crew training poses a major challenge as it requires more total flying hours and possibly less for each role.[280] Accordingly, RAF sources indicated in the early 1990s that not all Typhoon squadrons would be trained in more than one role, based partly on the realization that it is difficult to achieve high proficiency in all roles.[281] Besides multirole features, current Western aircraft programmes emphasised beyond-visual-range air combat performance and good manoeuvrability, combined with a night-vision capability for various roles such as aerial reconnaissance, target detection, and precision strikes.[282]

Russia has experienced parallel trends in combat aircraft development in recent years. The MiG-29 and the Sukhoi Su-27, originally conceived as air superiority fighters during the Cold War, have been further developed into true multirole aircraft with appropriate avionics and armament.[283] Disadvantages of integrating MiG-29 aircraft into a NATO air force are discussed in chapter 4, "German Air Force (Luftwaffe)."

The post–Cold War era has seen the emergence of various European weapons, significantly decreasing dependency upon the United States. The ubiquitous American short-range air-intercept missile AIM-9L Sidewinder, for instance, is being supplanted in many European air forces by the Infrared Imaging System—Tail/Thrust Vector-Controlled (IRIS-T) air-to-air missile (AAM). The IRIS-T is the outgrowth of a multinational European programme that began in 1997.[284] In the field of long- and medium-range AAMs, the Meteor missile is planned to offer an alternative to the American AIM-120 advanced medium-range air-to-air missile (AMRAAM) after 2010. Like IRIS-T, the Meteor programme is a European cooperative endeavour, with Great Britain being the lead nation.[285]

For air-to-ground weapons, the evolving strategic and political conditions of the post–Cold War era have rendered weapons particularly designed for Cold War scenarios obsolete, as was the case with the British JP233 airfield attack system. Originally optimised for the specific role of low-level counterair missions against Warsaw Pact airfields in Central Europe, JP233 provided, according to a high-ranking RAF officer, less than satisfactory outcomes during Operation Desert Storm, and eventually fell victim to the worldwide ban against anti-personnel mines.[286] On the other hand, precision air-to-ground weapons had a major impact upon aerial warfare. Though PGMs accounted for less than 10 per cent of the total air munitions tonnage dropped during Desert Storm, it was the first conflict to be fought with an extensive use of guided weapons. At the end of the 1990s, Operation Allied Force was planned to be fought almost exclusively with PGMs. However, the expansion of the air campaign made reserve stocks dwindle fast, which led to an increased use of unguided air-to-ground munitions. These campaigns clearly revealed the need for cheap all-weather munitions such as the Joint Direct Attack Munition (JDAM), a GPS-directed bomb.[287] Besides precision-guided bombs, various standoff cruise missiles are being added to the arsenals. European designs include the Franco-British-Italian Storm Shadow/Scalp or the German-Swedish Taurus KEPD 350 cruise missiles.[288]

The SEAD mission poses a particular challenge to Western air forces, as dedicated SEAD aircraft are scarce assets. Moreover, specific anti-radiation missiles such as the American high-speed anti-radiation missile (HARM) or the British air-launched anti-radiation missile (ALARM) are required.[289]

Air Mobility. Military airlift has basically been split into two categories—tactical airlift (intratheatre) and strategic airlift (intertheatre).[290] The former was typically covered by medium-sized aircraft, such as the C-130 Hercules or the C-160 Transall, capable of very good takeoff performance from austere airfields. These aircraft are capable of delivering maximum loads of 15 to 20 tons, including armoured vehicles. Significant limitations regarding width and height of the cargo bay, however, restrict carrying bulky items such as medium-sized helicopters.

Heavy strategic airlift, on the other hand, was the domain of the USAF and, to a lesser extent, of the former Soviet Air Force during the Cold War. It required the development of large aircraft such as the Lockheed C-5 Galaxy or the Antonov An-124 Condor, capable

C-5 Galaxy. A USAF ground and air crew loads a US Army CH-47D Chinook helicopter into a C-5 Galaxy aircraft at Joint Base Balad, Iraq, in August 2008.

of lifting nearly every bulky piece of equipment in service with the respective armed forces.[291] In contrast, Western European armed forces did not require these capabilities since they were supposed to fight "in place."[292]

Yet after the Cold War, with a significant increase in out-of-area operations, European air forces have suffered from a shortage in strategic airlift capacities. To partially bridge this capability gap, the ubiquitous C-130 and C-160 have been used in the inter-theatre airlift role. Even at the end of the 1990s, Western Europe did not own a single military wide-body, long-range strategic transport aircraft capable of lifting a main battle tank.[293]

Current European requirements actually blur the dividing lines between tactical and strategic airlift. The A400M programme aims at a military transport aircraft capable of lifting 30 tons for approximately 4,500 kilometers (km) and 20 tons for 6,400 km while at the same time offering similar tactical performance to a C-130 Hercules. Moreover, its cargo bay is expected to accommodate bulky vehicles such as medium-sized helicopters. Six European air forces originally ordered 170 aircraft.[294] Though critics argue that the A400M is not really a strategic transport aircraft, it is going to significantly enhance European intertheatre airlift capacities.[295]

Intratheatre airlift is also the domain of rotary-wing aircraft. Various helicopters have been built in Europe. In an era of deployed operations, an increasingly important role conducted by medium-sized helicopters is CSAR.[296] European helicopters such as the French Cougar have been specifically adapted for CSAR.[297]

A further specific element of air mobility is AAR. Tanker aircraft offer a deployed force not only extended range but also extended endurance over the theatre of operations. Air policing over Bosnia or Iraq in the 1990s, for instance, saw the extensive use of tankers.[298] Tanker aircraft are either derived from civilian airliners such as the DC-10 and the Airbus A310 or from military transport aircraft such as the C-130 Hercules. Whereas the US Air Force, Navy, and Marine Corps together operated more than 650 tanker aircraft of various types in 2006, the air forces of France, Germany, Italy, the Netherlands, Spain, Sweden, and the United Kingdom could only muster approximately 70 aircraft, with the UK and France providing the bulk of the European tanker fleet.[299] AAR is hence one of the major European shortcomings in deployed operations.

C4ISTAR. Carrying a long-range surveillance radar, the E-3 AWACS aircraft is capable of providing a comprehensive, real-time picture of air operations in a theatre. A British air power expert singled it out as the linchpin of the coalition's application of air power in Desert Shield and Desert Storm.[300] Through the NAEW&CF, European NATO members acquired this high-value asset in a cooperative approach. The E-8 Joint Surveillance Target Attack Radar System (JSTARS) is in effect an AWACS analogue for ground surveillance and showed an outstanding performance in Desert Storm. Capable of viewing the entire Kuwaiti theatre of operation in a single orbit, the aircraft proved key in thwarting the Iraqi attack against the Saudi coastal town of Al Khafji.[301] As was mentioned, European NATO members are planning to acquire a wide-area air-to-ground surveillance capability through the cooperative AGS Core project.

Another air power innovation of the post–Cold War era is the increasing use of remotely piloted aircraft. RPAs have proved to be an efficient supplement to manned aircraft in the role of intelligence, surveillance, target acquisition, and reconnaissance (ISTAR), where persistence over the battlespace is a key requirement. Only technological quantum leaps of the 1990s, such as improvement in computing and data-link technology, made the considerable potential of RPA systems a viable option for Western air forces. RPAs offer significant advantages over manned aircraft when it comes to missions involving high risk or extreme endurance. For the primary purpose of intelligence gathering, air forces have employed basically two types of RPAs—high-altitude, long-endurance and medium-altitude, long-endurance (MALE) systems.[302] The American HALE RPA Global Hawk, for instance, has a ceiling of 65,000 feet, an intercontinental range of 5,400 nautical miles, and an endurance of 32 hours.[303] Unmanned systems also offer significant advantages with regard to operational costs. While a pilot may need a minimum of approximately 150 hours of training per year, the RPA can remain stored in its hangar for most of that time, with the majority of RPA controller training conducted on simulators.[304] Despite these advantages, adequate concepts to fully exploit the potential of RPAs and to integrate them optimally into complex air campaigns still have to be developed.[305]

Further high-value air power assets in the domain of ISTAR are signals intelligence (SIGINT) aircraft. Capable of locating sources of radiated electromagnetic energy for the purpose of recognizing threats, targeting, and planning future operations, SIGINT aircraft such as the American RC-135 Rivet Joint are crucial in complex air operations.[306] Moreover, electronic standoff jamming aircraft are key assets in American-led air campaigns. As such, the American EA-6B Prowler was an integral part of US-led aircraft packages patrolling the southern and northern NFZs over Iraq.[307]

Data gathered by ISTAR assets has to be processed and disseminated in near real time in order to build the framework known as network-centric or network-enabled warfare. Complex and immense data loads have caused digital data links to become a necessity for conducting air operations. The NATO standard data link for air defence operations is the Link 16.[308] Link-16 data communication development began in 1975. The first Link-16 terminals were bulky and were installed only on AWACS and at US, UK, and NATO ground-control facilities. In the ensuing years, smaller terminals were developed for combat aircraft. Yet due to their high cost and relatively large size, only a limited number of Western fighter aircraft, such as the US Navy F-14Ds, a single squadron of

USAF F-15Cs, or the RAF's Tornado F3s, were equipped with these smaller terminals.[309]

To remedy this shortcoming in data-link terminals, the multi-functional information distribution system (MIDS) programme was launched in March 1994. The programme was aimed at developing a small, lightweight Link-16 terminal for US and Western European aircraft, including the F-15, F-16, F/A-18, Rafale, Mirage 2000, Tornado, and Eurofighter Typhoon. MIDS is a multinational programme. Partner nations are the United States, France, Germany, Italy, and Spain.[310] Because of programme delays, the first Link-16-capable F-16 squadron became operational only in the third quarter of 2003.[311]

The most important NATO alliance project in the field of C2 is the formation of a unified alliance air command and control system (ACCS). In the later stages of the Cold War, C2 was frag-

Courtesy GAF; photo by Peter Müller

GAF Patriot unit during live firings at the NATO Missile Firing Installation on the isle of Crete, Greece

mented into air defence and offensive air operations.[312] An ACCS is designed to provide seamless management of all types of air operations over European alliance territory and beyond out of fully integrated combined air operations centres (CAOC).[313]

Ground Based Air Defence. In the post–Cold War era, the perception of threat from theatre ballistic missiles—possibly loaded with WMDs—has increased considerably.[314] That the threat was real became particularly apparent during Desert Storm, when Iraq employed its Scud surface-to-surface missiles against Israel.[315] As early as 1990, NATO began studying defensive measures against theatre ballistic missiles and finally came up with the extended air defence (EAD) concept.[316] Separate from the American Strategic Defense Initiative that aims at intercepting strategic nuclear missiles, EAD was formulated by the NATO Air Defence Committee to include defence against combat aircraft, tactical ballistic missiles, cruise missiles, or any enemy aerial vehicle.[317]

In the Cold War, NATO GBAD was mainly deployed in fixed belts and later on in clusters along the inner German border.[318] Consisting of different layers, it included high- and medium-level SAMs and short-range air defence missiles as well as anti-aircraft artillery. Since the Cold War, improved technologies and the need for increased mobility of the systems have led to a blurring of distinct layers.[319]

Notes

1. Andreas Wenger and Doron Zimmermann, *International Relations: From the Cold War to the Globalized World* (London: Lynne Rienner Publishers, 2003), 238.

2. Ibid., 242.

3. Philip A. G. Sabin, "The Changing Face of Conflict," in *A Second Aerospace Century: Choices for the Smaller Nations*, ed. John Andreas Olsen (Trondheim, Norway: Royal Norwegian Air Force Academy, 2001), 91.

4. International Institute for Strategic Studies (IISS), *Strategic Survey 1999/2000* (Oxford, England: Oxford University Press for the IISS, May 2000), 56–57.

5. International Institute for Strategic Studies, *Strategic Survey 2001/2002* (Oxford, England: Oxford University Press for the IISS, May 2002), 31.

6. Christopher Daase, "Internationale Risikopolitik: Ein Forschungsprogramm für den sicherheitspolitischen Paradigmenwechsel," in *Internationale Risikopolitik: Der Umgang mit neuen Gefahren in den internationalen Beziehungen,*

ed. Christopher Daase, Susanne Feske, and Ingo Peters (Baden-Baden, Germany: Nomos Verlagsgesellschaft, 2002), 9.

7. International Institute for Strategic Studies, *Strategic Survey 1990/1991* (London: Brassey's for the IISS, May 1991), 135–48.

8. International Institute for Strategic Studies, *Strategic Survey 1991/1992* (London: Brassey's for the IISS, May 1992), 33–37; and International Institute for Strategic Studies, *Strategic Survey 1990/1991*, 49–91.

9. International Institute for Strategic Studies, *The Military Balance 1993–94* (London: Brassey's for the IISS, 1993), 33–37.

10. Franz-Josef Meiers, *Zu neuen Ufern? Die deutsche Sicherheits- und Verteidigungspolitik in einer Welt des Wandels 1990–2000* (Paderborn, Germany: Ferdinand Schöningh, 2006), 171.

11. Philip A. G. Sabin, "The Shifting Trade-Offs in UK Defence Planning," in *British Defence Choices for the Twenty-First Century*, ed. Michael Clarke and Philip A. G. Sabin (London: Brassey's, 1993), 151–71.

12. See Donald B. Rice, "Global Reach-Global Power," in *American Defense Policy*, 7th ed., ed. Peter L. Hays, Brenda J. Vallance, and Alan R. Van Tassel (London: The Johns Hopkins University Press, 1997); Robert L. Pfaltzgraff, Jr., "The United States as an Aerospace Power in the Emerging Security Environment," in *The Future of Air Power in the Aftermath of the Gulf War*, ed. Richard H. Schultz, Jr. and Robert L. Pfaltzgraff (Maxwell AFB, AL: Air University Press, 1992); and Eliot A. Cohen, "The Mystique of U.S. Air Power," *Foreign Affairs* 73, no. 1 (January/February 1994): 109–24.

13. Mason, *Air Power*, 243.

14. Office of the Under Secretary of Defense (Comptroller), *National Defense Budget Estimates for FY 2005* (Washington D.C.: Department of Defense, March 2004), 188–89, 194–95, 200–201.

15. International Institute for Strategic Studies, *The Military Balance 1990–91* (London: Brassey's for the IISS, 1990), 19, 22.

16. International Institute for Strategic Studies, *The Military Balance 2003–4* (Oxford, England: Oxford University Press for the IISS, 2003), 19–20.

17. Frans Osinga, "European Military Transformation and the Intangibles of Innovation," in *Royal Netherlands Air Force* [RNLAF] *Air Power Symposium* [19 November 2003] (The Hague: RNLAF, 2003), 36.

18. NATO Headquarters, *NATO Handbook* (Brussels: NATO Office of Information and Press, 2001), 99.

19. Ibid.

20. "Petersberg Tasks," *Europa Glossary*, accessed 26 February 2007, http://europa.eu/scadplus/glossary/petersberg_tasks_en.htm.

21. Meiers, *Zu neuen Ufern?*, 143.

22. Jean-Yves Haine, "ESDP: An Overview," European Union Institute for Security Studies, accessed 14 May 2005, http://www.iss-eu.org/esdp/01-jyh.pdf.

23. "Summary of the Declaration on European Defence," UK-French summit, St. Malo, 3–4 December 1998, in International Institute for Strategic Studies,

Strategic Survey 1998/99 (Oxford, England: Oxford University Press for IISS, May 1999), 37.

24. Haine, "ESDP: An Overview."

25. Meiers, *Zu neuen Ufern?*, 161.

26. Burkhard Theile, "Bridging the Gap: The Six Nation Framework Agreement and European Defence Technology," *Military Technology* 26, no. 2 (February 2002): 11; and Nicole Gnesotto et al., *European Defence: A Proposal for a White Paper*, report of an independent task force (Paris: European Union Institute for Security Studies, May 2004), 48.

27. Haine, "ESDP: An Overview."

28. Council of the European Union, *A Secure Europe in a Better World: European Security Strategy* (Brussels: European Union, 12 December 2003), 6–7, accessed 28 November 2005, http://ue.eu.int/uedocs/cmsUpload/78367.pdf.

29. Burkard Schmitt, *European Capabilities Action Plan (ECAP)* (Paris: European Union Institute for Security Studies, 2006), accessed 5 March 2007, http://www.iss-eu.org/esdp/06-bsecap.pdf.

30. Gerrard Quille, "'Battle Groups' to Strengthen EU Military Crisis Management?," *European Security Review*, no. 22 (April 2004): 1–2, http://www.isis-europe.org/pdf/esr_23.pdf.

31. European Union, *Headline Goal 2010* (Approved by General Affairs and External Relations Council, 17 May 2004, endorsed by the European Council of 17 and 18 June 2004), 1–3, http://www.consilium.europa.eu/uedocs/cmsUpload/2010%20Headline%20Goal.pdf.

32. European Union, "Declaration on European Military Capabilities," Military Capability Commitment Conference, Brussels (Approved by the General Affairs and External Relations Council, 22 November 2004), 3, http://www.consilium.europa.eu/uedocs/cmsUpload/MILITARY%20CAPABILITY%20COMMITMENT%20CONFERENCE%2022.11.04.pdf.

33. Quille, "'Battle Groups,'" 1.

34. Gnesotto et al., *European Defence*, 56–57.

35. Arita Eriksson, "Sweden and the Europeanisation of Security and Defence Policy," in *Strategic Yearbook 2004: The New Northern Security Agenda—Perspectives from Finland and Sweden*, ed. Bo Huldt et al. (Stockholm: Swedish National Defence College [SNDC], 2003), 126.

36. European Union, *Headline Goal 2010*, 4–5.

37. NATO Headquarters, *NATO Handbook*, 2001, 102.

38. Ibid., 44.

39. Ibid., 17.

40. NATO Ministerial Communiqués, "The Alliance's New Strategic Concept Agreed by the Heads of State and Government Participating in the Meeting of the North Atlantic Council in Rome on 7–8 Nov 1991," http://www.nato.int/docu/comm/49-95/c911107a.htm#I.

41. "Final Communiqué, Ministerial Meeting of the North Atlantic Council, NATO Headquarters, Brussels, Belgium," NATO Press Communiqué M-NAC-2

(92) 106, 17 December 1992, §4, http://wiretap.area.com/ftp.items/Gov/NATO /921217.ministers.

42. NATO Headquarters, *NATO Handbook*, 2001, 47–48.

43. Ibid., 48–49.

44. "Declaration of the Heads of State and Government Issued by the North Atlantic Council in Brussels, Belgium," NATO Press Communiqué M-1(94)3, 11 January 1994, 4–6, http://www.nato.int/docu/comm/49-95/c940111a.htm.

45. Shaun Gregory, *French Defence Policy into the Twenty-First Century* (London: MacMillan Press, 2000), 111.

46. Ibid., 113–14.

47. "Declaration of the Heads of State and Government," §14.

48. NATO Headquarters, *NATO Handbook*, 2001, 67.

49. NATO, "The Partnership for Peace," 24 November 2009, http://www .nato.int/cps/en/natolive/topics_50349.htm.

50. NATO Headquarters, *NATO Handbook*, 2001, 72.

51. Ibid., 61.

52. "The Alliance's Strategic Concept Approved by the Heads of State and Government Participating in the Meeting of the North Atlantic Council in Washington, D.C., on 23 and 24 April 1999," NATO Press Release NAC-S(99)65, 24 April 1999, §31, http://www.nato.int/docu/pr/1999/p99-065e.htm.

53. NATO, "The NATO Response Force: At the Centre of NATO Transformation," http://www.nato.int/cps/en/natolive/topics_49755.htm.

54. NATO, "The NATO Response Force: What Does This Mean in Practice?," http://www.nato.int/cps/en/natolive/topics_49755.htm.

55. NATO, "The NATO Response Force: How Did It Evolve?," http://www .nato.int/cps/en/natolive/topics_49755.htm.

56. North Atlantic Council, *The Prague Summit and NATO's Transformation: A Reader's Guide* (Brussels: NATO Public Diplomacy Division, 2003), 28.

57. NATO Headquarters, *NATO Handbook* (Brussels: Public Diplomacy Division, 2006), 88–89.

58. Gustav Lindstrom, *The Headline Goal* (Paris: European Union Institute for Security Studies, updated January 2007), 2, accessed 5 March 2007, http:// www.iss-eu.org/esdp/05-gl.pdf.

59. EU Council Secretariat, "Background—Development of European Military Capabilities," *The Force Catalogue 2006* (Brussels: Council of the European Union, November 2006), 3, http://www.consilium.europa.eu/ueDocs/cms_Data /docs/pressData/en/esdp/91704.pdf.

60. Lt Col Michael Trautermann, General Staff (GS), GAF, Ministry of Defence (MOD), Bonn, Germany, to the author, e-mail, 30 May 2006.

61. Wg Cdr Kraesten Arnold, European Union Military Staff Operations Directorate, Crisis Response Planning and Current Operations, Brussels, to the author, e-mail, 22 April 2010.

62. Jean-Laurent Nijean, "L'Armée de l'Air et l'Europe: enjeux et perspectives," *Air actualités*, no. 605 (October 2007): 26.

63. NATO, "NATO Airborne Early Warning and Control Force: E-3A Component," fact sheet, 2010, 1, accessed 30 September 2010, http://www.e3a .nato.int/html/media.htm.

64. Col Jean-Pierre Moulard, "L'organisation de la campagne aérienne dans le cadre de la NRF 5?," *Penser les ailes françaises*, no. 8 (January 2006): 42.

65. Col (GS) Lothar Schmidt, GAF (MOD, Bonn, Germany), interview by the author, 26 April 2006.

66. JAPCC, "History," accessed 23 March 2007, http://www.japcc.de/file admin/user_upload/History/JAPCC_History.pdf.

67. JAPCC, "Mission," http://www.japcc.de/mission.html.

68. European Air Group, "History of the EAG," accessed 15 May 2007, http://www.euroairgroup.org/index.php?s=history.

69. Gregory, *French Defence Policy*, 138.

70. European Air Group, "History of the EAG."

71. European Air Group, "Objectives of the European Air Group," accessed 2 March 2007, http://www.euroairgroup.org/objective.htm.

72. Ibid.

73. European Air Group, "Activities," accessed 2 March 2007, http://www .euroairgroup.org/act_FP.htm.

74. European Air Group, "Deployable Multinational Air Wing (DMAW) Project," accessed 2 March 2007, http://www.euroairgroup.org/act_DMAW.htm.

75. Wolfgang Lange, "Lufttransport—Ansätze zur Kompensation von Defiziten," *Europäische Sicherheit*, no. 10 (October 2002): 38–39.

76. Lt Col (GS) Jörg Lebert, "Einrichtung eines Europäischen Lufttransportkommandos," *Europäische Sicherheit*, no. 7 (July 2001): 20–22.

77. European Air Group, "European Airlift Centre (EAC)," accessed 2 March 2007, http://www.euroairgroup.org/act_EAC.htm.

78. Lt Col (GS) Armin Havenith, GAF (Federal MOD, Bonn, Germany), interview by the author, 26 April 2006.

79. Movement Coordination Centre, "Background of the MCCE," https:// www.mcce-mil.com.

80. European Union, *Headline Goal 2010*, 3.

81. Bundesministerium der Verteidigung (BMVg) (Federal Ministry of Defence), "Verlässlicher Zugriff auf 'fliegende Güterzüge,'" 23 March 2006, http:// www.bmvg.de.

82. Ibid.; and EU Council Secretariat, "Background," 3.

83. European Union, "Declaration on European Military Capabilities," 20.

84. Tony Mason, "The Air War in the Gulf," *Survival* 33, no. 3 (May/June 1991): 211.

85. Eliot A. Cohen, director, *Gulf Air Power Survey*, vol. 5, *A Statistical Compendium and Chronology* (Washington, D.C.: Dept. of the Air Force, 1993), 42.

86. Ibid., 42–45.

87. Ibid., 641.

88. Calculation of percentage is based upon figures from ibid., 232–33.

89. Thomas A. Keaney and Eliot A. Cohen, *Gulf War Air Power Survey: Summary Report* (Washington, D.C.: Dept. of the Air Force, 1993), 11–12.

90. Maj Gen David A. Deptula, "Effects-Based Operations: Change in the Nature of Warfare," in *Second Aerospace Century: Choices for the Smaller Nations*, ed. Olsen, 136–39.

91. Ibid., 146–47.

92. Keaney and Cohen, *Gulf War Air Power Survey*, 15.

93. Ibid., 21.

94. Benjamin S. Lambeth, *The Transformation of American Air Power* (Ithaca, NY: Cornell University Press, 2000), 121–24.

95. Mason, *Air Power*, 166.

96. Ibid., 8.

97. United States European Command, "Operation Northern Watch," 20 February 2004, accessed 3 November 2004, http://www.eucom.mil/Directorates /ECPA/Operations/onw/onw.htm (site discontinued).

98. Tim Ripley, *Air War Iraq* (Barnsley, South Yorkshire, England: Pen and Sword Aviation, 2004), 9.

99. GlobalSecurity.org, "Operation Southern Watch: 1993 Events," http:// www.globalsecurity.org/military/ops/southern_watch-1993.htm.

100. United States European Command, "Operation Northern Watch: Chronology of Significant Events," accessed 3 November 2004, http://www.global security.org/military/ops/northern_watch-1999.htm.

101. Ripley, *Air War Iraq*, 10.

102. Kenneth M. Pollack, *The Threatening Storm: The Case for Invading Iraq* (New York: Random House, 2002), 92–94.

103. Anthony H. Cordesman, *The Iraq War: Strategy, Tactics, and Military Lessons* (Westport, CT: Praeger, 2003), 253; GlobalSecurity.org, "Operation Southern Watch: 2001 Events," http://www.globalsecurity.org/military/ops /southern_watch-2001.htm"; and GlobalSecurity.org, "Operation Southern Watch: 2002 Events," http://www.globalsecurity.org/military/ops/southern _watch-2002.htm.

104. Walter J. Boyne, *Operation Iraqi Freedom: What Went Right, What Went Wrong, and Why* (New York: Tom Doherty Associates, 2003), 39.

105. David E. Johnson, *Learning Large Lessons: The Evolving Roles of Ground Power and Air Power in the Post–Cold War Era* (Santa Monica, CA: RAND, 2007), 106.

106. Boyne, *Operation Iraqi Freedom*, 34–35.

107. Cordesman, *Iraq War*, 256–57, 275; and Johnson, *Learning Large Lessons*, 113, 125.

108. Gen T. Michael Moseley, chief of staff, USAF, "Iraq 2003: Air Power Pointers for the Future" (speech, RAF Defence Studies Conference, RAF Museum, Hendon, England, 11 May 2004); and Boyne, *Operation Iraqi Freedom*, 90.

109. Robert A. Pape, "The True Worth of Air Power," *Foreign Affairs* 83, no. 2 (March/April 2004): 128.

110. Gen Tommy Franks, US Army, retired, keynote speech in *RNLAF Air Power Symposium*, 19 November 2003.

111. Boyne, *Operation Iraqi Freedom*, 35, 132.

112. Wg Cdr Richard Duance and Wg Cdr Pete York, "An Interview with Air Chief Marshal Sir Glenn Torpy, Chief of the Air Staff, Royal Air Force," *Transforming Joint Air Power: The Journal of the JAPCC* 3 (2006): 27.

113. Boyne, *Operation Iraqi Freedom*, 132.

114. Rebecca Grant, "'The Fallujah Model," air-force-magazine.com, February 2005, http://www.airforce-magazine.com/MagazineArchive/Pages/2005/February%202005/0205fallujah.aspx.

115. Lambeth, *Transformation of American Air Power*, 178.

116. Col Robert C. Owen, "Summary," in *Deliberate Force—A Case Study in Effective Air Campaigning: Final Report of the Air University Balkans Air Campaign Study*, ed. Col Robert C. Owen (Maxwell AFB, AL: Air University Press, 2000), 461.

117. Tim Ripley, *Air War Bosnia: UN and NATO Airpower* (Osceola, WI: Motorbooks International, 1996), 39.

118. Ibid., 88.

119. Mason, *Air Power*, 173.

120. Mark A. Bucknam, *Responsibility of Command: How UN and NATO Commanders Influenced Airpower over Bosnia* (Maxwell AFB, AL: Air University Press, 2003), 131–32.

121. Lambeth, *Transformation of American Air Power*, 178.

122. Owen, "Summary," 474–77.

123. Maj Gen Walter Jertz, *Im Dienste des Friedens: Tornados über dem Balkan*, 2d rev. ed. (Bonn, Germany: Bernard & Graefe Verlag, 2000), 61.

124. NATO Headquarters, *NATO Handbook*, 2001, 47–48; and Owen, "Summary," 483–84.

125. Ibid., 491, 502.

126. Lambeth, *Transformation of American Air Power*, 176.

127. Lt Col Richard L. Sargent, "Aircraft Used in Deliberate Force," in Owen, *Deliberate Force*, 204, 220–22.

128. Lt Col Richard L. Sargent, "Deliberate Force Combat Air Assessments," in Owen, *Deliberate Force*, 332.

129. Sargent, "Aircraft Used in Deliberate Force," 199; and Owen, "Summary," 485.

130. Calculation of percentage is based on figures from Keaney and Cohen, *Gulf War Air Power Survey*, 226.

131. Tim Ripley, *Operation Deliberate Force: The UN and NATO Campaign in Bosnia 1995* (Lancaster, England: Centre for Defence and International Security Studies, 1999), 250.

132. John Stone, "Air-Power, Land-Power and the Challenge of Ethnic Conflict," *Civil Wars* 2, no. 3 (Autumn 1999): 29–30.

133. Lt Gen André Soubirou, retired, French Army, "The Account of Lieutenant General (Ret) André Soubirou, Former Commanding General of the RRF

Multinational Brigade in Bosnia from July to October 1995," *Doctrine Special Issue*, February 2007, 23–29, http://www.cdef.terre.defense.gouv.fr/publications/doctrine/no_spe_chefs_francais/version_us/art07.pdf.

134. Ripley, *Air War Bosnia*, 9.

135. Ibid., 18.

136. Ibid., 28.

137. Benjamin S. Lambeth, *NATO's Air War for Kosovo: A Strategic and Operational Assessment* (Santa Monica, CA: RAND, 2001), xiii.

138. John E. Peters et al., *European Contributions to Operation Allied Force: Implications for Transatlantic Cooperation* (Santa Monica, CA: RAND, 2001), 37–38.

139. Ibid., 39.

140. Lambeth, *Transformation of American Air Power*, 182.

141. Ibid., 187–89.

142. Ivo H. Daalder and Michael E. O'Hanlon, *Winning Ugly: NATO's War to Save Kosovo* (Washington, D.C.: Brookings Institution Press, 2000), 221.

143. Lambeth, *NATO's Air War for Kosovo*, 27.

144. Ibid., xviii–xix.

145. Daalder and O'Hanlon, *Winning Ugly*, 5, 184, 199–200; Lambeth, *NATO's Air War for Kosovo*, 67–86; and Benjamin S. Lambeth, "Operation Allied Force: A Strategic Appraisal," in Olsen, *Second Aerospace Century*, 103–9.

146. Peters et al., *European Contributions to Operation Allied Force*, 52; and Lambeth, *NATO's Air War for Kosovo*, 33, 66.

147. Anthony H. Cordesman, *The Lessons and Non-Lessons of the Air and Missile War in Kosovo*, rev. ed. (Washington, D.C.: Center for Strategic and International Studies, 20 July 1999), 4–5.

148. Garden, "European Air Power," in Gray, *Air Power 21*, 114.

149. Jolyon Howorth and John T. S. Keeler, *Defending Europe: The EU, NATO and the Quest for European Autonomy* (Basingstoke, England: Palgrave Macmillan, 2003), 88–89; and Peters et al., *European Contributions to Operation Allied Force*, 20–24.

150. Peters et al., *European Contributions to Operation Allied Force*, 18–23.

151. Capt Mark E. Kosnik, "The Military Response to Terrorism," *Naval War College Review* 53, no. 2 (Spring 2000): 13–39, http://www.iwar.org.uk/cyberterror/resources/mil-response/response.htm.

152. Marvin Leibstone, "War against Terrorism and the Art of Restraint," *Military Technology* 25, no. 11 (November 2001): 19.

153. Stephen Biddle, *Afghanistan and the Future of Warfare: Implications for Army and Defense Policy* (Carlisle Barracks, PA: US Army War College, 2002), 44.

154. GlobalSecurity.org, "Operation Enduring Freedom—Afghanistan," http://www.globalsecurity.org/military/ops/enduring-freedom.htm.

155. Gp Capt Chris Finn, "The Employment of Air Power in Afghanistan and Beyond," *Royal Air Force Air Power Review* 5, no. 4 (Winter 2002): 6.

156. Benjamin S. Lambeth, *American Carrier Air Power at the Dawn of a New Century* (Santa Monica, CA: RAND, 2005), 17.

157. GlobalSecurity.org, "Operations in Somalia: Applying the Urban Operational Framework to Support and Stability," http://www.globalsecurity.org/military/library/policy/army/fm/3-06/appc.htm.

158. International Institute for Strategic Studies, *Strategic Survey 1994/95* (Oxford, England: Oxford University Press for the IISS, May 1995), 207–9.

159. Richard Connaughton, "The Mechanics and Nature of British Interventions into Sierra Leone (2000) and Afghanistan (2001–2002)," *Civil Wars* 5, no. 2 (Summer 2002): 84–85.

160. William Fowler, *Operation Barras—The SAS Rescue Mission: Sierra Leone 2000* (London: Cassell, 2004), 89.

161. GlobalSecurity.org, "Operation Uphold Democracy," accessed 18 November 2005, http://www.globalsecurity.org/military/ops/uphold_democracy.htm; and GlobalSecurity.org, "Operation Stabilise Timor Crisis—American Forces," accessed 26 November 2005, http://www.globalsecurity.org/military/ops/timor-orbat.htm.

162. AP 3000, *British Air Power Doctrine*, 2.8.8.

163. Ibid., 3.11.1.

164. Air Vice-Marshal Andrew Vallance, RAF (MOD, London), interview by the author, 24 January 2007.

165. See Col John A. Warden, USAF, retired, "Employing Air Power in the Twenty-First Century," in Schultz and Pfaltzgraff, *Future of Air Power*, 57–82; Edward Mann, "One Target, One Bomb: Is the Principle of Mass Dead?," *Airpower Journal* 7, no. 1 (Spring 1993): 35–43, http://www.airpower.maxwell.af.mil/airchronicles/apj/apj93/spr93/mann.htm; and Jason B. Barlow, "Strategic Paralysis: An Air Power Strategy for the Present," *Airpower Journal* 7, no. 4 (Winter 1993): 4–15, http://www.airpower.maxwell.af.mil/airchronicles/apj/apj93/win93/barlow.htm.

166. See Pape, "True Worth of Air Power," 117–19; and Robert A. Pape, *Bombing to Win: Air Power and Coercion in War* (Ithaca, NY: Cornell University Press, 1996).

167. See Mason, *Air Power*; and Lambeth, *Transformation of American Air Power*.

168. Deptula, "Effects-Based Operations," in Olsen, *Second Aerospace Century*, 147–52.

169. See Richard Lock-Pullan, "Redefining 'Strategic Effect' in British Air Power Doctrine," *Royal Air Force Air Power Review* 5, no. 3 (Autumn 2002): 60.

170. Ibid., 60–62.

171. Allied Joint Publication (AJP)-3.3, *Joint Air and Space Operations Doctrine*, July 2000, 4-9.

172. Tony Mason, "Rethinking the Conceptual Framework," in Gray, *Air Power 21*, 234–35.

173. AFM 1-1, *Basic Aerospace Doctrine of the United States Air Force*, vol. 1, March 1992, 5.

174. Deptula, "Effects-Based Operations," in Olsen, *Second Aerospace Century*, 135–73.

175. AFM 1-1, *Basic Aerospace Doctrine*, vol. 2, 147.

176. Lambeth, *Transformation of American Air Power*, 269.

177. AFM 1-1, *Basic Aerospace Doctrine*, vol. 1, 8–9.

178. Air Force Link, "Air Force Doctrine Center," accessed 21 January 2007, http://www.af.mil/factsheets/factsheet.asp?fsID=141.

179. AFDD 1, *Air Force Basic Doctrine*, September 1997, v.

180. Ibid., 51.

181. Owen, "Summary," 471, 511.

182. AFDD 1, *Air Force Basic Doctrine*, September 1997, 32.

183. AFDD 1, *Air Force Basic Doctrine*, November 2003, i.

184. Ibid., 75.

185. Air Vice-Marshal Andrew Vallance, RAF, MOD, London, to the author, e-mail, 15 January 2007.

186. Gp Capt Chris Finn, "British Thinking on Air Power—The Evolution of AP 3000," *Royal Air Force Air Power Review* 12, no. 1 (Spring 2009): 58.

187. AP 3000, *Royal Air Force Air Power Doctrine*, 1991, v.

188. Ibid., 2, 38–39, 42, 60–61.

189. Ibid., 59.

190. Ibid., v.

191. Vallance, interview.

192. AP 3000, *Royal Air Force Air Power Doctrine*, 2d ed., 1993, v.

193. Ibid., 18–20, 71.

194. Ibid., 50, 72.

195. AP 3000, *British Air and Space Power Doctrine*, 4th ed., 2009, 8.

196. AP 3000, *British Air Power Doctrine*.

197. Ibid., 2.7.1–2.9.9. See also Finn, "British Thinking on Air Power," 62.

198. AP 3000, *British Air Power Doctrine*, 2.6.1.

199. Finn, "British Thinking on Air Power," 63.

200. AP 3000, *British Air and Space Power Doctrine*, 8; and Finn, "British Thinking on Air Power," 63.

201. AP 3000, *British Air and Space Power Doctrine*, 8.

202. Finn, "British Thinking on Air Power," 64.

203. AP 3000, *British Air and Space Power Doctrine*, 39–40.

204. Air Power Development Centre, "About the Air Power Development Centre," http://airpower.airforce.gov.au/Contents/About-APDC/2/About-APDC.aspx.

205. Australian Air Publication (AAP) 1000, *Royal Australian Air Force Air Power Manual*, 1st ed., August 1990.

206. Ibid., 21.

207. AAP 1000, *Air Power Manual*, 2d ed., March 1994; AAP 1000, *Air Power Manual*, 3d ed., February 1998; and AAP 1000, *Fundamentals of Australian Aerospace Power*, 4th ed., August 2002.

208. Michael E. Thompson, *Political and Military Components of Air Force Doctrine in the Federal Republic of Germany and Their Implications for NATO Defense Policy Analysis* (Santa Monica, CA: RAND, September 1987), 3.

209. NATO, "Standardization Agreements," updated 21 May 2010, http://www.nato.int/docu/standard.htm.

210. See Allied Tactical Publication (ATP) 33(B), *NATO Tactical Air Doctrine*, November 1986, 2-4–2-5, 10-1.

211. AJP-3.3, *Joint Air and Space Operations Doctrine*, 7-1–7-9.

212. Ibid., 4-9.

213. Air University (AU), "SAASS Commandant's Welcome," accessed 24 January 2007, http://www.au.af.mil/au/saass/welcome.htm.

214. AU, "SAASS Historical Background," accessed 24 January 2007, http://www.au.af.mil/au/saass/background.htm.

215. AU, "SAASS Curriculum," accessed 24 January 2007, http://www.au.af.mil/au/saass/curriculum.htm.

216. AU, "SAASS: Results—the Bottom Line," accessed 24 January 2007, http://www.au.af.mil/au/saass/results.htm; and Philip S. Meilinger, ed., *The Paths of Heaven: The Evolution of Airpower Theory* (Maxwell AFB, AL: Air University Press, 1997).

217. AU, "SAASS: Results—the Bottom Line."

218. Gp Capt Chris Finn, RAF, retired (RAF College, Cranwell, England), interview by the author, 26 January 2007. See also http://www.airpowerstudies.co.uk/newmastersdegree.htm.

219. Finn, "British Thinking on Air Power," 63.

220. Air Power Development Centre, "Education and Courses," http://airpower.airforce.gov.au/Contents/Education-And-Courses/46/Education-And-Courses.aspx.

221. Air Power Development Centre, "Fellowships," accessed 24 January 2007, http://www.defence.gov.au/raaf.

222. "Welcome to Air & Space Power Journal," *Air and Space Power Journal*, http://www.airpower.maxwell.af.mil/airchronicles/Welcome.html.

223. Air Force Historical Foundation, "About Us," accessed 26 December 2007, http://afhistoricalfoundation.com.

224. Air Force Historical Foundation, "Publications," accessed 26 December 2007, http://www.afhistoricalfoundation.com.

225. Gp Capt Andrew Vallance, ed., *Air Power: Collected Essays on Doctrine* (London: MOD Director of Defence Studies, 1990).

226. Finn, "British Thinking on Air Power," 59–60; and Gp Capt Andrew Lambert and Gp Capt Arthur C. Williamson, eds., *The Dynamics of Air Power* (Bracknell, England: Royal Air Force Staff College, 1996), iv.

227. Lambert and Williamson, *Dynamics of Air Power*.

228. Stuart Peach, ed., *Perspectives on Air Power: Air Power in Its Wider Context* (London: The Stationery Office, 1998); Gray, *Air Power 21*; and Peter W. Gray, ed., *British Air Power* (London: The Stationery Office, 2003), iii.

229. Finn, "British Thinking on Air Power," 60, 62–63.

230. "The New Air Power Journal," *Royal Air Force Air Power Review* 1, no. 1 (1998).

231. "Contributions to RAF Air Power Review," *Royal Air Force Air Power Review* 1, no. 1 (1998).

232. Royal Air Force Centre for Air Power Studies, http://www.airpower studies.co.uk.

233. David R. Mets, "Bomber Barons, Bureaucrats, and Budgets: Your Professional Reading on the Theory of Strategic Air Attack," *Airpower Journal* 10, no. 2 (Summer 1996): 87, http://www.airpower.maxwell.af.mil/airchronicles /apj/apj96/sum96/mets1.pdf.

234. Owen, *Deliberate Force*.

235. Flt Lt David Tucker, "Preparing the Way for Cooperation in European Air Power," *Royal Air Force Air Power Review* 7, no. 1 (Spring 2004): 51.

236. Jürgen Groß, "Revision der Reform: Weiterentwicklung des Bundeswehrmodells '200F,'" in *Europäische Sicherheit und Zukunft der Bundeswehr: Analysen und Empfehlungen der Kommission am IFSH*, ed. Jürgen Groß (Baden-Baden, Germany: Nomos Verlagsgesellschaft, 2004), 118.

237. Michael Alexander and Timothy Garden, "The Arithmetic of Defence Policy," *International Affairs* 77, no. 3 (July 2001): 516.

238. See "Rüstung: Später fliegen," *Der Spiegel*, no. 49 (December 2006): 18.

239. Alexander and Garden, "Arithmetic of Defence Policy," 510.

240. See Rainer Hertrich, "Strategische Schlüsselindustrie," *Europäische Sicherheit*, no. 6 (June 2003): 35; and Hans-Christian Hagman, *European Crisis Management and Defence: The Search for Capabilities*, Adelphi Paper no. 353 (Oxford, England: Oxford University Press for IISS, 2002), 62.

241. See Christina Goulter, "Air Power and Expeditionary Warfare," in Gray, *Air Power 21*, 207.

242. Gates, "Air Power: The Instrument of Choice?," 26.

243. International Institute for Strategic Studies, *The Military Balance 2000–2001* (Oxford, England: Oxford University Press for the IISS, 2000), 28, 30–31.

244. Gates, "Air Power: The Instrument of Choice?," 28.

245. Jan Ahlgren, Anders Linnér, and Lars Wigert, *Gripen: The First Fourth Generation Fighter* (Sweden: Swedish Air Force, FMV and Saab Aerospace, 2002), 52–53.

246. Bitzinger, *Facing the Future*, 81.

247. International Institute for Strategic Studies, *The Military Balance 1998–99* (Oxford, England: Oxford University Press for the IISS, 1998), 34.

248. Susan Willett, Philip Gummet, and Michael Clarke, *Eurofighter 2000* (London: Brassey's for The Centre for Defence Studies, 1994), 2–5.

249. Mason, *Air Power*, 247.

250. See Hertrich, "Strategische Schlüsselindustrie," 38.

251. Dr. Hermann Hagena, brigadier general, GAF, retired, and Capt Hartwig Hagena, GAF, "Weichenstellung für das künftige europäische Transportflugzeug: Von der FIMA über EUROFLAG zum Joint Venture mit Antonow?,"*Europäische Sicherheit*, no. 4 (April 1998): 28–32.

252. Martin Agüera, "Zwischen Hoffnung und Verzweiflung: Strategischer Lufttransport für Europa," in *Europäische Sicherheit*, no. 6 (June 2002): 48–49.

253. Bernard Bombeau, "L'Armée de l'Air étudie l'achat de C-130J," *Air et Cosmos*, no. 2166 (April 2009): 33; and EADS, "A400M: Erfolgreicher Erstflug des weltweit modernsten Transportflugzeugs," 11 December 2009, accessed 4 January 2010, http://www.eads.net.

254. EADS, "EADS begrüsst Entscheidung der A400M–Kundennationen," 24 July 2009, http://www.eads.com/1024/de/investor/News_and_Events/news _ir/2009/2009/20090724_eads_a400m.html.

255. Mason, *Air Power*, 249.

256. Thompson, *Political and Military Components*, 22–23.

257. Ibid., 52.

258. Paul Dreger, "JSF Partnership Takes Shape: A Review of the JSF Participation by Australia, Canada, Denmark, Israel, Italy, the Netherlands, Norway, Singapore, Turkey and the UK," *Military Technology* 27, no. 4 (April 2003): 28–31.

259. MOD, UK, "Royal Air Force Gets a Lift from C-17 Deal," *Defence News*, 4 August 2006, accessed 15 August 2006, http://www.mod.uk.

260. Dr. Hermann Hagena, brigadier general, GAF, retired, "Charter oder Leasing? Zwischenlösung für neue Transportflugzeuge," *Europäische Sicherheit*, no. 4 (April 2002): 35.

261. Michael J. Gething, "Balancing the Pitfalls and Potential of Aircraft Upgrade," *Jane's International Defence Review* 31, no. 12 (28 November 1998): 37.

262. Timothy Garden, "The Future of European Military Power," in Olsen, *Second Aerospace Century*, 250.

263. Col (GS) Hans-Dieter Schön and Maj (GS) Markus Schetilin, "Material- und Ausrüstungsplanung der Luftwaffe," *Europäische Sicherheit*, no. 10 (October 2006): 45.

264. NATO, "NATO's Allied Ground Surveillance Programme Signature Finalised," 25 September 2009, accessed 4 January 2010, http://www.nato.int /cps/en.

265. NATO, "Strategic Airlift Capability: A Key Capability for the Alliance," 27 November 2008, accessed 2 January 2009, http://www.nato.int/issues/strategic -lift-air-sac/index.html; and NATO, "First C–17 Plane Welcomed at PAPA Airbase," 27 July 2009, http://www.nato.int/cps/en/natolive/news_56690.htm.

266. European Defence Agency, "EU Governments Launch New Plan to Build Defence Capabilities for Future ESDP Operations," 14 December 2006, http://www.eda.europa.eu/newsitem.aspx?id=56.

267. *Declaration of Intent between the Minister of Defence of the Kingdom of Belgium and the Minister of Defence of the Czech Republic and the Federal Minister of Defence of the Federal Republic of Germany and the Minister of National Defence of the Hellenic Republic and the Minister of Defence of the Kingdom of Spain and the Minister of Defence of the French Republic and the Minister of Defence of the Italian Republic and the Minister of Defence of the Grand-Duchy of Luxembourg and the Minister of Defence of the Kingdom of the Netherlands and the Minister of Defence of the Portuguese Republic and the Minister of De-*

fence of Romania and the Minister of Defence of the Slovak Republic Regarding the Establishment of a European Air Transport Fleet, Brussels, 10 November 2008.

268. NATO, "Defence Capabilities Initiative," Press Release NAC-S(99)69, 25 April 1999, §1, http://www.nato.int/docu/pr/1999/p99s069e.htm.

269. Lebert, "Einrichtung eines Europäischen Lufttransportkommandos," 20.

270. Hagman, *European Crisis Management and Defence*, 63–64.

271. NATO, "NATO after Prague," http://www.nato.int/docu/0211prague /after_prague.pdf, 3.

272. European Union, *Headline Goal 2010*, 2.

273. Schmitt, *European Capabilities Action Plan*.

274. European Union, *Headline Goal 2010*, 3.

275. *Declaration of Intent*, 2008.

276. James Elliot, "From Fighters to Fighter-Bombers," *Military Technology* 23, no. 8 (August 1999): 69–70.

277. Ibid., 73–76.

278. Ibid., 76.

279. Duance and York, "Interview with Chief of the Air Staff," 26.

280. Mason, *Air Power*, 267–68.

281. Clifford Beal, "The Multirole Fighter Debate: Accepting the Inevitable," *Jane's International Defence Review* 26, no. 6 (1 June 1993): 454.

282. Ibid.; and Mark Hewish and Joris Janssen Lok, "Turning the Night into Day: Night-Vision Systems Reveal a Whole New World for Pilots," *Jane's International Defence Review* 31 (1 August 1998): 26.

283. Elliot, "From Fighters to Fighter-Bombers," 75–76.

284. "Wirken gegen Ziele in der Luft," *Wehrtechnischer Report: Fähigkeiten und Ausrüstung der Luftwaffe*, no. 3 (2006): 44.

285. Ibid., 43–44.

286. Air Chief Marshal John Allison, commander in chief, Strike Command RAF, "The Future of Air Power—a European Perspective," *Military Technology* 23, no. 5 (May 1999): 10.

287. Gen Merrill A. McPeak, USAF, retired, former chief of staff, "Precision Strike—The Impact on the Battle Space," *Military Technology* 23, no. 5 (May 1999): 22–24.

288. Elliot, "From Fighters to Fighter-Bombers," 69, 73; Karl-Heinz Pitsch, "Modulare Abstandswaffe Taurus," *Europäische Sicherheit*, no. 10 (October 2001): 39; and *Armée de l'Air 2007: enjeux et perspectives* (Paris: Sirpa Air, 2007), 49.

289. Gert Kromhout, "From SEAD to DEAD: USAF implements New Defence Suppression Course," *Military Technology* 24, no. 12 (December 2000): 9–12.

290. AP 3000, *British Air Power Doctrine*, 2.8.3.

291. Sergio Coniglio, "A400M, An-70, C-130J, C-17: How Do They Stand? A Comparative Report of Military Transport Aircraft Programmes," *Military Technology* 27, no. 7 (July 2003): 51.

292. Howorth and Keeler, *Defending Europe*, 83–84.

293. Hagman, *European Crisis Management and Defence*, 22.

294. Airbus Military, "A400M," accessed 4 February 2008, http://www.air busmilitary.com.

295. Lawrence Freedman, "Can the EU Develop an Effective Military Doctrine?," in *A European Way of War*, ed. Steven Everts et al. (London: Centre for European Reform, 2004), 22–23; and Jan Joel Andersson, *Armed and Ready? The EU Battlegroup Concept and the Nordic Battlegroup*, report no. 2 (Stockholm: Swedish Institute for European Policy Studies, March 2006), 30–31.

296. See Lt Col Stefan Scheibl and Lt Col Wolfgang Schad, "Combat Search and Rescue—Bewaffnete Suche und Rettung," *Europäische Sicherheit*, no. 11 (November 2002): 42–43.

297. "EC 725 Cougar Medium Multimission Helicopter," airforce-technology .com, http://www.airforce-technology.com/projects/ec725.

298. Michael J. Gething and Bill Sweetman, "Air-to-Air Refuelling Provides a Force Multiplier for Expeditionary Warfare," *Jane's International Defence Review* 39 (February 2006): 43.

299. Ibid., 45.

300. Mason, *Air Power*, 155.

301. Lambeth, *Transformation of American Air Power*, 122–23.

302. Schön and Schetilin, "Material- und Ausrüstungsplanung der Luftwaffe," 44–45.

303. Gen Tom Hobbins, USAF, "Unmanned Aircraft Systems: Refocusing the Integration of Air & Space Power," *Transforming Joint Air Power: The Journal of the JAPCC*, no. 3 (2006): 6–7.

304. Andrea Nativi, "Manned vs Unmanned Aerial Systems—False Expectations, Real Opportunities & Future Challenges," *Transforming Joint Air Power: The Journal of the JAPCC*, no. 3 (2006): 39.

305. Hobbins, "Unmanned Aircraft Systems," 9.

306. Ripley, *Air War Iraq*, 120–24.

307. Ibid., 13.

308. Mark Hewish and Joris Janssen Lok, "Connecting Flights: Datalinks Essential for Air Operations," *Jane's International Defence Review* 31, no. 12 (28 November 1998): 42.

309. Myron Hura et al., *Interoperability: A Continuing Challenge in Coalition Air Operations*, RAND Monograph Report MR-1235-AF (Santa Monica, CA: RAND, 2000), 108, 111, http://www.rand.org/publications/MR/MR1235.

310. Hewish and Lok, "Connecting Flights," 42–45.

311. John Asenstorfer, Thomas Cox, and Darren Wilksch, *Tactical Data Link Systems and the Australian Defence Force (ADF): Technology Developments and Interoperability Issues*, rev. ed. (Edinburgh, South Australia: Defence Science & Technology, February 2004), 5, http://www.dsto.defence.gov.au/pub lications/2615.

312. Rainer Fiegle, "Neue Kommandostruktur für taktische Luftkriegsführung," *Europäische Sicherheit*, no. 3 (March 1995): 30–31.

313. NATO, "NATO Air Command and Control System," http://www.nato .int/issues/accs/index.html.

314. See Lt Gen B. A. C. Droste, commander in chief, RNLAF, "Shaping Allied TBM Defence," special issue, *NATO's Sixteen Nations and Partners for Peace* 42 (1997): 47.

315. Lambeth, *Transformation of American Air Power*, 145.

316. Droste, "Shaping Allied TBM Defence," 47.

317. Bernd Baron von Hoyer-Boot, "Erweiterte Luftverteidigung," *Europäische Sicherheit*, no. 9 (September 1992): 506.

318. Thompson, *Political and Military Components*, 107.

319. Trevor Nash, "Low Level Air Defence: Getting the Balance Right," *Armada International* 21, no. 2 (April/May 1997): 12.

Chapter 3

French Air Force (Armée de l'Air)

During WWI the French aircraft and aircraft engine industries were at the forefront in production and technical efficiency. By the last year of the war, the French air service was the second largest in the world; yet, only in 1933 did it become an independent service. However, Army officers—who dominated the high command of the armed forces—continued to provide strategic direction. This led to an inappropriate air power doctrine and contributed to France's defeat in 1940.[1]

After WWII and following two large-scale conflicts in Indochina and Algeria, a watershed in France's history was the collapse of the Fourth Republic in May 1958 and the establishment of the Fifth Republic under the leadership of Pres. Charles de Gaulle. This ushered in fundamental changes in the French force posture.[2] To guarantee French status, developing autonomous nuclear deterrence was considered indispensable. In parallel, de Gaulle pursued a path of gradually disengaging from NATO's integrated military command structure. These events culminated in a letter to the American president on 7 March 1966, where de Gaulle announced that French forces would no longer remain subject to NATO's integrated military command structure. On a political level, France continued to be a member of the alliance, and the French armed forces constituted an operational reserve in NATO's central region. A certain level of operational cooperation was retained between NATO and the French Air Force. In 1970 France and NATO signed an MOU regarding the exchange of recognised air picture data.[3]

From 1959 onwards, budgetary and doctrinal priority had been placed upon strategic nuclear forces, resulting in the first French nuclear bomb test on 13 February 1960. In October 1964, the FAF with its Mirage IV bombers became the first operational pillar of the strategic forces. The FAF also acquired responsibility for ground-launched strategic ballistic missiles, which reached operational status in 1971. The nuclear strike role was considered to give the FAF a truly independent status.[4]

Mirage 2000N. In October 1964, the FAF with its Mirage IV bombers became the first operational pillar of the strategic nuclear forces. The airborne nuclear strike role was later transferred to the Mirage 2000N, shown here armed with a mock-up *air-sol moyenne portée* (ASMP), or medium range air-to-ground weapon, a standoff nuclear missile.

For conventional materiel, France pursued a path towards autarchy. Whereas the FAF was dependent on foreign aircraft immediately after WWII, it relied almost exclusively on a national industrial base from the 1960s onwards. Yet, emphasis on nuclear forces rendered conventional forces secondary.[5] Moreover, staying outside NATO's integrated military command structure led to a reduced level of interoperability with NATO partners.

The ensuing shortfalls in the FAF's conventional force posture became particularly apparent during Operation Desert Storm. Hence, the FAF aspired to significantly improve its conventional capabilities and to enhance interoperability with its NATO partners. This did not imply that France would forego its particular national status in multinational air campaigns, an issue that became particularly apparent during Operation Allied Force in 1999. When French decision makers didn't consent to certain target sets, their American counterparts viewed them as non-cooperative. This led to the perception, particularly in the Anglo-Saxon world, that France had effectively hindered a more effective air campaign.[6] The image of France as an obstinate ally was again reinforced in the course of the Iraq crisis in early 2003. In contrast to this impression, the FAF has made valuable contributions in combined operations. For instance, the FAF was the first European air force to deploy combat aircraft for Operation Enduring Freedom.

How Has the French Air Force Adapted to the Uncertainties Created by Shifting Defence and Alliance Policies?

This section first analyses France's post–Cold War defence policy and its influence on the FAF's evolution. France's alliance policies are described next, along with a view towards how they have affected its Air Force.

Defence Policy

In the post–Cold War era, French defence policy has been articulated primarily through three key documents—the 1994 *Defence White Book*, published under the late presidency of François Mitterrand; Jacques Chirac's reform, *A New Defence 1997–2015*; and the 2008 *Defence White Book* published in the early presidency of Nicolas Sarkozy. While the 1994 *Defence White Book* represented a cautious adjustment to the post–Cold War environment, *A New Defence* initiated far-reaching reforms in the second half of the 1990s. The 2008 *Defence White Book* ushered an era of a renewed relationship between France and the transatlantic alliance.

The 1994 *White Book* primarily described the emerging geostrategic environment immediately after the end of the Cold War and laid out broad directions for the French defence architecture. Conspicuously, it avoided referring to critical issues such as the future composition of the nuclear forces.[7] *A New Defence,* on the other hand, was regarded as arguably the most substantial revision of French defence policy since the 1966 withdrawal from NATO's integrated military command structure.[8] It set the background for a major military reform, the Armed Forces Model 2015.[9] This long-term plan was then broken down into so-called military planning laws, each covering a five-year period and providing a framework for force structuring and defence spending. In his foreword to the 2008 *White Book*, Sarkozy, while acknowledging the importance of the previous reforms, stated that the Armed Forces Model 2015 had been too ambitious and that a reform was necessary.[10]

Threat and Risk Perception. In 1990 French decision makers only hesitantly modified their threat perception, particularly compared to their counterparts in the United States and the United Kingdom. According to the French view, although the threat of a massive surprise attack seemed to have vanished, the Soviet Union still posed a threat due to its vast arsenal of conventional and unconventional weapons. Only after the failed coup in Moscow in mid-1991 did French officials seriously start to acknowledge the demise of the Soviet/Russian threat.[11]

With the 1994 *Defence White Book*, French threat and risk perception had further adjusted to post–Cold War realities. Regional conflicts were considered to destabilise global stability, and the fragmentation of the Soviet Union was observed with great concern, particularly regarding nuclear proliferation. Moreover, nonmilitary threats and risks received considerable appreciation in the *White Book*. Among these, terrorism was singled out as the most substantial.[12]

The view that the distinction between internal and external security had blurred was accentuated in the wake of 11 September. According to the security assessment as laid out in the military planning law for 2003–8, a new form of terrorism had emerged. Consequently, the vulnerability of French society had to be taken into account in all its dimensions. Moreover, it was assumed that the threat potential of failing states and WMD proliferation had increased further.[13]

The 2008 *White Book* expanded these security concerns by emphasising the phenomenon of globalisation, ushering in an era of strategic uncertainty. While acknowledging the positive aspects of globalisation, the *White Book* underlines several security concerns caused by a more globalised environment. It stresses the issue of energy and resource security and identifies increased military spending—particularly in India and China, as well as a reasserting Russia—as a potentially destabilising factor.[14] Given all these developments, the *White Book* concludes that France and Europe are in a more vulnerable situation now than at the end of the Cold War.[15]

Tasks of the Armed Forces. The end of the Cold War made a conceptual realignment of the tasks of the French armed forces necessary. The 1994 *Defence White Book* lays out three objectives for French defence policy—defending and safeguarding French interests, contributing to a common European defence and to international stability, and embracing a broad approach to defence that encompasses both internal and external security.[16] These objectives are translated into four missions for the French armed forces:

- Preserve vital French interests against any form of aggression.

- Contribute to the security and defence of Europe and the Mediterranean space and build a common European defence.

- Help to strengthen peace and respect for international law.

- Support civil authorities through subsidiary missions.[17]

The *White Book* further states that nuclear deterrence remains an indispensable pillar of French military strategy. In contrast to the Cold War, however, it is argued that a new balance between nuclear and conventional forces must be achieved. Moreover, France has to be capable of defending its interests, both autonomously and in a coalition. While conventional forces would allow France to intervene in regional conflicts, nuclear forces must be capable of permanently fulfilling two key functions: to inflict unacceptable damage upon any opponent in a second strike and to conduct limited strikes against military objectives.[18] The latter function represents a novelty and is a response to WMD proliferation concerns.[19]

A major shift in paradigms occurred at the conventional force level. During the Cold War, the preeminence of nuclear doctrine rendered conventional forces secondary, as they were largely embedded into nuclear strategy.[20] Yet, the 1994 *Defence White Book* conceives their primary employment outside the nuclear context. In the new strategic environment, conventional forces were to act autonomously according to three main functions—prevention, action, and protection—with deterrence being the fourth function and relegated to the strategic forces. In the context of prevention, diplomatic, economic, and military means were to be coordinated, with the armed forces detecting potential crises and preventing them from unfolding. In case of failure, action would be required. This function primarily foresaw combined and joint crisis management operations. The third function, protection, was chiefly related to the defence of national territory.[21]

With the publication of *A New Defence* in 1996, Chirac basically reiterates these functions, though slightly adapted. The function of *action* was translated into *projection*. Chirac explicitly argues that the first line of defence is no longer close to the national border but often far away from French territory. Projection became the first priority for French conventional forces.[22]

In the wake of 11 September, the 2003–8 military planning law again confirms the four strategic functions of deterrence, prevention, projection-action, and protection. It also underlines that preemptive action could be considered given an explicit and imminent threat.[23] Given Iran's nuclear activities, for instance, President Chirac announced in January 2006 that states supporting WMD terrorism should be aware of exposing themselves to a potentially firm response.[24]

The 2008 *White Book* introduces a new and fifth strategic core function: cognition and anticipation. In an environment of strategic uncertainty, it is considered to be the most important. By reiterating the original four functions, albeit with a slightly different wording—*intervention* replaced *projection-action*—Sarkozy provides French defence policy in the post–Cold War era with continuity.[25] In accordance with the new emphasis upon cognition and anticipation, the ability of gathering information autarchically was deemed of utmost importance to guarantee France's strategic autonomy. Regarding intervention, the *White Book* explicitly acknowledges the predominantly multinational character of deployed operations.[26]

Towards Professional Armed Forces. The 1990–91 Gulf War showed serious shortfalls in French power projection capabilities. Different from small-scale deployments to Africa during the 1980s, a visible contribution to the Gulf War required materiel and personnel on a much larger scale. Despite an overall strength of more than 450,000 personnel, the French armed forces at first were able to deploy only around 5,000 troops to the Middle East. Finally, the French contingent was augmented to 19,000 personnel, including a reserve force of 3,400 in Djibouti. Nevertheless, it was dwarfed by the US deployment of more than half a million troops and was significantly weaker than the UK contingent of 36,000 troops.[27]

Even with the increasing demands for power projection, the 1994 *Defence White Book* reconfirms the adherence to conscription, arguing that it remains an important link between the armed forces and society, plays a crucial role in the defence of French territory, and is indispensable for the reconstitution of the French armed forces. Nevertheless, it identifies a need to augment professional personnel.[28] The FAF itself had always had a high degree of professional personnel—it was 62 per cent in 1993.[29] However, this did not prevent shortcomings in the FAF's projection capability during the Gulf War.[30]

In line with his major defence reform, Chirac argued in the mid-1990s that power projection requirements rendered conscription obsolete.[31] As a consequence, the French armed forces were planned to be converted into a fully professional force reduced to 350,000 troops. This was supposed to allow power projection of between 50,000 and 60,000 personnel beyond the national borders.[32] Particularly, shortcomings against the backdrop of Operation Desert Storm were to be the driving factor behind Chirac's reform.[33] In this thrust towards fully professional armed forces, the FAF was to undergo a 29 per cent cut in personnel from 89,200 to 63,000 in the second half of the 1990s.[34]

Compared to previous decisions, the 2008 *White Book* reduced quantitative projection ambitions to 30,000 deployed ground troops for a major operation.[35] In the same vein, the military planning law for 2009–14 announced a reduction of military and civil Air Force personnel, with a force goal of 50,000 by 2015.[36]

Defence Policy and Its Impact upon the French Air Force. The evolution of the FAF in the post–Cold War era has primarily been guided by three directives—the Armed Forces 2000 project, the directive for the reorganisation of the FAF command structure as of 1994, and the Armed Forces Model 2015. Each step attempted to provide an organisational response to the shifting defence policy paradigms.

The Armed Forces 2000 project was presented for the first time in June 1989, four months prior to the fall of the Berlin Wall. Two years later, the modalities to implement the Armed Forces 2000 project were laid out. Its main objective was to improve military crisis management capabilities by enhancing power projection.[37] A reform was considered urgent due to telling shortfalls in the

FAF's conventional force posture, as revealed during the Gulf crisis, and due to French overemphasis on nuclear forces. Throughout 1990 and 1991, for instance, FAF conventional forces were not even allocated enough funds to attain 180 flying hours per pilot per year.[38] Accordingly, a 1992 RAND report concludes that the FAF faced the largest modernisation and replacement requirements of any air force in NATO's central region.[39]

Against the backdrop of the reform, it was decided to render the FAF leaner by focusing on quality versus quantity. Plans were to reduce the combat aircraft fleet from 450 to between 390 and 350 aircraft by 1994.[40] The Armed Forces 2000 project's major shortfall, however, was its being conceptually grounded in the Cold War. Despite its focus on crisis management, it did little to prepare the French armed forces for the upheavals during the first half of the 1990s.[41]

Adapting the FAF's force structure to the new geostrategic context was indispensable. In 1994 a new command structure was introduced, distinguishing between operational and organic commands. Prior to this reform, the FAF had largely been organised according to the structure of 1960. The implementation of that structure was at the time the consequence of the adoption of the nuclear task. The structure itself consisted of four operational commands—Strategic Air Command, Tactical Air Command, Air Defence Command, and Airlift Command. These four commands were complemented by three support commands responsible for training, signals, and engineering work.[42]

To respond to the new requirements, the structural reform of 1994 had to improve the flexible employment of conventional air assets. The result was the creation of the Air Defence and Operations Command. This command was charged with planning and conducting all conventional operations, both offensive and defensive, and hence enabled centralised command.[43] Alongside Strategic Air Command, Air Defence and Operations Command was henceforth the only operational command. This significantly streamlined the command structure of the FAF and reduced unnecessary redundancies. Apart from these two operational commands, the 1994 FAF structure included five organic commands in charge of the combat aircraft fleet, the airlift fleet, the ground environment and signals, ground protection, and training.[44]

The reorganisation of the FAF's command structure represented a direct response to the shifting geostrategic context. Since a major clash within Europe was no longer an immediate threat, large command structures and forces preassigned for specific tasks had become obsolete. Instead, continuous operations abroad and the thrust towards combined and joint operations required a flexible approach. The structure was fully in line with the 1994 *White Book*, which largely decouples conventional forces from nuclear strategy and puts a premium upon power projection and operations in multinational frameworks.[45] Three years later, under Chirac's presidency, military planning law 1997–2002 confirmed the reform of the command structure.[46]

The military planning law for 1997–2002 was supposed to be the first out of three to achieve the goals set by the Armed Forces Model 2015. It initiated a streamlining process of the aircraft fleet, as depicted in the following table:

Table 1. Armed Forces Model 2015 (FAF)

1996	*2002*	*2015*
405 combat aircraft (Mirage F1, Mirage 2000, Jaguar, Mirage IV)	360 combat aircraft (Mirage F1, Jaguar, Mirage IV, Mirage 2000)	300 modern combat aircraft
80 tactical transport aircraft	80 tactical transport aircraft	50 modern transport aircraft (A400Ms)
11 AAR aircraft	14 AAR aircraft	16 AAR aircraft
6 specialised C-160 Transall	6 specialised C-160 Transall	2 specialised C-160 Transall
4 AWACS	4 AWACS	4 AWACS
1 DC 8 Sarigue (electronic intelligence)	1 DC 8 Sarigue	1 DC 8 Sarigue
101 helicopters	86 helicopters	84 helicopters

Reprinted from Loi de Programmation Militaire (LPM) [French military programme bill of law] 1997–2002, *Rapport fait au nom de la commission des Affaires étrangères, de la défense et des forces armées (1) sur le projet de loi, adopté par l'Assemblée Nationale après déclaration d'urgence, relatif à la programmation militaire pour les années 1997 à 2002* [Report on behalf of the commission of foreign and defence affairs and the armed forces on the military programme bill of law 1997–2002, approved by the national assembly after a declaration of urgency], no. 427 (Paris: Sénat, 1996), 132.

Yet, the delay of several procurement programmes due to financial, technological, or industrial reasons prevented the FAF from attaining the original goals of Armed Forces Model 2015. For instance, as of 2003, only 150 Rafale combat aircraft at most were expected to be delivered by 2015 out of a total of 234 aircraft destined for the FAF.[47]

Courtesy FAF, Sirpa Air

Rafale, the latest French combat aircraft

To support the primary drive towards enhancing power projection, the number of AAR aircraft was to be increased. Against the projection logic, however, was the planned reduction of the airlift fleet. A reason for this cutback lies in the aging of available airframes and delays in their replacement. With power projection identified by Chirac's reform as the primary mission of conventional forces, a major goal set for the FAF has been the capability to deploy up to 100 combat aircraft in an out-of-area operation.[48] Given delays in the A400M programme and an aging tanker aircraft fleet, this goal proved too ambitious for operations beyond the close vicinity of Europe.[49] Chirac's defence reform also foresaw a simplification and reduction of the nuclear forces by abandoning the terrestrial component. Henceforth, nuclear deterrence has been maintained by submarine- and air-launched nuclear missiles.[50]

To further adapt the FAF to the new environment and enhance its effectiveness, a new Air Force structure—Air 2010—was envisaged in 2003. In the summer of 2005, Air 2010 was initiated by establishing the Support Command in Bordeaux.[51] While the two operational commands, Strategic Air Command and Air Defence and Operations Command, were retained, the organic commands were reduced to two. Air Forces Command, essentially a force provider, prepares the various FAF units while Support Command provides operational support.[52] As such, Air 2010 resulted in a further streamlining of the FAF's command structure.

In 2008 Sarkozy declared that the Armed Forces Model 2015 had been too ambitious and proposed a reform.[53] In line with the new strategic core function, cognition and anticipation, the 2008 *White Book* prioritises the modernisation of the ground-based air-surveillance architecture.[54] Yet, the military planning law for 2009–14 significantly reduced the total force goal of modern combat aircraft, Rafales and Mirage 2000Ds, to 300 for both the FAF and the French Navy.[55]

Emphasis upon Air Power. The French defence budget not only contains the single service budgets of the Army, Navy, and Air Force but also expenditures for the national military police force, the administration, and others. When only comparing the three service budgets of 2002 with each other, the Army receives 41.6 per cent, the Navy 28.95 per cent, and the FAF 29.45 per cent.[56] Yet, air power is not relegated to the FAF. Both the Army and Navy operate a considerable number of air assets.

In the early 1990s, the Army air wing (*Aviation légère de l'armée de terre*) contained slightly more than 700 utility, transport, and combat helicopters as well as a couple of fixed-wing aircraft. Additionally, the Army operated more than 1,200 air defence guns as well as 345 SAM systems, including man-portable air defence systems (MANPADS) and a considerable number of Roland I/II and Hawk systems. With its organic air arm, the Navy operated 156 fixed-wing and 52 rotary-wing aircraft in the early 1990s.[57] A decade later, the Army's air assets had been significantly reduced to slightly more than 400 aircraft. The naval air arm's inventory, on the other hand, remained stable with 123 fixed-wing and 78 rotary-wing aircraft.[58] The military planning law for 2003–8 projects the modernisation of the Army air wing and the naval air arm. The latter was planned to be equipped with 60 Rafale multirole combat aircraft.[59] The military planning law for 2009–14 announces the procurement of 80 Tigre combat helicopters and 130 modern transport helicopters for the Army air wing.[60] These figures reveal that the French armed forces emphasise air power far beyond the slightly less than 30 per cent FAF budget allocation.

With regard to the relevance of the FAF in real operations, interesting evolutions have taken place in the post–Cold War era. During the 1990–91 Gulf crisis, the FAF's mission was originally conceived to be purely defensive. As well as contributing to the defence of Saudi airspace, its main mission was to cover the action of the French Army division Daguet.[61] Hence, air power was considered to be a mere adjunct to the manoeuvre of ground forces. Only later did Mitterrand decide that the FAF was to take on its share of the coalition air campaign. In the Balkans, the weight given to air versus land power was more balanced. While the Army excelled by its robust commitment to ground operations in Bosnia—making France the political and military leader of the United Nations Protection Force (UNPROFOR)—the FAF took on a major European role in NATO's air campaigns over Bosnia.[62] The forward-deployed forces in Africa have been inherently integrated.[63] With the Army providing the bulk of personnel, the FAF has been a crucial strategic enabler, given the large distances in Africa. As portrayed in the next section, French contributions at the upper force spectrum in Kosovo and later in Afghanistan primarily relied upon air power. The FAF has gradually gained importance throughout the post–Cold War era and now can be regarded on a par with its sister services in deployed operations.

Regarding nuclear deterrence, while the maritime component with its nuclear submarines is the most powerful and survivable component, the strategic air component is considered useful for political signalling due to its visible and reversible deployment.[64] Hence, the FAF continues to play an important role in this arena.

Alliance Context

France's alliance policy of the post–Cold War era has been considerably marked by its Gaullist heritage. Autonomy and independence in decision making have remained cornerstones of French foreign and defence policy. Yet the end of bipolarity, which had allowed France a certain degree of freedom of manoeuvre between the two superpowers, and the new security challenges meant that France had to embark upon a path of rapprochement towards its NATO partners.

Reasons for rapprochement were manifold. A crucial one was France's active involvement in UN operations. While France's rank and status were secured by an independent nuclear force during the Cold War era, this formula was only partially valid in the post–Cold War era. Particularly in the early 1990s, France sought to secure its international status and permanent seat in the UN Security Council through a decisive role and enhanced visibility in peace support operations. As the conduct of these operations was largely shaped by NATO councils and committees, participation in the decision-making process was considered necessary.[65]

Only by embedding its actions into a multilateral framework was France able to defend its interests and to influence international affairs. In the first half of the 1990s, France undertook a whole series of initiatives within the WEU, EU, or NATO aimed at establishing an autonomous European defence and security architecture.[66] For instance, the setting up of a Franco-German army corps in 1992, later to become known as Eurocorps, was conceived to be the nucleus of a European army.[67] In short, Europe had become France's new frontier, and French officials perceived it as their sphere of influence. Particularly through the construction of a common European defence, French status would be fostered.[68] However, not being able to prise its European partners away from the comfort of an American commitment to European security, France had to attain reform from within NATO to foster a more autonomous European defence.[69]

Already in 1994, the *Defence White Book* conveyed a certain shift in French paradigms. On the one hand, it argued that NATO must become an arena where the European security and defence identity could establish itself and that the adaptation of NATO to the new strategic conditions must be resolutely pursued. On the other, the *White Book* clearly restated the particular French status within NATO. It argued that the principles laid down in 1966, including non-participation in NATO's integrated military structure, would continue to guide French relations with the organisation.[70]

Franco-NATO rapprochement was further strengthened when Jacques Chirac came to office in spring 1995. The new president aimed at exploiting the ongoing evolutions within NATO to restore France as a major player within the alliance and to Europeanise the alliance from within. However, a reintegration of France into NATO's military command structure was not conceivable. The dominance of the United States and the automaticity of the integrated command structure were considered to be incompatible with France's key principles.[71]

The most tangible sign of French rapprochement towards NATO was the increasing participation of French officials in its councils and committees. In September 1994, France was for the first time since 1966 represented in an informal meeting of NATO defence ministers in Seville, Spain.[72] In 1995 French participation became more formalised. After the French chief of staff had attended a meeting of NATO's Military Committee, it was declared that France would henceforth participate fully in the Military Committee and that French defence ministers would participate in the North Atlantic Council meetings on a regular basis.[73]

France's rapprochement towards NATO in the mid-1990s did not imply acceptance of the US-dominated status quo. It was French pragmatism that accelerated Franco-NATO rapprochement. The setting up of an autonomous European defence architecture remained the overarching goal of French defence policy.[74] Recognising the dominance and importance of NATO in the medium term, French decision makers conceded to the necessity of building a European defence within the alliance to the point where European strategic autonomy would become a reality.[75]

However, French decision makers did not succeed in transforming NATO's command structure according to their views in the mid-1990s. With American preponderance at the highest level of the NATO command structure—Allied Command Europe and ACLANT—the French first sought to install a European SACEUR commanding ACE to shape the alliance in line with the European Security and Defence Identity, which was embraced by the alliance during the January 1994 Brussels summit. After this attempt had failed, French decision makers tried to make adjustments at the next lower command level. In particular, they pushed for a European general commanding Allied Forces Southern Europe (AFSOUTH) in Naples. Yet this proposal was not only rejected by the Americans in late 1996 but also by Germany, France's close ally in European defence matters. As a result of these developments, Paris decided to slow down rapprochement. This state of affairs made any further Europeanisation of NATO less likely and threatened the core objective France herself had sought within NATO.[76]

After a period of stagnation in setting up a European defence pillar, new momentum was gained at the Franco-British meeting of December 1998 in St. Malo, France. As discussed in chap. 2, both countries agreed that the EU must have the capacity for autonomous military action. An agreement between Britain and France was particularly important as both countries were considered to possess the most relevant armed forces in Europe.[77] The overcoming of obstacles between France and Britain laid the foundation for the European Security and Defence Policy.

Moreover, despite tensions over the Iraq crisis in 2003, France has reiterated its policy of rapprochement towards NATO and has made significant contributions to the newly set up NATO Response Force, with the logic that only a strong French commitment can contribute to the alliance becoming more European.[78]

The 2008 *White Book* announced France's intention to renew the transatlantic partnership.[79] In 2009 France finally returned to NATO's integrated military command structure. The reintegration was underlined by France receiving two of NATO's top military positions—the post of Supreme Allied Commander Transformation (SACT) and the command of the Joint Command Lisbon.[80] Yet on numerous occasions, the president and French officials made it clear that French autonomy, particularly in the domain of the country's nuclear deterrent, would not be compromised by this move, thereby retaining a certain degree of continuity of Gaullist paradigms.[81]

France's return to NATO's integrated military command structure did not mean that the country would no longer put a premium upon the European Union as a major actor in the domain of security. Both the 2008 *White Book* and the military planning law 2009–14 declare that the EU has far-reaching international responsibilities, including military ones. As such, its crisis management capabilities were to be further reinforced to achieve a better task sharing in security matters between Americans and Europeans.[82]

Alliance Policy and Its Impact upon the French Air Force

In the post–Cold War era, France has come to realise that no European nation can afford and sustain the costs of an air force that is capable of intervening independently in major crises.[83] Thus, France's approach on a grand-strategic level has been directly translated onto an air power level. As the FAF has a limited but balanced force structure, the European context has been perceived to enhance its operational capability.[84] To build European defence, the FAF has supported and promoted the harmonisation of procedures and structures.[85] While a certain degree of independence continues to be regarded as essential, the FAF has sought to integrate European air power under the aegis of the EU and NATO as well as through bilateral and multilateral cooperation.[86]

ESDP Context. As discussed, the FAF, together with the German Air Force, has been the lead air force in setting up the EU Air RRC. In the context of the Helsinki Headline Goal, France, similarly to Britain and Germany, had committed itself to provide 20 per cent of the air assets put at the disposal of the EU. A potential French contribution would theoretically have consisted of a maximum of 80 combat and reconnaissance aircraft and 30 transport aircraft with the corresponding support materiel.[87] In 2007 the contribution was slightly reduced to 75 combat aircraft for potential EU operations, with another 25 retained for national deployed operations.[88] Given shortfalls in the FAF's airlift capacity, as is shown later, this goal was too ambitious for operations beyond Europe's vicinity. Thus, Sarkozy's reform of 2008 made adjustments and reduced the overall number of deployable FAF and French Navy combat aircraft to 70.[89] However, the number of aircraft assigned to potential EU or NATO operations was not specified in the new *Defence White Book*.

NATO Context. The prospect of advancing the European pillar within the Atlantic Alliance has certainly featured prominently in French calculations of its significant contributions to the NRF. But France approved the NRF concept only under certain conditions. For instance, the employment of French armed forces on behalf of the European Union must not be affected by NRF commitments.[90] Despite these restrictions, the FAF has been a key player within the NRF. While France was in charge of the NRF 5

French Mirage 2000 and Belgian F16s

CJFAC during the second half of 2005, command of the NRF 6 CJFAC was passed on to the RAF at the end of the year.[91] Together with the RAF, the FAF provided the main bulk of air assets, and a French general acted as the NRF 5 CJFACC.[92] Approximately 1,500 French personnel were put on standby. Regarding air assets, the French contribution consisted of 24 combat aircraft, six transport and four tanker aircraft, as well as one AWACS aircraft.[93] Together, the RAF and the FAF provided 80 per cent of the CJFACC's staff and 40 per cent of the air assets, with other NATO countries supplying the remaining personnel and materiel. Cooperation between the two air forces generated synergies. French structures and national procedures were fully interoperable with NATO and hence allowed smooth integration of allied air forces.[94] Also, both air forces were cooperating closely in ground support. Together, they provided three deployable air bases for NRF 5 and NRF 6.[95] France and Great Britain were at the time the only European nations able to provide such a capability on a national basis.[96]

Bilateral and Multilateral Organisational Relationships. The FAF has also been a key player in establishing bilateral and multilateral organisational relationships in the post–Cold War era. The European Air Chiefs Conference (EURAC), for example, was launched on a French initiative in 1993. It brings together at least once a year (until 2004, two annual formal meetings) the chiefs of the Air Staffs of several European air forces to reflect on common problems.[97] Through the EURAC, the FAF has been able to promote a number of projects aimed at cooperation, such as the Advanced European Jet Pilot Training that was conceived to pool European fighter pilot training. The FAF intends to play an important role in this field and to offer its infrastructure and assets.[98] A milestone was reached in September 2004 when the Franco-Belgian Advanced Jet Training School was established.[99]

As mentioned, France and Great Britain were also in the lead in establishing a common approach to European air power by inaugurating the Franco-British European Air Group in High Wycombe, UK, on 30 October 1995.[100] The FBEAG was later transformed into the EAG, taking on board other European nations. Given the professionalism and experience of the British armed forces, Great Britain has always been an important partner for France to develop meaningful approaches to cooperation.

On the continent, it has largely been the German Air Force with which the FAF has sought to deepen its ties. This basically reflects the Franco-German partnership at the grand-strategic level in promoting European defence. A few months after the creation of the FBEAG, the FAF set up an exchange system for airlift with the German Air Force. Henceforth, the system allowed each nation to utilise each other's transport aircraft.[101] France and Germany have also been the lead nations in developing a European air transport command, as is explained later.

In the wake of 11 September, the FAF has mainly been pursuing European cooperation in the area of air policing. According to the view of the chief of the Air Staff, multilateral cooperation and efforts to integrate national airspaces were to be identified and air policing procedures to be harmonised.[102] By early 2006, the FAF had formalised or was in the process of formalising cooperation with Spain, Switzerland, Belgium, Italy, Germany, the United Kingdom, and Brazil.[103] Furthermore, Luxembourg contemplated the FAF controlling its airspace.[104]

Cooperation on an Operational Level. In the post–Cold War era, the FAF has gained a high level of experience in combined air campaigns. In Operation Allied Force, Mirage 2000Ds were often air refuelled by British tankers, escorted by Dutch F-16s, and supported by German electronic combat/reconnaissance (ECR) Tornados.[105] The Europeanisation of air power was particularly visible when a French Mirage 2000D detachment to Central Asia was superseded by a trinational air component consisting of Danish, Dutch, and Norwegian F-16s. Against the backdrop of Operation Enduring Freedom, this component took over from the FAF in late 2002. During the transitional phase, the airlift assets of the involved European nations were coordinated through the EACC.[106]

Moreover, to build European defence, integrated exercises and exchange of pilots have been considered of major importance. Integrated exercises in the framework of NATO have allowed the

Courtesy Gripen International; photo by FMV

A FAF Boeing C-135FR tanker aircraft refuelling SwAF Gripen combat aircraft

FAF to test, train, and improve its expertise in terms of operational planning and the conduct of combined operations and can hence be regarded as an important enabler of French high-level contributions to the NRF air component.[107] In the post–Cold War era, the thrust towards combined and joint exercises has increased. The FAF has not only participated in exercises with NATO partners, such as Red Flag, Maple Flag, or Anatolian Eagle, but has also taken part in exercises in Brazil and the Arab Emirates.[108]

How Has the French Air Force Responded to the Challenges of Real Operations?

In the wake of the Algerian War, France adopted a new military strategy to protect its remaining overseas interests. This strategy hinged upon light and rapidly deployable units as well as on theatre reception bases, which allowed for interventions on a much smaller scale but not necessarily at the lower end of the spectrum of military force. Particularly under the presidency of Valéry Giscard d'Estaing (1974–81), French overseas interventions experienced a drastic rise against the backdrop of growing Soviet-Cuban involvement in Africa.[109]

Gradually, reliance upon the FAF to project power grew. This trend is illustrated by the composition of French deployments to Chad between 1968 and 1987. The first operations hinged upon light and mobile infantry forces, with air power playing a secondary role and being primarily relegated to inter- and intratheatre airlift. In the late 1970s, French Jaguar aircraft took on an increasingly important role by delivering concentrated firepower.[110] In the first half of the 1980s, the increasing Libyan air threat for the first time made the deployment of French air defence assets such as the Mirage F1 and Crotale SAMs necessary.[111] The final major deployment to Chad, lasting from early 1986 to late 1987, reversed the original composition of French intervention forces. For the first time, an Air Force officer was appointed as the joint force commander. The operation's aim was to deter enemy air and land incursions. For this purpose, the FAF executed two major offensive counterair missions deep inside enemy-held territory and engaged Libyan aircraft at least twice in combat.[112]

FAF C-160 Transall operating from a gravel strip in Chad

The FAF was also engaged in counterinsurgency operations. During Operation Lamantin in late 1977 and early 1978, for instance, up to eight French Jaguar combat aircraft conducted decisive strikes over two days against Polisario rebels raiding Mauritanian government installations, thereby contributing to the establishment of a ceasefire. They were supported by tanker aircraft and guided by French Navy Atlantic aircraft that served as command and SIGINT platforms. French forces never exceeded 350 personnel.[113] This experience, spanning the entire spectrum from counterinsurgency operations to offensive counterair missions, rendered the FAF one of the most combat experienced Western air forces in the late 1980s.[114] Nonetheless, operations in Africa remained secondary to the FAF's nuclear role.[115]

After decades of autonomous operations, interoperability became a particular challenge for the FAF at the outset of the post–Cold War era. Within a decade, the FAF came a long way. Nowadays, it is capable of rapidly integrating itself into major combined air operations and accomplishing leadership functions, as the French commitment to the NRF air component conveys. Moreover, the FAF has retained its ability to operate autonomously, particularly in African theatres. It offers French decision makers a broad range of options across the entire spectrum of force.

French Contributions to Western Air Campaigns

Desert Storm constituted a watershed for French defence policy in general and for the FAF in particular. Politically, French participation in Operation Desert Storm aroused tensions within the republic's leadership. While François Mitterrand was basically the only one to support a troop deployment from the very beginning, the minister of the interior, for instance, warned of military engagement in an American-led operation and its consequences, particularly with regard to Franco-Arab relations.[116] At the special request of French defence minister Jean-Pierre Chevènement, French combat aircraft were not based with American forces to prevent the FAF from being too closely linked with the coalition. Sympathetic to the Iraqi leadership, the defence minister wanted to relegate French operations to purely defensive actions. In the course of the campaign, he accordingly opposed strikes against targets on Iraqi territory but was overruled by Mitterrand. On 27 January 1991, Chevènement finally resigned from his post.[117]

The altered geostrategic environment and experience of Desert Storm caused France to gradually abandon this isolationist stance. There was a realisation that no European state could sustain any longer the costs of an air force capable of intervening independently in major crises. The FAF was from the very beginning closely integrated in the air campaign over the Balkans and was able to integrate itself swiftly into the American-led air campaign over Afghanistan in early 2002. This change in paradigm is also reflected in France's involvement in cooperative endeavours, which aspire—amongst others—to harmonise European air power.[118]

In 1990 political reluctance to subordinate French forces to the joint force commander during the force buildup led to a military marginalisation of the FAF contingent.[119] Besides support aircraft, the deployment contained 14 Jaguar, 10 Mirage 2000, and six Mirage F1 CR reconnaissance aircraft.[120] At dawn on 17 January, French Jaguar aircraft conducted low-level attacks against an Iraqi air base in Kuwait. Though four aircraft were hit by ground fire, they accomplished their mission and returned safely.[121] After these offensive counterair strikes, the FAF targeted artillery sites in Kuwait and Republican Guard units in southern Iraq. French Jaguar aircraft conducted precision strikes by means of AS-30-L laser-guided air-to-ground missiles and Mirage F1 CR photo- and radar-reconnaissance missions.[122] Though French aircraft flew 1,237 combat and a significant number of tanker and airlift sorties, French participation represented only 2 per cent of the total sorties flown. This was partly due to tight political control. French decision makers demanded to be informed of potential targets 48 hours in advance; this proved to be incompatible with the high tempo of the air campaign.[123] Moreover, the reason for this relatively limited contribution was related to materiel. French aircraft lacked compatible identification, friend or foe (IFF) equipment, preventing them from working closely with the allies. French Jaguar aircraft, despite their precision strike capability, lacked night-sight capabilities.[124] In addition, the employment of French Mirage F1 CR aircraft was considerably restricted to avoid confusion with Iraqi Mirage F1.[125]

In July 1991, an official French report concluded that the FAF's aircraft fleet generally lacked the capability of immediately operating in an advanced technological environment.[126] Desert Storm proved at the time too big, too technically advanced, and too Anglo-Saxon for the FAF.[127] This experience became a catalyst for improvements in France's power projection capabilities and conventional force renewal.[128] As is discussed later, it was also at the root of FAF efforts to acquire a computerised and interoperable C2 architecture. Despite autonomous deployed air operations in the 1970s and 1980s, Desert Storm was a wake-up call. If the FAF was to play a major role on the international stage, significant reforms had to be initiated. Desert Storm had considerable corollaries both on a political and military level.

Immediately after Desert Storm, the implementation of the NFZ over northern and southern Iraq represented the first opportunities to enhance operational experience in a multinational environment. On 25 July 1991, eight Mirage F1 CR reconnaissance aircraft and a C-135FR tanker were dispatched to Incirlik Air Base in Turkey. Operation Aconit was the French contribution to Operation Provide Comfort, aimed at protecting the Kurdish population in northern Iraq. Prior to the arrival of RAF Jaguars in late September, French Mirage aircraft were the only tactical reconnaissance assets of the allied contingent. The FAF regarded Opera-

Courtesy FAF, Sirpa Air

FAF Mirage F1 CR reconnaissance aircraft

tion Provide Comfort as a valuable experience in adapting to allied procedures and using English for tactical communication.[129]

For Operation Southern Watch, the first French contingent dispatched on 1 September 1992 and consisted of eight Mirage 2000Cs, one C-135FR tanker, and transport aircraft.[130] It was a force primarily designed for defensive purposes, consistent with French containment policy. France did not support operations of a predominantly coercive nature.[131] For instance, the French refused to back an extension of the southern NFZ in 1996. They also withdrew their forces from Turkey when Operation Northern

Watch superseded Operation Provide Comfort in January 1997.[132] The different approach to dealing with Iraq was also apparent in mid-1998, when France dispatched a Mirage IV strategic reconnaissance aircraft to the Middle East to provide UNSCOM with a second intelligence source independent of and complementary to American sources.[133]

The FAF not only contributed to crisis management in the Middle East but also in the Balkans from the early 1990s.[134] In late August 1995, Deliberate Force superseded Operation Deny Flight. On the first day of operations, hostile GBAD shot down a French Mirage 2000N from Strategic Air Command. Only after three months of captivity did the Bosnian Serbs release the pilot and his navigator.[135] This incident, however, did not prevent the FAF from making a critical European contribution to the operation. Though the United States carried out the bulk of the operation, France was third after the United States and the UK in sorties flown (see chap. 2, "Western Air Operations over the Balkans"). Moreover, the French Army provided most of the heavy artillery of the multinational brigade inserted into Bosnia, pinning down Bosnian Serb military movement around Sarajevo.[136]

In 1999 the French contribution to Operation Allied Force was the second largest in both sorties flown and dispatched aircraft, with a total of approximately 100 including French naval aircraft. The FAF deployed 20 strike aircraft, including Mirage 2000Ds and Jaguars, with an additional eight Mirage 2000Cs in the air-to-air role as well as three Mirage IVPs and six Mirage F1 CRs in the reconnaissance role. With these capacities in place, French forces flew most of the European strike sorties.[137] They dropped 582 laser-guided bombs (LGB), more than double the amount UK forces dropped, accounting for approximately 7.2 per cent of PGMs employed during the campaign.[138] This significant achievement (in European terms) is largely because all of the FAF's offensive aircraft were capable of conducting precision strikes by day, with the Mirage 2000D extending this capability to nighttime.[139] Compared to Deliberate Force, when French aircraft released merely 14 PGMs that accounted for approximately 2 per cent of all PGMs released, this represented a significant improvement.[140] Furthermore, French aircraft flew 21 per cent of all reconnaissance missions in Allied Force, and the FAF contributed key assets such as two E-3F AWACS and up to 10 C-135FR tanker aircraft.[141]

Aware of the relative magnitude of their contribution, French decision makers were able to influence the air campaign without—according to their view—negatively influencing the course of the campaign or jeopardizing alliance cohesion.[142] Contrarily, American decision makers in the Pentagon took the view that European partners had exerted excessive restrictions on the Americans, particularly against the backdrop of European air power shortfalls.[143] In a hearing of the Senate Committee on Armed Services, Lt Gen Michael Short particularly singled out France as the key country in restricting targets.[144] In fact, the issue was more complex—severe tensions existed between Gen Wesley Clark, SACEUR, and the air component commander about the most effective target sets. Whilst Clark identified Serbian ground forces in Kosovo as the centre of gravity of the campaign, General Short put a premium upon leadership-related targets in Belgrade, Serbia.[145]

Besides President Clinton, only Tony Blair and Jacques Chirac were in a position to review and veto possible targets. Chirac particularly expressed concern about targets with the potential of causing significant collateral damage and targets in Montenegro, including SAM sites and an airfield.[146] As such, he vetoed for a time the bombing of two television towers and key bridges crossing the Danube in Belgrade, which led to the above American accusations. Yet in their study on the Kosovo conflict, Daalder and O'Hanlon point out that "the strategy of gradual escalation and the decision to publicly rule out the deployment of ground forces came out of Washington."[147] In essence, the debate involved a generic issue about gradualism versus shock and awe or, in USAF parlance, parallel warfare. Given the intra-American disputes, it can be argued that General Short used France as a convenient scapegoat.

Despite the alliance frictions of 1999, the FAF was the first European air force to engage targets in Afghanistan. From 23 September 2001, a French SIGINT aircraft gathered information over Afghanistan. This effort was augmented by the deployment of two Mirage IVPs together with two tanker aircraft to the Arab Emirates to deliver important imagery intelligence. The French used a two-pronged approach in deploying combat forces. The first was dispatching French Super Etendards from the aircraft carrier

Charles de Gaulle to fly missions over Afghanistan from December 2001. The second was sending six Mirage 2000Ds and two tanker aircraft from France to Manas, Kyrgyzstan, on 26 and 27 February 2002. On 2 March, these aircraft took part in Operation Anaconda.[148] The American air component commander, Gen T. Michael Moseley, argued that given the ferocity of the fighting on the ground, he immediately had to employ the French Mirage aircraft without giving them time to acclimatise. The French detachment was the first to be based at Manas, and General Moseley acknowledged France's role in establishing a new front for operations over Afghanistan. In June 2002, the French Mirage aircraft in Manas were joined by USMC F/A-18D Hornets.

Courtesy FAF, Sirpa Air

FAF deployment. Six Mirage 2000D combat aircraft and two tanker aircraft deployed from France to Manas, Kyrgyzstan, on 26 and 27 February 2002. On 2 March, these aircraft took part in Operation Anaconda.

Both in the air and on the ground, cooperation between the two contingents was very tight. For instance, the American fighter aircraft were air-to-air refuelled by the French tanker aircraft based in Manas.[149] Within seven months, French Mirages logged 4,500 flying hours and 900 sorties, destroying or neutralizing 32 targets.[150] The French contingent at Manas significantly reduced the logistical footprint by substantially drawing on regional supply, quite in contrast to the American approach.[151]

French participation in the early stages of Operation Enduring Freedom was not limited to combat, reconnaissance, and tanker aircraft. The French transport command very early established an air bridge to Central Asia and was amongst the first to conduct night flights into Mazar-i-Sharif under extremely adverse circumstances.[152]

In the ensuing years, the FAF has continued to deploy combat aircraft to Asia for both Operation Enduring Freedom and the International Security Assistance Force (ISAF). In August 2007, the FAF Central Asia detachment, Serpentaire, began redeployment from Dushanbe, Tajikistan, to Kandahar to reduce transit time and to increase NATO's firepower in the southern Afghan provinces. The detachment has usually been comprised of a total of six combat aircraft, normally combining three Mirage 2000Ds with either three Mirage F1s or three Rafales. The Air Force's Rafales, in concert with the French Navy's Rafales operating from the *Charles de Gaulle*, saw action over Afghanistan in March 2007 for the first time.[153]

French Interventions in Africa

As during the Cold War, French operations in Africa have generally been autonomous national operations. In many cases, French forces have intervened to temporarily stabilize hotspots or to evacuate Western citizens. As these operations have required quick reaction, power projection by air has proved indispensable for mission success. Besides airlift, the FAF has often provided combat aircraft, giving the lean French ground force deployments a decisive edge in firepower.

Already during the Gulf crisis in 1990–91, French troops were simultaneously involved in crisis management operations in Rwanda.[154] This can be considered a precursor to Operation Tur-

quoise, which lasted from 22 June until 22 August 1994. Operation Turquoise was aimed at stopping genocide and at establishing
a safe haven. Since Rwanda was a landlocked country and more
than 8,000 kilometres away from France, rapid deployment could
only be executed by airlift. The air bridge was considerably augmented by chartered Russian wide-body transport aircraft. A total
of 3,000 personnel as well as 700 vehicles and additional cargo
were moved during the operation. Firepower was delivered by
1,200 French frontline troops supported by 12 combat aircraft.[155]
Jaguar and Mirage F1 combat aircraft supported by a C-135FR
tanker aircraft were stationed in neighbouring Kisangani and
Goma, from where they were capable of delivering air cover and
executing low-level flights aimed at deterring combatants.[156] Further major joint interventions in Africa were conducted in the
Central African Republic (1996), the Congo (1997), and the Ivory
Coast beginning in late 2002. Forward-deployed bases and troops
proved to be an essential key to success.[157]

The French armed forces not only conducted national operations in Africa but also multinational ones. The most prominent
of these have been the French contribution to the UN operations
in Somalia in the early 1990s and in the Democratic Republic of
Congo in mid-2003. Whereas the former were American-led, the
latter was conducted under the aegis of the EU, with France being
the lead nation. The French president's positive response to the
UN secretary-general's call for assistance to stabilise the crisis in
the Ituri district of the Congo paved the way for the EU's first autonomous military operation outside Europe.[158] Out of 1,860
troops, France provided 1,660, by far the main bulk for Operation
Artemis.[159] The Ugandan capital of Entebbe served as the forward
operational base. During the operation, French Mirage aircraft
provided CAS and reconnaissance.[160]

In 2006 the second autonomous EU military operation after
Artemis took place in the Democratic Republic of Congo. The European Union Force (EUFOR) aimed at providing security during
the elections in conjunction with other civilian crisis management
cells of the EU. Towards this end, several European countries dispatched airlift assets, and Belgium deployed Hunter RPAs. The
FAF added a crucial contribution by making available forward-
deployed bases and assets both in Chad and Gabon and by provid-

Operation Artemis. Operation Artemis in 2003 was the EU's first autonomous military operation outside Europe. The FAF rapidly deployed a camp into the crisis area.

ing a contingent of Mirage F1s, which appeared to have a deterring effect at potential trouble spots.[161] France's military experience on the African continent also played a crucial role in the joint and multinational operation EUFOR in Chad and the Central African Republic. The EU operation, lasting from 28 January 2008 to 15 March 2009, took place against the backdrop of the crisis in Darfur. A key objective for the 3,700 troops deployed to Chad and to the Central African Republic was to protect civilians, particularly refugees.[162]

While retaining a sufficient presence to facilitate rapid joint military interventions, Sarkozy reduced the French military footprint on the African continent. The emphasis shifted towards the major strategic axis of the Atlantic-Mediterranean-Indian Ocean, as identified by the 2008 *Defence White Book*.[163] In particular, a permanent French military presence was built up in the United Arab Emirates.[164]

Humanitarian Operations outside Africa

While Africa has remained a major hotspot for French stabilization and humanitarian operations since the days of the Cold War, the focus has become broader in the post–Cold War era. Cambodia and Bosnia in particular have attracted French assistance. Moreover, the FAF has assisted civil authorities in emergency situations such as floods within France itself.[165]

In 1992 the FAF dispatched personnel and materiel under the UN banner to Cambodia in support of democratic elections. The deployment contained three Transall transport aircraft, air controllers, and ground personnel.[166] The FAF was also at the forefront in the relief effort of Sarajevo. On 29 June 1992, a French Transall aircraft conducted the first supply flight into Sarajevo and thereby opened the longest lasting military air bridge in history. In its course, the FAF delivered 161,700 tons of supplies to Sarajevo and evacuated 1,100 wounded and sick. French ground personnel and air controllers played a crucial role in keeping the Sarajevo airport running. FAF personnel also proved indispensable when it came to repairing devastated runways, as was the case in Mostar in December 1995.[167]

When in early 1993 the USAF had embarked upon airdrop missions in Eastern Bosnia, the FAF joined in this humanitarian effort. On 27 March 1993, the first French Transall airdropped its cargo over Bosnia. French participation lasted until 31 March 1994, with 1,447 tons of food and medicine delivered by French aircraft.[168]

Air Policing

The 11 September tragedy had a major impact upon air policing in France. Immediately after the attacks, the FAF augmented the number of aircraft on high-readiness alert from two to 12, and the interception time over French territory was lowered from 30 to 10 minutes. Armed helicopters, AWACS and tanker aircraft, mobile ground radars, and GBAD supplemented these efforts.[169]

The FAF has also provided for security in the air during important events both in France and abroad, particularly during Franco-African summits.[170] Moreover, international cooperation has been extended within Continental Europe. During the Group of Eight (G8) summit in 2003, for example, the FAF closely cooperated with the Swiss Air Force to establish a NFZ.[171] Since April 2007, the FAF has regularly contributed to NATO's air policing and quick reaction alert (QRA) efforts in the Baltic by deploying Mirage 2000 interceptors.[172]

How Has the French Air Force Responded to the New Intellectualism in Air Power Thinking and Doctrine?

French airmen in the interwar years had a culture of in-depth debate on air power theory, which found expression in several air power journals.[173] To underline the importance of an independent air force, for instance, the concepts of the Italian general Giulio Douhet were discussed widely.[174]

Following WWII, in 1947 the FAF published its first air power doctrine, the *Instruction provisoire pour l'emploi des forces aériennes* (*Provisional Instruction for the Employment of the Air Force*). Though the publication had only provisional status, it caused tensions with the Army and with the joint chief of staff, who was not informed before its release. Moreover, putting a premium upon the distinct use of air power, it stood in stark contrast to the FAF's operations in Indochina and Algeria, where it primarily played a supporting role.[175] The postwar years were further characterised by a lively debate on air power, which found expression in periodicals such as *Forces aériennes françaises* (*French Aerial Forces*), published from 1946 until 1973 and serving as a platform for doctrinal ideas and views upon military strategy.[176] This conventional air power debate was largely superseded by nuclear doctrine from the late 1950s onwards.[177]

In the post–Cold War era, a renaissance of the conventional air power debate has been taking place. It is this debate, rather than formal air power doctrine, which characterises French air power thinking.

Air Power Doctrine

Since the end of the Cold War, there have been two failed attempts at formalising an official French air power doctrine—the first in 1998 and the second in 2003. A 2006 French study argues that these failures reveal the lack of an institutionalised doctrinal process within the FAF.[178] A third attempt resulted in an air power doctrine, the *Concept de l'Armée de l'Air* (*Concept of the FAF*), published in September 2008.

The first attempt in 1998 can be related to the *Concept d'emploi des forces* (*Concept for the Employment of the Armed Forces*), published in 1997. This joint document establishes the purpose of the French armed forces according to the four strategic core functions as identified in the 1994 *Defence White Book*—deterrence, prevention, projection, and protection.[179] According to the 2006 study, this first attempt represents the breaking down of the *Concept d'emploi des forces* onto a FAF level and was hence entitled the *Doctrine d'emploi de l'Armée de l'Air* (Doctrine for the employment of the FAF). The writing of the document seems to have been restricted to a small nucleus consisting of three colonels and a major general. For unknown reasons, however, this air power doctrine was not published.[180]

The second attempt in 2003 can also be related to the publication of a major joint document, the *Doctrine interarmées d'emploi des forces en opération* (*Joint Doctrine for the Employment of the Armed Forces in Operations*). It comes next in the doctrine hierarchy after the 1997 *Concept d'emploi des forces* and sets guidelines for the mode of employment of the French armed forces.[181] More concrete than the 1997 concept, this joint doctrine devotes single sections to certain key air power roles and characteristics but does not come up with distinct French air power doctrine features.[182] As a response to this joint doctrine, FAF officers attending the Joint Forces Defence College (Collège interarmées de défense [CID]) in Paris were tasked with writing air power doctrine. However, the hastily written document was not approved within the wider FAF community, mainly due to concerns that an immature document, compared to joint doctrine documents or foreign air power doctrine manuals, might be detrimental for the FAF.[183]

The 1998 and 2003 attempts at formalising French air power doctrine were primarily driven by joint doctrine documents and hence lacked sufficient internal motivation. There has, in fact, existed an aversion towards formal doctrine as well as concerns, raised by a former chief of the Air Staff and other officers during interviews, that formal doctrine might be received as dogmatic and thus inhibit innovative thinking and limit the room for manoeuvre regarding both employment and resource allocation.[184]

Despite, or perhaps because of, the lack of a formal air power doctrine, it is interesting to note that in the air power periodical

Penser les ailes françaises (*To Think of French Wings*), a broader doctrinal debate was launched bottom-up. In October 2005, a colonel published a short draft version of a basic air power doctrine to gather ideas from the wider Air Force community for use in formalizing a doctrine.[185] Though the colonel found an echo from his fellow officers, he retired from the Air Force prior to completing this endeavour. The project was not continued, again confirming the lack of an institutionalised doctrinal process.[186]

In 2007 the chief of the Air Staff, Stéphane Abrial, ordered a formal air power doctrine to be produced.[187] This top-down approach resulted in the air power doctrine *Concept de l'Armée de l'Air*, published in September 2008. Released shortly after the 2008 *Defence White Book*, the hallmarks of the latter are clearly visible and were deliberately incorporated. Moreover, in light of the concerns above, the chief of the Air Staff made it clear that the document was to be adapted on a regular basis to respond to future developments.[188] With slightly more than 20 pages, the doctrine appeared to be deliberately kept short, probably to avoid excessive detail. Its conciseness was also supposed to encourage reading, as the document aimed at sharing the airman's view on air power both nationally and internationally.[189]

After introducing some general aspects on the third dimension and its military exploitation, the doctrine was closely related to the 2008 *White Book*'s threat and risk spectrum and to its five strategic core functions—cognition-anticipation, prevention, deterrence, protection, and intervention.[190] In the same vein, it also reflects other key themes of the *White Book* such as building European security as well as a European aerospace industry.[191]

Conceptually, the doctrine developed an artificial dichotomy between air power in the service of political authorities and air power in the service of militaries, thereby seeming to partly put into question the Clausewitzian notion that all form of military action is the continuation of policy by other means. On the positive side, this approach does not preclude gradualist campaigns as in Kosovo or punctual engagements. Both forms are seen as instrumental in supporting coercive diplomacy.[192] In the service of militaries, air power's ability to conduct parallel warfare and to supposedly produce strategic paralysis was highlighted.[193] The doctrine concluded by providing an outlook into the future, anticipating joint warfare in an urban environment as a future trend.[194]

Prior to 2008, against the backdrop of a lack of strategic air power doctrine, the FAF had drawn upon various joint doctrine manuals.[195] On an operational and tactical level, the Joint Staff and the Army have published manuals regarding the conduct of air campaigns, airborne operations, CSAR, or ground based air defence.[196]

A number of other interesting innovations have occurred in recent years. For instance, in 2004–5 two new organisations were established: on a joint level, the Centre interarmées de concepts, de doctrines et d'expérimentations (CICDE), or the Joint Forces Centre for Concept Development, Doctrine, and Experimentation; and on an air force level, the Cellule études et stratégie aérienne militaire (CESAM), or air power strategy and studies cell. At the time of writing, the CESAM was still very young and consisted of a nucleus of three officers. Their work focused on issues of major relevance to the current environment, such as CAS in urban terrain.[197] Again, it can be argued that the establishment of the CESAM was a FAF response to the establishment of the CICDE on a joint level. Nevertheless, this expresses a more institutionalised approach towards air power thinking. Moreover, the Centre d'études stratégiques aérospatiales (CESA), or Centre of Strategic Aerospace Studies (formerly Centre d'enseignement supérieur aérien)—primarily responsible for the education of midcareer officers—has recently stepped up its research activities in the domain of air power, with a focus upon low-intensity conflicts and asymmetrical threats.[198] On a NATO level, the FAF has also sent representatives to the recently established Joint Air Power Competence Centre in Germany.[199]

The above-mentioned French study on air power doctrine concluded that the recent thrust towards more institutionalised thinking on air power could partly be seen in the context of the FAF's increasing involvement within NATO and the ESDP.[200] In fact, if the FAF aspires to play a major part in European air power, it also has to shape the doctrinal landscape. Moreover, it might have been felt necessary to strengthen the FAF's position regarding the formulation of joint doctrine, as the Army had clearly taken the lead.[201]

Teaching of Air Power Thought

As regards higher education of FAF officers, the two major innovations in the post–Cold War era have been an increasing thrust towards joint and, later on, towards combined teaching. Since 1993 higher military education has been delivered at the CID. It basically replaced single-service education, which in the case of the FAF was provided by the École supérieure de guerre aérienne (ESGA), or Advanced School of Air Warfare.[202] At the CID, selected midcareer officers are educated for 10 months.[203] The syllabus emphasises operational aspects such as campaign planning in joint and multinational scenarios.[204]

Throughout the year, French CID course participants are exposed to an international environment, reflecting the growing importance of combined operations. Out of 66 Air Force participants in the 13th course, 22 were foreigners.[205] Concurrently, five to six Air Force officers are sent each year to higher staff colleges in the UK, Germany, the United States, or Italy.[206] This push towards multinational education has been complemented since 2001 by the Combined Joint European Exercise (CJEX) in which the staff colleges of France, the UK, Germany, Italy, and Spain participate. For one week, these colleges exchange students to conduct a multinational joint exercise.[207]

Since 2002 the CID has strived to encourage thinking on air power theory and doctrine. The participants in the ninth course conducted various studies on air power strategy, and the results were assembled in a booklet. The project's goal was to gradually promote an intellectual debate on air power theory and strategy in a broader circle.[208] This effort was reiterated by the following course. Topics ranged from specific air power topics such as targeting or reconnaissance to broader topics such as the evolution of the employment of air power.[209] In the meantime, the recently set up air power periodical *Penser les ailes françaises* has provided the CID course participants with a new platform for disseminating their thoughts on various aspects, such as the fundamentals of air and space power, command in air operations, a critical view upon the FAF's participation in Desert Storm, or the use of cruise missiles.[210] These developments reveal a growing interest in the intellectual aspects of the employment of air power. Yet so far, CID course participants have not been exposed to an intense reading programme of various air power literature, including the latest Anglo-Saxon air power thinking.[211]

Dissemination of Air Power Thought

In parallel with the increased publishing activities of the CID course participants, two series of air power conferences have been created within the FAF since 2004. It was primarily the CESA that stepped up its efforts to establish a broad air power debate. It organises the Ateliers de l'Armée de l'Air (Air Force conference) and the Ateliers du CESA (CESA conference). The former is an annual air power conference, which took place for the first time on 29 June 2004 at the special request of the chief of the Air Staff. These conferences are organised around several roundtables aimed at gathering industrialists, researchers, politicians, journalists, and militaries.[212] The first conference was devoted to reflections upon the FAF's raison d'être and its prospects, particularly its purpose and role in a common European defence architecture.[213] In his keynote address, the chief of the Air Staff underlined the new approach the FAF took with that conference. To open the debate, civilians were invited to take charge of the various roundtables and discussions.[214]

The second conference series the CESA organises is the Ateliers du CESA. Several take place throughout the year. As in the case of the Ateliers de l'Armée de l'Air, these conferences aim at gathering militaries and civilians to discuss aerospace-relevant topics. The first conference, "Du Vietnam à l'Irak: réflexions sur quarante ans d'engagements aériens" ("From Vietnam to Iraq: Reflections on Forty Years of Air Campaigning"), took place on 18 July 2005.[215] The second Ateliers du CESA was devoted to Douhet and his influence upon strategic thinking.[216] After these rather historical beginnings, ensuing Ateliers du CESA turned to more topical issues such as air operations in Afghanistan, air power in urban warfare, air power's role in low-intensity conflicts, or reflections on joint warfare. Some of the conferences have an international dimension, extending invitations to high-ranking air force officers from around the world, particularly from the USAF and the RAF. The conference proceedings are published in *Penser les ailes françaises*.[217]

Whereas the Ateliers du CESA have a more conceptual and doctrinal focus, the Ateliers de l'Armée de l'Air have more of a politico-military flavour. In their own ways, both intend to foster air power thinking within and beyond the Air Force. Since September 2006, these conferences have been augmented by the Rencontres air et espace du CESA (aerospace meetings). These serve as a platform from which to engage the views of an outstanding personality on a particular aerospace topic. The speakers are often retired or active general officers and scholars.[218]

While air power colloquia as such are not a new phenomenon within the FAF, they have never been so large and frequent. The historical service of the FAF has organised international colloquia since the days of the Cold War. The audience has, however, been primarily an academic one.[219]

Parallel to these developments within the Air Force, we also find efforts to encourage aerospace power thinking outside, though not separate from, the Air Force. One of the most prominent figures in this domain has been Prof. Pierre Pascallon, who was a member of the National Defence Commission at the National Assembly. Since 1996 he has organised several colloquia dealing with aerospace issues.[220] His most encompassing air power colloquium, entitled "L'Armée de l'Air: les armées françaises à l'aube du XXIᵉ siècle" ("The French Air Force: The French Armed Forces at the Dawn of the Twenty-first Century"), took place in February 2003. The seminar provided an opportunity to bring together a wide range of officers, industrialists, academics, and politicians to discuss current and future concerns of French air power, with an emphasis upon force structuring. Among the speakers were a significant number of Air Force generals, revealing the interest on the part of the FAF.[221]

Not only conferences but also the launching of the periodical *Penser les ailes françaises* in 2003 reveals a keen interest in encouraging a common reflection upon air power within the FAF.[222] The journal is in the tradition of the review *Forces aériennes françaises,* published from 1946 until 1973.[223] As mentioned above, *Penser les ailes françaises* has provided CID course participants with a platform for expressing their thoughts on air power. Moreover, the periodical serves the explicit purpose of disseminating and fostering French air power thinking.[224] Prior to *Penser les ailes françaises*, ideas and views on air power had to be expressed in journals outside the Air Force, particularly in *Défense nationale (National Defence)*.[225] With the introduction of *Penser les ailes françaises* in 2003, the FAF again offers its own platform for debating aerospace issues.

These days, the most prolific French writer in the domain of air power is Patrick Facon, the director of the air branch of the armed forces' historical service. He has written extensively on air power in World War II and has published various articles in defence journals.[226] At the time of this writing, Facon and his team were undertaking several book projects, such as the translation of Douhet's *The Command of the Air* and Gen William Mitchell's *Skyways: A Book on Modern Aeronautics*.[227] Though this historical approach has its proper value, a French colonel publicly remarked in 2005 that publications on modern air operations had hardly been produced in France, particularly in contrast to the United Kingdom. He added that *Penser les ailes françaises* and the recently set up air power conferences were intended to make a first step towards remedying this situation.[228]

On the positive side, FAF officers have also started to write on doctrinal aspects outside the Air Force. In 2004 Col Régis Chamagne published his book *L'art de la guerre aérienne* (*The Art of Air War*). Although the book was a personal initiative, its publication received great attention by the FAF and was offered to all Air Force officers attending the CID.[229] It was referred to as the first modern piece by a French author scrutinising the complexities of air war in their entirety and was distinguished by the Académie des sciences morales et politiques (Academy of Moral and Political Sciences).[230] However, the book does not develop entirely new concepts. In fact, it is strongly influenced by John Warden's concepts and explicitly refers to them.[231] As such, the term *strategic* is very much linked to independent air strikes, and the author rather thinks in terms of target sets than strategic effect.[232] Accordingly, he views strikes against leadership targets in Baghdad during Operation Desert Storm as strategic and strikes against Iraqi ground forces in the Kuwaiti theatre of operation as attritional, with the former supposedly generating more coercive effect.[233]

Another French officer who saw the need to promote thinking on air power in France was Col Philippe Steininger. He translated

Warden's book *The Air Campaign* and wrote the preface to it. The French translation appeared in 1998.[234] His personal writing has also been profoundly influenced by Warden's concepts.[235]

This overemphasis upon Warden's theories was also remarked upon in the recent French study on air power doctrine.[236] In fact, the reemerging French air power debate seemed to favour theories on the distinct use of air power, hence the still keen interest in Douhet's theories. As was noted, one of the very first Ateliers du CESA was devoted to him, and the translation of *The Command of the Air* is amongst the projects of the armed forces' historical service. French air power thinking thus reiterated an interwar preoccupation by underlining the issue of an independent mission. As evidenced by the CESA's air power conferences, the focus has turned to more topical aerospace power issues.

Research in the field of air power history and theory has also been encouraged by the highest authorities. The defence minister introduced in 2004 the Prix de l'Armée de l'Air (prize of the FAF). It honours outstanding achievements of doctoral theses or master's dissertations or the equivalent.[237] The prize represents a tightening link between the military and academia.

How Have French Defence Planners Attempted to Maintain a Relevant Air Force in Light of Escalating Costs and Advanced Technologies?

Right after WWII, French decision makers aspired to reestablish a potent indigenous aircraft industry. The emerging postwar FAF soon contained three components—a NATO component furnished with modern American materiel, a national component gradually being equipped by the national postwar aircraft industry, and an overseas component in Indochina that basically consisted of second-rate American materiel. From the 1960s onwards, however, the FAF has been primarily equipped with indigenous materiel. At the same time, it started to support the French industry in its thrust towards increased exports by providing training for foreign pilots.[238] Sales abroad were crucial to reduce the unit costs for the FAF's own equipment.[239] Yet, the primacy of nuclear doctrine and deterrence rendered French conventional forces secondary.[240] Substantial parts of the French conventional force inventory were second-rate by the end of the Cold War.[241]

One French commentator argued that the shortfalls the FAF encountered during Desert Storm seriously put into question previous procurement priorities. As discussed, French aircraft were devoid of appropriate IFF systems, night vision capability, adequate electronic self-protection suites, and sufficient navigation systems. Furthermore, the FAF did not possess the sophisticated C2 systems which proved indispensable for an air campaign of this scale and complexity.[242]

After Operation Allied Force, the importance of interoperable C2, identification, and air-to-air refuelling systems was reemphasised. That a meaningful contribution to an American-led campaign required a certain level of sophistication since technological shortfalls could lead to isolation was also stressed.[243]

Renewal of French conventional forces has not been an easy task due to budget constraints. For industrial survival, all major procurement programmes were to be maintained in the early 1990s. This could only be achieved by delaying certain programmes and by reducing the number of units to be purchased. Politico-industrial concerns were clearly given priority over military operational concerns. Moreover, unlike most of France's allies, the French armed forces more or less retained their Cold War size-up to the mid-1990s, which—in light of budget constraints—negatively affected the modernisation of the conventional forces. This situation also resulted in considerable delays for the various programmes. Interim solutions were therefore required to bridge the most significant operational shortfalls. Paradoxically, investments in interim solutions resulted in a diversion of funds originally assigned to major weapons programmes and caused further delays.[244] International observers concluded in 2003 that although the FAF was in the process of being equipped with a number of new assets, these were actually long overdue.[245]

An important feature of French procurement policy was its emphasis upon national autarchy. Since the days of de Gaulle's rise to power, the French armed forces have been primarily equipped with French materiel. In the aftermath of the Cold War, this policy was reiterated by successive defence ministers.[246] French officials have been conscious that independence does not allow for short-term bargains but requires significant long-term investments. Yet autarchy in the field of defence procurement has acquired an extended meaning. Gradually, French decision makers have shifted their rhetoric from the notion of national autonomy to the notion of European autonomy.[247] Accordingly, both the 1994 and the 2008 *Defence White Books* underline the importance of preserving national competencies in the defence industrial sector and of European cooperation.[248] Given this European focus, the FAF puts a premium upon the ESDP and its ventures in the field of defence planning and procurement.[249]

Combat Aircraft

In 1985 France decided to go ahead with its own Rafale project rather than to remain in the Eurofighter programme.[250] The FAF opted for a multirole fighter that could replace several aircraft types in its inventory. This was deemed a necessity both for operational and financial reasons.[251] Slight adjustments regarding the Rafale order were made in the wake of Operation Desert Storm, when dual-seat aircraft were deemed more appropriate for air-to-ground missions than single-seat aircraft, particularly for long-range penetration missions.[252] To respect the financial framework and due to the higher costs of dual-seat aircraft, the number of planned units was reduced from 250 to 234 in 1991. Out of these 234 units, 95 were planned to be single-seat and 139 dual-seat aircraft.[253] The dual-seat version was conceived to exploit to the maximum Rafale's versatility, and crews were to be trained with complementary skills.[254] Yet in 2004, while still acknowledging the advantages of dual-seat aircraft on long-range missions, the chief of the Air Staff decided to emphasize production of the single-seat version.[255] Improvements in the man-machine interface might have led to this revision.

Unlike multinational programmes, the Rafale programme was conducted by a lean organisation, reportedly resulting in reduced costs and leading Dassault Aviation's chief executive officer to maintain that the 294 Rafales the FAF (234) and French Navy (60) planned to acquire would cost less than Germany's 180 Eurofighters. Moreover, the FAF actively contributed to reducing costs by preventing excessively perfect solutions in areas considered less relevant in today's environment. Avionics and weapons became the focus of development efforts, while plans for a more powerful engine were put on hold. A major focus was also directed towards the use of all-European technology.[256] Since a number of parameters are confidential, it is difficult to compare the latest European combat aircraft. Whether the Rafale is actually much cheaper than the Eurofighter is debatable. With the French armed forces so far being the only customer, exports have not yet contributed to economies of scale. In late 2008, overall planning figures were reduced from 294 Rafale combat aircraft to 286. While the projected number of Rafales for the FAF was reduced from 234 to 228, current plans foresee 58 Navy Rafales—a reduction of two aircraft.[257]

Delays of the programme caused serious problems for defence planners. While the first Rafales were originally expected to be delivered in 1992, French naval Rafale aircraft entered service in 2001, and the first FAF Rafale squadron only became operational in mid-2006.[258] Delays were primarily due to funding problems as well as to the reorganisation of the French aerospace industry in the early and mid-1990s.[259]

Interim solutions were required. In particular, two aircraft—the Mirage 2000D for attack missions and the Mirage 2000-5 air superiority fighter—have bridged the capability gap. The Mirage 2000D is the FAF's first conventional all-weather attack aircraft.[260] Its importance in bridging the capability gap pending introduction of the Rafale was formally acknowledged in the military planning law for 1997–2002. A total of 86 units were planned to be procured.[261] Prior to the introduction of the Mirage 2000D, only the Mirage 2000Ns assigned to Strategic Air Command were all-weather capable, highlighting French primacy of nuclear over conventional forces during the Cold War.[262]

The Mirage 2000-5 originally did not respond to operational requirements; the motivations for its development were rather industrial.[263] The decision to set up two Mirage 2000-5 squadrons in January 1993 was at the time controversial but was justified by the delay of the Rafale programme due to technical and financial reasons.[264] The Mirage 2000-5 is basically a modified Mirage 2000C with improved air-to-air radar and fire-and-forget Matra MICA (interception and aerial combat) missiles. The first of 37 converted aircraft was delivered in December 1997.[265] Though the Mirage 2000-5 and the Mirage 2000D were pragmatic interim solutions, they could only provide a relatively lean interim force of advanced weapon systems.

At the outset of the post–Cold War era, the FAF's principal precision-guided air-to-ground weapon was the laser-guided AS-30-L missile, widely used by French Jaguars during Desert Storm. Besides the AS-30-L, the FAF could draw upon the BGL, an indigenous LGB. The weapon was used during the conflicts in Bosnia and Kosovo but was withdrawn from service thereafter as it was more expensive than standard LGBs of US origin.[266] Aside from the BGL, the American GBU-12 Paveway was introduced into the FAF's inventory in 1995. At the time of this writing, French GBU-12s were being improved to enhanced Paveway standard with GPS guidance. To deliver these laser-guided weapons, the FAF has introduced a number of indigenous laser-designator pods since the mid-1980s. While the early pods only allowed for daylight operations, more sophisticated pods, introduced in 1993 and in 1999, offered a nighttime capability.[267]

Operation Allied Force made the FAF realize that the relationship between operational requirements and the industrial capacity to produce PGMs had to be readjusted.[268] Given the importance of all-weather PGMs, the 2003–8 military planning law announced the introduction of a new type of weapon, the indigenous *armement air-sol modulaire*.[269] The AASM is essentially an all-weather PGM propelled by a rocket booster. Depending on its release altitude, it can engage targets at close or medium ranges exceeding 50 km with various options of terminal impact angles. Currently, two guidance systems are available; the more sophisticated one integrates an infrared imager seeker with a combined inertial measuring unit/GPS receiver navigation kit. Delivery of a third guidance system specifically adapted for engaging mobile targets is expected for 2012. The new system integrates a semiactive laser with the combined inertial measuring unit/GPS receiver. Further sophisticated guidance kits, including a millimetric-wave radar seeker, are likely to be developed. In April 2008, Rafales engaged Taliban positions with AASMs for the first time, with the target coordinates provided by a Canadian forward air controller.[270] The weapon represents a low-cost complement to the more complex cruise missiles that the FAF commenced to acquire in the late 1990s.

Rafale flying over Afghanistan armed with AASM precision-guided munitions

Delivery of the Apache cruise missile started in 1998. It is a standoff weapon carrying submunitions for use against area targets such as airfields. The Scalp, a derivative of the Apache, was at the time being developed.[271] It is conceived for attacks against point targets at a distance exceeding 200 km.[272] Scalp was planned to become operational with the Mirage 2000D squadrons in 2003 and with the first Rafale squadron in 2006.[273] The 2003–8 military planning law foresaw the purchase of 450 Scalps for the FAF, with an additional 50 for naval aviation.[274]

In the domain of air-to-air missiles, a limited number of the indigenous MICA fire-and-forget radar-guided missiles were ordered in the 1990s to equip the new Mirage 2000-5 squadrons.[275] The MICA is available with either a radar or an infrared seeker and has become the standard AAM for the Rafale fleet.[276] Moreover, France has been participating in the multilateral development of the Meteor AAM.[277]

Since available funds were primarily concentrated upon the Rafale programme, France bailed out of a bilateral anti-radar missile programme with Germany.[278] With the decommissioning of its Martel anti-radiation missile-armed Jaguars in the latter half of the

1990s, the FAF basically gave up its dedicated SEAD capability.[279] The FAF has instead been relying upon its allies for this particular capability in combined air operations.[280] Yet, French lessons learned from Operation Allied Force stressed the need to reacquire a SEAD capability.[281] In the longer term, an anti-radar seeker might be developed for the modular AASM.[282]

In the years to come, manned aircraft assets are planned to be complemented by remotely piloted combat aircraft as well as by a limited ground based TBMD capability. The military planning law for 2003–8 announced that by 2010 an initial TBMD capability should be in place.[283] The 2008 *Defence White Book* highlights the growing threat posed by ballistic and cruise missiles against Europe. In 2009 the FAF received its first *sol-air moyenne portée terrestre* (SAMP-T), or land-based medium range surface-to-air missile, fire unit. Dubbed "Mamba," it provides a TBMD capability against missiles in the Scud category.[284] Dassault has taken on design leadership of the Neuron remotely piloted combat aircraft programme. By mid-2007, Greece, Italy, Spain, Sweden, and Switzerland participated in the project, with France supplying half of the programme's budget.[285]

Air Mobility

Throughout the second half of the Cold War, the C-160 Transall was the FAF's principal transport aircraft. The C-160 was the result of a joint Franco-German requirement statement issued in the late 1950s. It was conceived as a tactical transport aircraft in the same class as the C-130 Hercules. Yet, its performance fell short of that of the Hercules and remained constrained.[286] For this reason, the C-160 Transall fleet was complemented in 1987 by 12 C-130 Hercules aircraft for missions up to 5,000 km. Yet despite this measure, the FAF's airlift capacity has suffered from a chronic weakness, incompatible with the missions assigned to the French armed forces.[287] Reliance on the support of large US transport aircraft was therefore necessary in several operations during the Cold War.[288]

As in the case of combat aircraft, the FAF has been seeking interim solutions to mitigate the strain on the transport fleet. Amongst others, 20 Casa CN-235 transport aircraft of limited size and range were procured.[289] Likewise, two Airbus A310-300s were purchased in the early 1990s to replace two aging DC-8Fs.[290] The last two DC-8Fs were replaced by two Airbus A340 TLRAs (*très long rayon d'action* or very long range) aircraft in 2006–7, providing extended reach. The FAF leased the two aircraft with the option to buy at the end of the leasing period.[291] Regarding the much larger C-160 Transall fleet, a major life extension and upgrade programme was undertaken in the post–Cold War era.[292] Despite additional improvements in the areas of self-protection and night-flight capability, the lifespan of the Transall aircraft is limited.[293] The 1997–2002 military planning law foresaw their retirement from 2003 onwards, but that time frame was deferred to bridge expected shortfalls in airlift capacity.[294]

Courtesy Airbus Military; photo by L. Olivas

A400M on its first flight. France plans to procure 50 A400M transport aircraft.

In 2003 delivery of the first A400M for the FAF was scheduled for October 2009. Yet due to serious delays in the programme, the earliest possible deliveries are not expected to take place before late 2012. As a result, French aviators expressed an interest in potentially acquiring up to 12 American C-130J Hercules aircraft as a further interim solution.[295]

In the post–Cold War era, AAR has become an ever more important element of air mobility. The FAF's AAR capacity is closely linked to its airborne nuclear component and reaches back to the

1960s, when the FAF ordered 12 American Boeing C-135FR aerial tankers to support its Mirage IV strategic bomber squadrons.[296] From 1985 onwards, the C-135FRs received new engines to improve their performance. This, however, only slightly eased French shortfalls in AAR, which became evident during Operation Desert Storm and again during Operation Allied Force.[297] Moreover, the availability rate has decreased due to the high degree of maintenance required for the aging C-135FR fleet. This problem also affected the C-160 Transall fleet and early versions of the Mirage 2000—their availability rate fell at times to between 50 to 60 per cent. Emergency sums for maintenance were disbursed in 2002 in an attempt to alleviate the deficit.[298] As in the case of the combat aircraft fleet, the FAF introduced various interim solutions to maintain its AAR capacities at a sufficient level. For instance, a number of C-135FRs were equipped with additional wing-mounted refuelling pods. Moreover, the purchase of three additional secondhand C-135s was announced in the military planning law for 1997–2002.[299] Beyond 2015 the aging French tanker fleet is planned to be renewed by the acquisition of 14 Airbus multirole transport tanker (MRTT) aircraft.[300]

Another specific element of air mobility is CSAR. The FAF has undertaken particular efforts in this field in recent years. During the Balkans crisis, the FAF provided the only European CSAR capability alongside the Americans.[301] A CSAR helicopter squadron consisting of nine specially equipped Puma helicopters was formed for this purpose in 1994.[302] To enhance its CSAR capability, the FAF ordered the EC 725 Caracal helicopters conceived for the CSAR task. The FAF views CSAR as a key capability that can also provide leverage in coalition operations.[303]

C4ISTAR

Shortfalls in the area of C4ISTAR proved to be serious during Operation Desert Storm. The lack of electronic C2 systems particularly constrained the FAF in its freedom of action.[304] Efforts to improve the situation followed throughout the 1990s.

A quantum leap was made at the outset of the post–Cold War era by the procurement of four Boeing E-3 AWACS aircraft. In a national scenario, they significantly complemented ground-based detectors. In an out-of-area scenario, they provided an unprece-

Courtesy FAF, Sirpa Air

The FAF ordered EC 725 Caracal helicopters to enhance its CSAR capability.

dented early warning capability.[305] In the field of computer-aided C2 systems, however, the FAF was still facing a major shortfall in the early 1990s, as it still hinged upon a 30-year-old system. Improvements were expected through the introduction of the successor system, the SCCOA (*système de commandement et de conduite des opérations aériennes* or air operations command and control system). This system was conceived to automate most C2 and detection functions as well as to facilitate the conduct of defensive and offensive missions by fusing data of various sensors.[306] The SCCOA encompasses a deployable component and is interoperable with its NATO equivalent. The deployable component of SCCOA can support a JFACC in deployed combined operations, a capability which allows France to take on lead-nation status.[307]

Interoperability is further facilitated by the introduction of Link-16 terminals.[308] Aside from the Rafale, the Mirage 2000-5 and Mirage 2000D were planned to be retrofitted with MIDS Link-16 terminals in 2008 and 2009, respectively.[309] France has played a particularly prominent role in the multinational MIDS development programme by covering 26.5 per cent of the costs involved.[310]

With regard to ISTAR, the FAF relied upon a broad range of sensors throughout the post–Cold War era. For instance, it used to operate two types of SIGINT aircraft, the C-160NG Gabriel and the DC-8 Sarigue. In 1999 the Sarigue aircraft was replaced by another DC-8 Sarigue NG (*nouvelle génération* or new generation) with enhanced systems.[311] Yet due to exceedingly high maintenance costs, the Sarigue NG was retired in 2004.[312] On the positive side, the 2003–8 military planning law announced development programmes in the field of electronic warfare, particularly standoff jamming—if possible, through European cooperation.[313] These steps are aimed to remedy a European shortfall that had become particularly apparent against the backdrop of Operation Allied Force.

For tactical reconnaissance, the FAF's Mirage F1 CR proved to be a versatile aircraft due to the fusion of radar, infrared, and conventional wet film systems.[314] In the wake of the decommissioning of the Mirage IV, used as a strategic reconnaissance aircraft in the post–Cold War era, the FAF introduced new reconnaissance pods for the Mirage F1 CR, followed by a new system for the Rafale, the Reco-NG, from 2009 onwards. The Reco-NG integrates a data link capable of real-time imagery.[315]

In the field of RPAs, the FAF has taken a gradual approach that has been marked by a number of setbacks in the post–Cold War era. Beginning in 1995, four Hunter systems were procured from Israel to jointly evaluate performances and requirements for RPAs. In 2000 the Hunter systems were put under the authority of the FAF, which had employed them in Kosovo. The systems were retired in September 2004 and intended to be replaced by an interim system, the Harfang SIDM (*système intérimaire de drone MALE* or interim MALE RPA system).[316] The Harfang SIDM is based upon an Israeli cell, with the avionics originally developed jointly by France and the Netherlands. Due to delays, however, the Netherlands pulled out of the programme.[317] Although the Harfang SIDM was originally scheduled for 2003, it only entered service in 2008—a five-year delay.[318] For a potential successor system, various cooperative European MALE RPA ventures and off-the-shelf procurements are being considered, including purchasing the American Reaper as a further interim solution.[319]

Overall, concerns over European autarchy and politico-industrial aspects have tended to outweigh concerns over operational needs.

Conclusion

In line with its international status, France has continued to pursue a proactive defence policy in the post–Cold War era. Significant French commitments to operations in the former Yugoslavia, combined with lessons learned from participation in Desert Storm, made a major defence reform in the mid-1990s inevitable. Moreover, France's emphasis upon nuclear doctrine had to be reevaluated in the context of deployed multinational operations. The new strategic environment required a thorough reorganisation and modernisation programme of the FAF's conventional forces. The 1994 command structure reform allowed for a flexible use of all available air assets by streamlining the command echelons. Constant adaptation has allowed the FAF to contribute to all strategic core functions of French defence policy. The transition into a fully professional force occurred against the backdrop of an increased emphasis upon projection. Since French defence policy puts a premium upon national autarchy, the FAF has retained a coherent force structure covering a broad array of air power roles. Financial constraints, however, have forced the FAF to forego certain specialised roles such as a dedicated SEAD capability, and its RPA programmes have also suffered from several setbacks. In the context of the Armed Forces Model 2015, two aircraft types were given priority, the Rafale and the A400M, leading to a homogenous force structure. Despite the cancellation of various minor procurement programmes, major goals of the Model 2015 could not be achieved in time, and plans proved to be over-ambitious. President Sarkozy embarked upon a reform in 2008, reducing the force goal of modern combat aircraft to 300 for both the FAF and the French Navy together, 70 of which are planned to be simultaneously deployable.

In the domain of alliance policy, France's ambition has been to establish an autonomous European defence that it could actively shape and also take on a centre stage position. Since most European countries have proved reluctant to forego the American security umbrella, France had to seek the strengthening of the European defence pillar from within the NATO alliance. This paradoxically led to an increased engagement with NATO that culminated in France's 2009 return to NATO's integrated military command structure. In return, France received two of NATO's top military positions—the post of SACT and the command of Joint Command Lisbon—while retaining a degree of national autonomy, particularly in the domain of the country's nuclear deterrent. This reintegration into NATO did not mean, however, that France would no longer attach considerable importance to the European Union as a major actor in the domain of security.

Along with France's increasing participation in NATO and its lead role in the ESDP, the FAF has been shaped accordingly. As such, it is now capable of providing the nucleus of a larger multinational force. The FAF has played a significant role in the NRF air component and other major European cooperative ventures such as the EAG, the EATC, or the EU Air RRC. Through integration, the FAF is able both to overcome its shortcomings, such as the lack of a dedicated SEAD capability, and to gain leverage through the provision of key capabilities, such as C2 systems and deployable air bases that still represent critical bottlenecks in European air power. In short, integration has allowed the FAF to enhance its operational effectiveness and to raise its international silhouette.

Throughout the post–Cold War era, the FAF has stood up to the challenge of real operations across a broad range of military engagements, from humanitarian enterprises to high-intensity warfare. Apart from Operations Northern Watch, Desert Fox, and Iraqi Freedom, it has participated in all major Western air campaigns and has simultaneously been able to conduct smaller, autonomous operations in Africa. To remedy significant deficiencies in interoperability as experienced in Desert Storm, the FAF has consistently pursued a path towards operating in a combined mode and is nowadays able to swiftly integrate itself into a coalition or to take on a lead role in multinational operations. As regards the African theatre, the French armed forces retained their expertise in managing regional crisis quickly with lightly equipped, joint contingents. Airlift and firepower delivered by air have proven critical in these operations. The ability to rapidly deploy air assets over long distances and to operate them from bare bases was also critical in the deployment of French aircraft to Central Asia in early 2002, making the FAF the first European air force to engage targets in Afghanistan with fighter-bombers. The French contingent paved the way for a European F-16 component. Despite these achievements, Chirac's goal—to deploy up to 100 combat aircraft or the FAF's theoretical contribution of up to 80 combat aircraft with additional support aircraft to potential deployed EU operations—proved too ambitious given the shortfalls in strategic airlift or AAR. So far, the FAF has never deployed simultaneously more than 50 combat aircraft to a particular post–Cold War campaign, with Operation Allied Force witnessing the largest deployment. While the deployment to Manas in early 2002 can be considered an achievement in terms of rapid power projection, it nevertheless has to be pointed out that the detachment, including six Mirage 2000Ds, was far below the quantitative goals above.

Notwithstanding the FAF's increasing involvement in combined operations, only in 2008 did it issue an air power doctrine. Released the same year as Sarkozy's *Defence White Paper*, the hallmarks of the latter are clearly visible. Amongst others, it implicitly introduces the concept of parallel warfare, which supposedly produces strategic paralysis.

Prior to the 2008 air power doctrine, French air power thinking rather found expression in a bottom-up approach. This is particularly demonstrated by FAF officers expressing their views in *Penser les ailes françaises* and elsewhere. Moreover, through the establishment of two air power conference series in 2004–5, the FAF invited scholars, industrialists, and others to share their views on French air power, representing a significant step towards critical thinking.

Given that French military thinking was long overshadowed by nuclear doctrine, modern thinking on conventional air power is relatively young. While the reemerging French air power debate focused on historical issues and the idea of an independent mission—as exemplified by the keen interest in Douhet's and Warden's theories—the focus has turned to current challenges in joint warfare.

Before 2008 the FAF preferred to hinge upon implicit air power doctrine rather than to publish an official strategic air power doctrine. The highest echelons of the FAF considered formal doctrine as dogmatic and inhibiting flexible thinking. Yet, given France's aspirations to lead-nation status in European defence matters, the FAF was actually obliged to shape the European doctrinal landscape and, hence, to come up with its own formal air power doctrine.

Furthermore, although Joint Forces Defence College participants increasingly express their views in *Penser les ailes françaises*, the college syllabus appears to focus upon staff work rather than air power theory. They have not been exposed to an intense reading programme of various air power literature. On the positive side, the French were amongst the first to establish a joint staff college in the post–Cold War era, taking account of the integrated and complex nature of modern military operations.

Operation Desert Storm proved to be a watershed for the FAF in the field of force structuring, as it blatantly disclosed conventional shortfalls preventing France from making a contribution congruent with its political objectives. Given France's aspiration to lead-nation status in coalition warfare, French officials and defence planners realised that the FAF's conventional equipment had to be state of the art and fully interoperable with allied air forces. In this respect, the FAF has put a premium upon the introduction of NATO interoperable C2 systems and upon the buildup of key capabilities such as CSAR. Moreover, France's lead role in the development of the MIDS Link-16 terminal is fully in line with national strategic ambitions. Significant delays of major aircraft programmes, however, have considerably constrained the FAF's reform. Interim solutions have mitigated capability gaps; however, along with the growing maintenance costs of increasingly aging materiel, they have also diverted resources from major programmes. Gap-filling programmes such as the Mirage 2000-5 and the Mirage 2000D had to be kept at the lowest possible level. Pending introduction of the Rafale, the FAF had to rely upon a very lean fleet of conventional state-of-the art combat aircraft throughout the 1990s. A limited force of Mirage 2000D and Mirage 2000-5 aircraft arguably represented a subcritical conventional interim force in light of French ambitions. Yet, less sophisticated aircraft such as the Jaguar or the Mirage F1 proved indispensable throughout the 1990s in the African theatre and elsewhere. Also, given the rather permissive environment of Western air campaigns in the post–Cold War era, French Jaguars performed effectively in air-to-ground precision strike missions.

In the Gaullist tradition, French defence planners continued to pursue a path of autarchy. Yet the unilateral Rafale programme has left the FAF little room for manoeuvre regarding the development of other capabilities. For instance, France had to bail out of a bilateral Franco-German SEAD armament programme. In the medium term, however, the combination of a versatile platform combined with sophisticated sensors and France's latest air-to-ground weapon, the AASM, is expected to make a dedicated SEAD capability no longer necessary. Besides the Rafale, major national programmes have been conducted in the field of computerised C2 systems as well as air-to-air and air-to-ground armament. Only in cases where a national programme would have been excessively expensive, such as in the case of AWACS, did French defence planners buy assets of American origin. Yet an escalation in costs and advanced technologies made French decision makers shift their rhetoric from national to European autonomy. As such, major programmes such as the A400M or a future remotely piloted combat aircraft are European cooperative ventures. Yet, these politico-industrial aspirations have not allowed for short-term bargains. Capability shortfalls—especially in the area of airlift—have developed pending delivery of national or European assets. It can be concluded that politico-industrial concerns have taken precedence over operational concerns. At the same time, capability shortfalls have reinforced the FAF's thrust towards closer European cooperation.

The 1990s and the early years of the twenty-first century represented for the FAF a particularly critical period, as it had to undergo a considerable transition from a force primarily geared up for nuclear war to a force that aspired to be at the forefront of conventional air power in deployed operations. As the FAF's commitment to the NRF air component or to deployed multinational operations conveys, this transition has been successful. Given significant delays in major procurement programmes, however, many of the FAF capabilities currently being introduced are overdue.

Notes

1. James Corum, "Airpower Thought in Continental Europe between the Wars," in Meilinger, *Paths of Heaven*, 151–53.

2. Mark A. Lorell, *Airpower in Peripheral Conflict: The French Experience in Africa* (Santa Monica, CA: RAND, 1989), 2.

3. André Martel, "Conduire la défense: les institutions, les forces, les alliances," in *Histoire militaire de la France*, Tome 4, *De 1940 à nos jours,* ed. André Martel, director (Paris: Presses Universitaires de France, 1994), 402–4.

4. Ibid., 365; and Claude Carlier, "L'aéronautique et l'espace, 1945–1993," chap. 10, in Martel, *Histoire militaire de la France*, Tome 4, 456–59, 462.

5. Carlier, "L'aéronautique et l'espace," 452–53, 463, 465.

6. See Greg Seigle, "USA Claims France Hindered Raids," *Jane's Defence Weekly* 32, no. 17 (27 October 1999): 3.

7. Defence Minister Charles Millon, "Vers une Défense nouvelle," *Défense nationale*, no. 7 (July 1996): 13.

8. Gregory, *French Defence Policy*, 29.

9. "Une Défense nouvelle 1997–2015," hors série [special edition], *Armées d'aujourd'hui*, no. 208 (March 1996): 9.

10. Nicolas Sarkozy, president of the French Republic, "Foreword," in Ministère de la Défense [Ministry of Defence] (MOD), France, *Défense et Sécurité nationale: le Livre blanc* (Paris: Odile Jacob/La Documentation française, June 2008), 9.

11. Louis Gautier, *Mitterrand et son armée 1990–1995* (Paris: Grasset, 1999), 43–44.

12. MOD, France, *Livre blanc sur la défense* [White book on defence] (Paris: MOD, 1994), 12–13, 22, http://lesrapports.ladocumentationfrancaise.fr/BRP /944048700/0000.pdf.

13. Loi de programmation militaire (LPM) [French military programme bill of law] 2003–2008, *La politique de défense* [Defence policy] (Paris: Journaux officiels [Official journals of the French Republic], May 2003), 14–15.

14. MOD, France, *Défense et Sécurité nationale: le Livre blanc*, 19–30, 37–39.

15. Ibid., 43.

16. MOD, France, *Livre blanc sur la défense*, 23–24.

17. Ibid., 73–74.

18. Ibid., 49–50, 58.

19. Gautier, *Mitterrand et son armée 1990–1995*, 89, 95–96.

20. Carlier, "L'aéronautique et l'espace," 465.

21. MOD, France, *Livre blanc sur la défense*, 59–62.

22. "Une Défense nouvelle 1997–2015," 7, 13.

23. LPM 2003–2008, 19–20.

24. Jacques Chirac, "Dissuasion nucléaire—'garantie fondamentale de notre sécurité,' "*Air actualités*, no. 589 (March 2006): 17.

25. MOD, France, *Défense et Sécurité nationale: le Livre blanc*, 65.

26. Ibid., 66, 71.

27. Gregory, *French Defence Policy*, 44–45.

28. MOD, France, *Livre blanc sur la défense*, 95–96.

29. Ibid., table between 114 and 115.

30. Gautier, *Mitterrand et son armée 1990–1995*, 162.

31. "Une Défense nouvelle 1997–2015," 13, 20.

32. Ibid., 4.

33. Meiers, *Zu neuen Ufern?*, 152.

34. Gregory, *French Defence Policy*, 94.

35. MOD, France, *Défense et Sécurité nationale: le Livre blanc*, 211.

36. LPM 2009–2014, *Projet de loi relatif à la programmation militaire pour les années 2009 à 2014 et portant sur diverses dispositions concernant la défense* [Law project pertaining to military programme bill of law 2009–2014 and to different directives regarding defence], Assemblée Nationale [National Assembly], 29 October 2008, rapport annexé [annexed report], 21.

37. André Martel, "Conclusion: Armées 2000; un système de transition," in Martel, *Histoire militaire de la France,* 597–98.

38. Gregory, *French Defence Policy*, 45; and Carlier, "L'aéronautique et l'espace," 474–75.

39. Mark A. Lorell, *The Future of Allied Tactical Fighter Forces in NATO's Central Region* (Santa Monica, CA: RAND, 1992), 24.

40. Carlier, "L'aéronautique et l'espace," 473.

41. Gregory, *French Defence Policy*, 36.

42. Carlier, "L'aéronautique et l'espace," 461–62.

43. Jacques Patoz and Jean-Michel Saint-Ouen, *L'Armée de l'Air: survol illustré dans les turbulences du siècle* (Paris: Editions méréal, 1999), 138–39.

44. Gert Kromhout, "The New Armée de l'Air," *Air Forces Monthly*, no. 118 (January 1998): 42.

45. MOD, France, *Livre blanc sur la défense*, 59.

46. LPM 1997–2002, *Rapport fait au nom de la commission des Affaires étrangères, de la défense et des forces armées (1) sur le projet de loi, adopté par l'Assemblée Nationale après déclaration d'urgence, relatif à la programmation militaire pour les années 1997 à 2002* [Report on behalf of the commission of foreign and defence affairs and the armed forces on the military programme bill of law 1997–2002, approved by the national assembly after a declaration of urgency], no. 427 (Paris: Sénat, 1996), 131.

47. Gen François Bourdilleau, "Evolution de l'Armée de l'Air vers le modèle Air 2015," in *L'Armée de l'Air: les armées françaises à l'aube du XXIe siècle*, Tome 2, ed. Pierre Pascallon (Paris: L'Harmattan, 2003), 255.

48. "Une Défense nouvelle 1997–2015," 18.

49. Ibid., 121.

50. Ibid., 7.

51. "Air 2010: simplification et cohérence—feu vert pour la deuxième étape," Armée de l'Air website, 24 March 2006, http://www.defense.gouv.fr/sites/air /base/breves/2006/mars/220306_air_2010_simplification_et_coherence.

52. *Charte de fonctionnement de l'Armée de l'Air*, no. 196/DEF/MGAA, 16 September 2008, 5, 19–23.

53. Sarkozy, "Foreword," 9–10.

54. MOD, France, *Défense et Sécurité nationale: le Livre blanc*, 178.

55. LPM 2009–2014, rapport annexé, 21.

56. Calculation of percentage is based upon figures from Jean-Luc Mathieu, *La Défense nationale* (Paris: Presses Universitaires de France, 2003), 54.

57. International Institute for Strategic Studies, *The Military Balance 1991–92* (London: Brassey's for the IISS, 1991), 56–57.

58. International Institute for Strategic Studies, *The Military Balance 2002–03* (Oxford, England: Oxford University Press for the IISS, 2002), 40–41.

59. LPM 2003–2008, 27–28.

60. LPM 2009–2014, rapport annexé, 14–15, 21.

61. Cdt José Souvignet and Lt Col Stéphane Virem, "L'Armée de l'Air dans la tourmente: la campagne aérienne du Golfe," *Penser les ailes françaises*, no. 9 (February 2006): 94.

62. Rachel Utley, *The Case for Coalition: Motivation and Prospects: French Military Intervention in the 1990s*, The Occasional, no. 41 (New Baskerville, England: Strategic and Combat Studies Institute, 2001), 22.

63. François Ouisse, "Présence française en Centrafrique: l'Armée de l'Air sans frontières," *Air actualités*, no. 493 (June 1996): 50–51.

64. "Une Défense nouvelle 1997–2015," 14; and Gen Michel Mathe, "Garantir la permanence et la crédibilité de la composante aérienne de dissuasion: les Forces Aériennes Stratégiques," in Pascallon, *L'Armée de l'Air*, 156.

65. Kai Burmester, "Atlantische Annäherung—Frankreichs Politik gegenüber der NATO und den USA," in *Die verhinderte Grossmacht: Frankreichs Sicherheitspolitik nach dem Ende des Ost-West-Konflikts*, ed. Hanns W. Maull, Michael Meimeth, and Christoph Nesshöver (Opladen, Germany: Leske & Budrich, 1997), 100–101.

66. Gautier, *Mitterrand et son armée 1990–1995*, 64, 69.

67. Burmester, "Atlantische Annäherung," 105.

68. Gautier, *Mitterrand et son armée 1990–1995*, 75.

69. Utley, *Case for Coalition*, 38.

70. MOD, France, *Livre blanc sur la défense*, 35–37.

71. Gregory, *French Defence Policy*, 106–7.

72. Gautier, *Mitterrand et son armée 1990–1995*, 81.

73. Gregory, *French Defence Policy*, 109–10.

74. Burmester, "Atlantische Annäherung," 112.

75. Gregory, *French Defence Policy*, 125–27.

76. Ibid., 112–18.

77. Charles Grant, "Conclusion: The Significance of European Defence," in Everts et al., *European Way of War*, 73–74.

78. Ibid., 68–69.

79. MOD, France, *Défense et Sécurité nationale: le Livre blanc*, 99.

80. NATO, "France Moves on Reintegrating NATO's Military Structure," SHAPE news release, 27 July 2009, http://www.nato.int/shape/news/2009/07/090727a.html.

81. Sarkozy, "Foreword," 9; LPM 2009–2014, rapport annexé, 4; and British Broadcasting Corporation, "France Ends Four-Decade NATO Rift," 3 December 2009, http://news.bbc.co.uk/2/hi/europe/7937666.stm.

82. MOD, France, *Défense et Sécurité nationale: le Livre blanc*, 101; and LPM 2009–2014, rapport annexé, 4.

83. Gen Richard Wolsztynski, chief of the Air Staff, "La contribution de l'Armée de l'Air à la construction de l'Europe de la défense," *Défense nationale*, no. 7 (July 2004): 11.

84. Ibid., 7.

85. Gen Patrick de Rousiers, "Contribution de l'Armée de l'Air à la construction de la défense européenne: nature, perspectives, importance," in Pascallon, *L'Armée de l'Air*, 545.

86. Wolsztynski, "La contribution de l'Armée de l'Air," 11–12.

87. Ibid., 11.

88. "La contribution de l'Armée de l'Air à l'outil de défense," *Armée de l'Air 2007: enjeux et perspectives* (Paris: Sirpa Air, 2007), 14.

89. MOD, France, *Défense et Sécurité nationale: le Livre blanc*, 213.

90. De Rousiers, "Contribution de l'Armée de l'Air," 542–43.

91. Moulard, "L'organisation de la campagne aérienne," 39–40, 42.

92. "NRF: The French Air Force in Control," *Air actualités*, no. 583 (July/August 2005): 41.

93. Lt Col Emmanuel Maisonnet, FAF, to the author, e-mail, subject: Lessons Learned/Doctrine Division A7, 14 March 2006.

94. "NRF," 43.

95. Moulard, "L'organisation de la campagne aérienne," 40.

96. "NRF," 43.

97. De Rousiers, "Contribution de l'Armée de l'Air," 544.

98. Wolsztynski, "La contribution de l'Armée de l'Air," 8–9.

99. Jean-Laurent Nijean, "L'Armée de l'Air et l'Europe," *Air actualitiés*, no. 605 (October 2007): 20.

100. European Air Group, "History of the EAG."

101. Gregory, *French Defence Policy*, 138.

102. Wolsztynski, "La contribution de l'Armée de l'Air," 8.

103. MOD, France, *Defence against Terrorism: A Top-Priority of the Ministry of Defence* (Paris: Délégation à l'information et à la communication de la défense, April 2006), 17.

104. Col Marc Weber (FAF, Air Staff, Paris), interview by the author, 11 October 2005.

105. De Rousiers, "Contribution de l'Armée de l'Air," 540.

106. Eugénie Baldes, "Les Mirages français passent le relais," *Air actualités*, no. 556 (November 2002): 34–36.

107. De Rousiers, "Contribution de l'Armée de l'Air," 541, 544–45.

108. Weber, interview.

109. Lorell, *Airpower in Peripheral Conflict*, 22, 75–76.

110. Ibid., 67–68.

111. Jérôme de Lespinois, "Emploi de la force aérienne: Tschad 1969–1987," *Penser les ailes françaises*, no. 6 (2005): 70.

112. Lorell, *Airpower in Peripheral Conflict*, 68.

113. Patoz and Saint-Ouen, *L'Armée de l'Air*, 276–77.

114. Lorell, *Airpower in Peripheral Conflict*, 2.

115. Etienne de Durand and Bastien Irondelle, *Stratégie aérienne comparée: France, Etats-Unis, Royaume-Uni* (Paris: Centre d'études en sciences sociales de la défense, 2006), 141.

116. Jérôme de Lespinois, "La participation française à la campagne aérienne de la guerre de libération du Koweït (1991): prolégomènes politico-diplomatiques," *Penser les ailes françaises*, no. 7 (Octobre 2005): 68.

117. Souvignet and Virem, "L'Armée de l'Air dans la tourmente," 93; and de Lespinois, "La participation française," 72–73.

118. Souvignet and Virem, "L'Armée de l'Air dans la tourmente," 92–94.

119. Wg Cdr K. J. Baldwin, "Can Europe Project Air Power without the Support of the United States?," *Royal Air Force Air Power Review* 4, no. 1 (Spring 2001): 67.

120. Patoz and Saint-Ouen, *L'Armée de l'Air*, 288; and Lieutenant General Lemieux, "Moyen-Orient: premiers enseignements," *Air actualités*, no. 441 (April 1991): 14.

121. Patoz and Saint-Ouen, *L'Armée de l'Air*, 19–20.

122. Lt Col Henri Guyot, "30 jours de guerre aérienne," *Air actualités*, no. 440 (March 1991): 8–9.

123. Lt Col Henri Guyot, "Moyen-Orient: 43 jours de guerre aérienne," *Air actualités*, no. 441 (April 1991): 12; and Souvignet and Virem, "L'Armée de l'Air dans la tourmente," 92, 95.

124. Gregory, *French Defence Policy*, 45; and Utley, *Case for Coalition*, 15.

125. Durand and Irondelle, *Stratégie aérienne comparée*, 143.

126. *Rapport no. 32/DEF/CGA du 19 juillet au Contrôleur des armées Hervouet concernant la participation de l'Armée de l'Air aux opérations de libération du Koweït et de la couverture des Émirates*, 1991, quoted in Durand and Irondelle, *Stratégie aérienne comparée*, 143.

127. Souvignet and Virem, "L'Armée de l'Air dans la tourmente," 92.

128. Ibid., 84–85.

129. "Opération Aconit: huit Mirage F1 CR en Turquie," *Air actualités*, no. 447 (November/December 1991): 11–13.

130. Patoz and Saint-Ouen, *L'Armée de l'Air*, 197.

131. Gen Jean-Pierre Job, "La projection aérienne au service de la diplomatie préventive," *Défense nationale*, no. 6 (June 1999): 95.

132. Ripley, *Air War Iraq*, 10; and United States European Command, "Operation Northern Watch," 20 February 2004, http://www.eucom.mil.

133. Job, "La projection aérienne," 95.

134. Ripley, *Air War Bosnia*, 57.

135. Ouisse, "104 jours d'angoisse," *Air actualités*, no. 488 (January 1996): 5.

136. Soubirou, "Account of Lieutenant General (Ret) André Soubirou," 26; see also Ripley, *Operation Deliberate Force*, 175.

137. Peters et al., *European Contributions to Operation Allied Force*, 18–21, 30; Durand and Irondelle, *Stratégie aérienne comparée*, 159; and Sénateur Xavier de Villepin, *Rapport d'information fait au nom de la commission des Affaires étrangères, de la défense et des forces Armées (1) sur les premiers enseignements de l'opération "force alliée" en Yougoslavie: quels enjeux diplomatiques et militaires?* [Information report on behalf of the commission of foreign and defence affairs and the armed forces on the first lessons learned of Operation Allied Force in Yugoslavia: what kind of diplomatic and military challenges?], no. 464 (Paris: Sénat, 30 June 1999), http://www.senat.fr/rap/r98-464/r98-464_mono.html.

138. Garden, "European Air Power," 114. Calculation of percentage is based upon figures from Cordesman, *Lessons and Non-Lessons*, 4.

139. David S. Yost, "The U.S.-European Capabilities Gap and the Prospects for ESDP," in Howorth and Keeler, *Defending Europe*, 88.

140. Calculation for percentage is based upon figures from Lt Col Richard L. Sargent, "Weapons Used in Deliberate Force," in Owen, *Deliberate Force*, 258.

141. Peters et al., *European Contributions to Operation Allied Force*, 20–21; and Yost, "U.S.-European Capabilities Gap," 89.

142. MOD, France, *Les enseignements du Kosovo: analyses et références* (Paris: MOD, November 1999), 11–12.

143. Gnesotto et al., *European Defence*, 44.

144. Gen Wesley K. Clark, supreme allied commander, Europe, and Lt Gen Michael Short, testimony before the Senate Subcommittee on Armed Services, "Lessons Learned from Military Operations and Relief Efforts in Kosovo," 21 October 1999, 12–14.

145. Ibid., 11; Wesley K. Clark, *Waging Modern War: Bosnia, Kosovo, and the Future of Combat* (Oxford, England: PublicAffairs, 2001), 278, 458; Lambeth, "Operation Allied Force," 121–22; and Daalder and O'Hanlon, *Winning Ugly*, 198.

146. Daalder and O'Hanlon, *Winning Ugly*, 222; and Clark and Short, testimony before the Senate Subcommittee on Armed Services, 14.

147. Daalder and O'Hanlon, *Winning Ugly*, 221.

148. Gen Patrice Klein, "Bilan et enseignements des opérations Héraclès et Ammonite," in Pascallon, *L'Armée de l'Air*, 203; and "Destination Kirghizistan," *Air actualités*, no. 550 (April 2002): 4.

149. Frédéric Castel, "Sept mois de coopération exemplaire entre la France et les États-Unis," *Air actualités*, no. 556 (November 2002): 30–32.

150. Baldes, "Les Mirages français passent le relais," 34–35; and MOD, France, *Defence against Terrorism*, 24.

151. Jean-Patrice Le Saint (commandant, FAF, air power strategy and studies cell [CESAM], Paris), interview by the author, 11 October 2005.

152. Corinne Micelli, "Le transport en tête de ligne," *Air actualités*, no. 549 (February–March 2002): 4–7.

153. Stephan de Bruijn, "Rafales in Afghanistan," *Air Forces Monthly*, August 2008, 58–60.

154. Utley, *Case for Coalition*, 31.

155. Adm Jacques Lanxade, "L'opération Turquoise," *Défense nationale*, no. 2 (February 1995): 9–11.

156. "Bilan 1994 des actions extérieures," *Air actualités*, no. 479 (February 1995): 35; and Patoz and Saint-Ouen, *L'Armée de l'Air*, 192.

157. François Ouisse, "Bangui: les soldats français éteignent l'incendie," *Air actualités*, no. 493 (June 1996): 47–49; Ouisse, "Présence française en Centrafrique," 50–51; Patoz and Saint-Ouen, *L'Armée de l'Air*, 142; and Député Marc Joulaud, *Avis présenté au nom de la commission de la défense nationale et des forces armées, sur le projet de loi de finances rectificative pour 2003*, no. 1267 (Paris: Assemblée Nationale, 2003), 19, http://www.assemblee-nationale.fr/12/pdf/rapports/r1267.pdf.

158. Olivier Harmange, "Opération Oryx," *Air actualités*, no. 459 (February 1993): 20–21; and United Nations, Peacekeeping Best Practices Unit, Military Division, *Operation Artemis: The Lessons of the Interim Emergency Multinational Force* (UN: October 2004), 3–4.

159. Joulaud, *Avis présenté au nom de la commission de la défense nationale*, 20.

160. United Nations, *Operation Artemis*, 12.

161. Jean-Laurent Nijean, "L'Armée de l'Air et l'Europe: opération EUFOR RDC," *Air actualités*, no. 605 (October 2007): 34–35.

162. European Union, European Security and Defence Policy, "EU Military Operation in Eastern Chad and North Eastern Central African Republic," updated March 2009, http://www.consilium.europa.eu/uedocs/cmsUpload/Final_FACTSHEET_EUFOR_TCHAD-RCA-version9_EN.pdf.

163. MOD, France, *Défense et Sécurité nationale: le Livre blanc*, 72–73, 77; and LPM 2009–2014, rapport annexé, 9.

164. MOD, France, "Inauguration de l'implantation militaire française aux Emirats arabes unis," 26 May 2009, http://www.defense.gouv.fr.

165. Patoz and Saint-Ouen, *L'Armée de l'Air*, 225.

166. Ibid., 266; and Jérôme Billère, "Dans le ciel de Phnom Penh," *Air actualités*, no. 459 (1993): 19.

167. Patoz and Saint-Ouen, *L'Armée de l'Air*, 155, 294.

168. Ibid., 77.

169. Gen Patrick Thouverez, commander, CDAOA (Air Defence and Operations Command), "La contribution de l'Armée de l'Air à la posture permanente de sûreté," in Pascallon, *L'Armée de l'Air*, 138–39.

170. Ibid., 170.

171. Formerly the G6, France created the group in 1975 for the governments of France, Germany, Italy, Japan, the United Kingdom, and the United States. Canada joined the group in 1976, and Russia joined in 1997. Aldo Wicki, "Sicherheit durch Kooperation," *Schweizer Luftwaffe*, 2004, 13–15.

172. Craig Hoyle, "Baltic Exchange," *Flight International* 175, no. 5191 (2–8 June 2009): 33; and "France Takes Over Baltic Air Policing," *Air Forces Monthly*, March 2010, 12.

173. These journals include *Les ailes, Revue de l'Armée de l'Air, Revue de l'aéronautique militaire*, and *Revue des forces aériennes*. See also Centre of Strategic Aerospace Studies (CESA), "Bibliographie thématique: stratégie aérienne," accessed 7 April 2006, http://www.cesa.air.defense.gouv.fr/IMG/pdf/BIBTHEMA.pdf (site discontinued).

174. Corum, "Airpower Thought in Continental Europe," 153.

175. Durand and Irondelle, *Stratégie aérienne comparée*, 137.

176. Patrick Facon (director, AIR, historical service of the armed forces, Château de Vincennes, Paris), interview by the author, 2 March 2006; and Patrick Facon, to the author, e-mail, 19 April 2006. See also CESA, "Bibliographie thématique."

177. Durand and Irondelle, *Stratégie aérienne comparée*, 138.

178. Ibid., 133.

179. PIA [Joint publication]-00.100, *Concept d'emploi des forces* [Force employment concept] (Paris: Etat-major des armées, 23 July 1997), 10–12.

180. Durand and Irondelle, *Stratégie aérienne comparée*, 146–48.

181. PIA-00.200, *Doctrine interarmées d'emploi des forces en opération* [Joint doctrine for force employment in operations] (Paris: Etat-major des armées, division emploi, September 2003), preamble.

182. Ibid., chap. 5, 25–28, and chap. 8, 10–15.

183. Durand and Irondelle, *Stratégie aérienne comparée*, 149–50.

184. See also ibid., 153.

185. Col Marc Weber, "Vers un concept d'emploi de l'arme aérienne," *Penser les ailes françaises*, no. 7 (October 2005): 3–10.

186. Brig Gen Michel de Lisi, director, CESA, "Éditorial," *Penser les ailes françaises*, no. 8 (January 2006): 3; and Col Jean-Christophe Noël, FAF, director, CESAM, Paris, to the author, e-mail, 8 May 2006.

187. "Wird sich stark verändern," interview of Gen Stéphane Abrial, chief of the Air Staff, FAF, conducted by *Y. Magazin, Bundeswehr*, 8 February 2007, http://www.luftwaffe.de.

188. *Concept de l'Armée de l'Air* (Paris: Sirpa Air, September 2008), 3.

189. Ibid.

190. Ibid., 4–8, 11–12.

191. Ibid., 14–15.

192. Ibid., 14.

193. Ibid., 16–17.

194. Ibid., 20–21.

195. "Wird sich stark verändern."

196. *La campagne aérienne* (Paris: Etat-major des armées, 1999); PIA 03.122, *Concept interarmées de défense surface-air* [Joint concept for surface to air defense] (Paris: Etat-major des armées, 18 April 2000); *Concept national de recherche et de sauvetage au combat—RESCO* [National concept for CSAR] (Paris: Etat-major des armées, August 2000); and Instruction 3400, *Concept national des opérations aéroportées* [National concept for airborne operations] (Paris: Armée de terre, 26 April 2002).

197. Col Jean-Christophe Noël (FAF, director, CESAM, Paris) and Cdt Jean-Patrice Le Saint (FAF, CESAM, Paris), interviews by the author, 27 February 2006.

198. Brig Gen Michel de Lisi, FAF, director, CESA, to the author, e-mail, 21 March 2006 (forwarded by his deputy, Col Denis Gayno).

199. Lt Col Yvan Sais (FAF, Joint Air Power Competence Centre, Kalkar, Germany), interview by the author, 5 May 2006.

200. Durand and Irondelle, *Stratégie aérienne comparée*, 164.

201. Ibid., 149.

202. CESA, "Historique du CESA," http://www.cesa.air.defense.gouv.fr/article.php3?id_article=7.

203. "Présentation de l'enseignement militaire supérieur air et du colloque," *Penser les ailes françaises*, spécial colloque, no. 4 (September 2004): 3.

204. CID, "Programme," accessed 22 February 2006, http://www.college.interarmees.defense.gouv.fr.

205. CID, "Physionomie," accessed 10 March 2006, http://www.college.inter armees.defense.gouv.fr.

206. Col Gilles Lemoine (FAF, chief, Air Power Studies, CID, Paris), interview by the author, 1 March 2006.

207. CID, "Enseignement opérationnel: Euro exercice CJEX," accessed 28 March 2006, http://www.college.interarmees.defense.gouv.fr.

208. CID, "Stratégie aérienne: préface du Colonel Michel de Lisi," accessed 22 February 2006, http://www.college.interarmees.defense.gouv.fr.

209. CID, "Stratégie aérienne: 19 articles sur la stratégie aérienne de la 10e promotion," accessed 22 February 2006, http://www.college.interarmees.defense.gouv.fr.

210. Air Working Group No. 2, 10th course of the CID, "Puissance aérienne et spatial," *Penser les ailes françaises*, no. 1 (July 2003): 43–55; Air Working Group No. 4, 10th course of the CID, "Fondamentaux de la puissance aérienne," *Penser les ailes françaises*, no. 1 (July 2003): 9–20; Air Working Group No. 1, 10th course of the CID, "Fondamentaux de la puissance aérienne et enseignements," *Penser les ailes françaises*, no. 2 (February 2004): 3–37; Air Working Group No. 8, 12th course of the CID, "Gains et contraintes liés à l'utilisation des missiles croisière: l'exemple du SCALP," *Penser les ailes françaises*, no. 9 (February 2006): 78–83; Cdt Laurent Charrier, "Le commandement des opérations aéroportées: un défi interarmées," *Penser les ailes françaises*, no. 8 (January 2006): 61–65; Lt Col Dominique Colin, "Vers un modèle de processus décisionnel itératif de haut niveau," *Penser les ailes françaises*, no. 9 (February 2006): 97–104; Colonel Lefebvre and CID course members, "Quelle Armée de l'Air pour notre défense?," *Penser les ailes françaises*, no. 5 (February 2005): 3–11; and Souvignet and Virem, "L'Armée de l'Air dans la tourmente."

211. Col Gilles Lemoine, chief, Air Power Studies, CID, Paris, to the author, e-mail, 29 March 2006.

212. CESA, "Ateliers de l'Armée de l'Air," http://www.cesa.air.defense.gouv.fr/rubrique_tousarticles.php3?id_rubrique=110.

213. See September 2004 issue, *Penser les ailes françaises*, spécial colloque, no. 4, which is entirely devoted to the first Ateliers du CESA.

214. Gen Richard Wolsztynski, chief of the Air Staff, "Allocution d'ouverture du colloque 'L'Armée de l'Air, enjeux et perspectives,'" *Penser les ailes françaises*, spécial colloque, no. 4 (September 2004): 6.

215. Brig Gen Michel de Lisi, director, CESA, "Présentation des 'Ateliers du CESA,'" *Penser les ailes françaises*, no. 8 (2006): 5.

216. *Penser les ailes françaises*, no. 9 (2006): 5.

217. CESA, "Ateliers du CESA," http://www.cesa.air.defense.gouv.fr/rubrique _tousarticles.php3?id_rubrique=107.

218. CESA, "Les Rencontres air et espace du CESA," http://www.cesa.air .defense.gouv.fr/rubrique_tousarticles.php3?id_rubrique=108.

219. Facon to the author, e-mail. The air historical branch organised these seminars: "L'adaptation de l'arme aérienne aux conflits contemporains" [The adjustment of the air arm in contemporary conflicts] (Paris, 1984), "Histoire de la guerre aérienne" [History of air warfare] (Paris, 1987), "Penseurs et prophètes de la puissance aérienne" [Air power thinkers and prophets] (Paris, 1990), "L'aviation a-t-elle gagné la guerre en 1944?" [Did aviation win the war in 1944?] (Paris, 1995), and "Cent ans d'aviation militaire" [100 years of military aviation] (Paris, 1999).

220. Prof. Pierre Pascallon, editor, *L'Armée de l'Air*, to the author, e-mail, 21 March 2006.

221. Pascallon, *L'Armée de l'Air*, 615–23.

222. *Penser les ailes françaises*, no. 1 (2003): 2.

223. Facon to the author, e-mail.

224. *Penser les ailes françaises*, no. 9 (2006): 3.

225. Examples include *Défense nationale*, special edition, no. 6, 1999; Gen Pierre-Marie Gallois, "La guerre du Golfe: missiles et anti-missiles," *Défense nationale*, no. 3 (March 1991): 9–14; and Gen Jean Saulnier, "La guerre du Golfe: cas d'espèce ou modèle reproductible," *Défense nationale*, no. 6 (June 1991): 11–21.

226. Examples include Patrick Facon, *La bataille d'Angleterre (1940): la bataille aérienne décisive de l'histoire* (Paris: Economica, 1992); Patrick Facon, *L'Armée de l'Air dans la tourmente: la bataille de France 1939–1940* (Paris: Economica, 1997); Patrick Facon, *La guerre aérienne, 1933–1945* (Clichy, France: Larivière, 2003); Patrick Facon, *L'histoire de l'Armée de l'Air: une jeunesse tumultueuse 1880–1945* (Clichy, France: Larivière, 2004); and Facon to the author, e-mail.

227. Facon to the author, e-mail.

228. Col Marc Weber, "Introduction aux Ateliers du CESA," *Penser les ailes françaises*, no. 8 (January 2006): 6.

229. Col Régis Chamagne, FAF, retired, to the author, e-mail, 27 Mar 2006.

230. La bibliothèque de l'aéronautique, "L'art de la guerre aérienne," http:// www.aerostories.org; and CESA, "Prix pour l'art de la guerre aérienne," http:// www.cesa.air.defense.gouv.fr.

231. Col Régis Chamagne, FAF, retired, *L'art de la guerre aérienne* (Fontenay-aux-Roses, France: L'esprit du livre éditions, 2004), 197–98.

232. Ibid., 118.

233. Ibid., 85–86.

234. John A. Warden, *La campagne aérienne: planification en vue du combat*, trans. Philippe Steininger (Paris: Economica & Institut de Stratégie Comparée, 1998).

235. Col Philippe Steininger, "La paralysie stratégique selon Boyd et Warden," *Penser les ailes françaises*, no. 9 (February 2006): 50–62.

236. Durand and Irondelle, *Stratégie aérienne comparée*, 145.

237. CESA, "Le prix Armée de l'Air 2005," http://www.cesa.air.defense.gouv.fr/article.php3?id_article=106.

238. Carlier, "L'aéronautique et l'espace," 449, 452–53, 463.

239. Utley, *Case for Coalition*, 14.

240. Carlier, "L'aéronautique et l'espace," 465.

241. Gregory, *French Defence Policy*, 49.

242. Gautier, *Mitterrand et son armée 1990–1995*, 163–65.

243. MOD, France, *Les enseignements du Kosovo*, 15, 23–24.

244. Gautier, *Mitterrand et son armée 1990–1995*, 114, 348, 363–64.

245. Joris Janssen Lok and J. A. C. Lewis, "New French Air Power," *Jane's International Defence Review* 36 (June 2003): 41.

246. Gautier, *Mitterrand et son armée 1990–1995*, 345.

247. Gen Vincent Lanata, "Faire face: l'ère des nouveaux défis," *Défense nationale*, no. 8/9 (August/Septembre 1993): 16–17; and LPM 2003–2008, 17.

248. MOD, France, *Livre blanc sur la défense* 115, 120; and MOD, France, *Défense et Sécurité nationale*, 266–68.

249. Bourdilleau, "Evolution de l'Armée de l'Air," 258.

250. Willett, Gummet, and Clarke, *Eurofighter 2000*, 5.

251. Carlier, "L'aéronautique et l'espace," 471.

252. Ibid., 472; and Elliot, "From Fighters to Fighter-Bombers," 73.

253. Carlier, "L'aéronautique et l'espace," 472.

254. Bill Sweetman, "Pragmatic Rafale: A Study in French Philosophy," *Jane's International Defence Review* 38 (June 2005): 67.

255. Armée de l'Air, "Audition du général Wolsztynski, Commission de la Défense nationale et des forces armées," 22 October 2004, accessed 19 May 2007, http://www.defense.gouv.fr/air/archives/22_10_04_audition_du_general_wolsztynski.

256. Sweetman, "Pragmatic Rafale," 64–65, 67.

257. Xavier Pintat and Daniel Reiner, *Avis présenté au nom de la commission des Affaires étrangères, de la défense et des forces armées (1) sur le projet de loi de finances pour 2009, adopté par l'Assemblée Nationale*, Tome 5, *défense—équipement des forces*, no. 102 (Paris: Sénat, 20 November 2008), chap. 2, sec. 4, "Engagement et combat, B. Frapper à distance," http://www.senat.fr/rap/a08-102-5/a08-102-5.html.

258. LPM 1997–2002, 133; and J. A. C. Lewis, "French Connection: A New President and Preparations for a New Defence White Paper Promise Changes for the French Military," *Jane's Defence Weekly* 44, no. 27 (4 July 2007): 27.

259. "European Military Aircraft Programmes Revisited," *Military Technology*, no. 6 (1997): 8.

260. Col Jean-François Louvion, "Armée de l'Air et projection," *Défense nationale*, no. 7 (July 1995): 63.

261. LPM 1997–2002, 133.

262. Durand and Irondelle, *Stratégie aérienne comparée*, 140.

263. Gautier, *Mitterrand et son armée 1990–1995*, 363.

264. Ibid., 168.

265. Kromhout, "New Armée de l'Air," 42.

266. Col Denis Mercier, FAF, Air Staff, Paris, to the author, e-mail, 1 September 2006; and De Villepin, *Rapport d'information fait au nom de la commission des Affaires étrangères.*

267. Mercier to the author, e-mail.

268. MOD, France, *Les enseignements du Kosovo*, 17.

269. LPM 2003–2008, 40.

270. Henri-Pierre Grolleau, "French Precision: Rafale's Lethal Punch," *Air International* 76, no. 4 (April 2009): 21–22, 25; and Armée de l'Air, "Commande de 680 kits d'AASM," February 2010, http://www.defense.gouv.fr/air/base/breves/2010/fevrier/commande_de_680_kits_d_aasm.

271. Kromhout, "New Armée de l'Air," 44.

272. LPM 1997–2002, 140–41.

273. LPM 2003–2008, 40.

274. Ibid., 28.

275. Kromhout, "New Armée de l'Air," 44.

276. *Armée de l'Air 2007*, 48.

277. Armée de l'Air, "Le CEMAA aux rencontres parlementaires," 26 January 2006, accessed 20 May 2007, http://www.defense.gouv.fr/air/archives/26_01_06_le_cemaa_aux_rencontres_parlementaires (site discontinued).

278. Kromhout, "New Armée de l'Air," 42; Mathias Lehmann, "The Further Development of Luftwaffe's Equipment," *Military Technology*, special issue, no. 2 (1999): 53–54.

279. Janssen Lok and Lewis, "New French Air Power," 41–42.

280. Kromhout, "New Armée de l'Air," 42.

281. De Villepin, *Rapport d'information fait au nom de la commission des Affaires étrangères.*

282. Grolleau, "French Precision," 25.

283. LPM 2003–2008, 44.

284. MOD, France, *Défense et Sécurité nationale: le Livre blanc*, 50; and Délégation générale pour l'armement, "La DGA livre le premier système sol-air SAMP/T de série à l'Armée de l'Air," http://www.defense.gouv.fr/defense/votre_espace/journalistes/communiques/communiques_du_ministere_de_la

_defense/la_dga_livre_le_premier_systeme_sol_air_samp_t_de_serie_a_l
_armee_de_l_air.

285. Maryse Bergé-Lavigne and Philippe Nogrix, *Rapport d'information fait au nom de la commission des Affaires étrangères, de la défense et des forces armées (1) à la suite d'une mission sur le rôle des drones dans les armées*, no. 215 (Paris: Sénat, 22 February 2006), http://www.senat.fr/rap/r05-215/r05-2151.pdf, 35.

286. Lorell, *Airpower in Peripheral Conflict*, 11.

287. General Perret, "La force aérienne de projection: un défi à relever pour la fin du XXe siècle," *Air actualités*, hors série no. 1 (July/August 1997): 6.

288. Carlier, "L'aéronautique et l'espace," 478.

289. Ibid., 478; and International Institute for Strategic Studies, *The Military Balance 2004/05* (Oxford, England: Oxford University Press for the IISS, 2004), 50.

290. Lanata, "Faire face," 16.

291. Armée de l'Air, "L'Airbus A340 TLRA arrive en France," 19 May 2006, http://www.defense.gouv.fr/air/base/breves/2006/mai_06/19_05_06_l_airbus_a_340_tlra_arrive_en_france.

292. Kromhout, "New Armée de l'Air," 45.

293. Lanata, "Faire face," 16.

294. LPM 1997–2002, 137; Armée de l'Air, "Audition du général Wolsztynski"; and Bourdilleau, "Evolution de l'Armée de l'Air," 256.

295. Bombeau, "L'Armée de l'Air," 32–33; and EADS, "A400M."

296. Lorell, *Airpower in Peripheral Conflict*, 17–18.

297. Carlier, "L'aéronautique et l'espace," 479; and MOD, France, *Les enseignements du Kosovo*, 20–21.

298. Janssen Lok and Lewis, "New French Air Power," 41.

299. LPM 1997–2002, 135–36.

300. LPM 2009–2014, rapport annexé, 20.

301. MOD, France, *Les enseignements du Kosovo*, 18; and Kromhout, "New Armée de l'Air," 46.

302. Patoz and Saint-Ouen, *L'Armée de l'Air*, 284–85.

303. Gen Michel Fouquet, "L'hélicoptère et le combat SAR [Search and rescue]," in Pascallon, *L'Armée de l'Air*, 173.

304. Gautier, *Mitterrand et son armée 1990–1995*, 164.

305. Patoz and Saint-Ouen, *L'Armée de l'Air*, 125.

306. Gen François Vallat, "La conduite des opérations aériennes dans l'infosphère," *Défense nationale*, no. 8/9 (August/Septembre 1993): 51.

307. Bourdilleau, "Evolution de l'Armée de l'Air," 248.

308. Ibid., 257.

309. Armée de l'Air, "Audition du général Wolsztynski."

310. Asenstorfer, Cox, and Wilksch, *Tactical Data Link Systems*, 7.

311. Kromhout, "New Armée de l'Air," 42, 46–47.

312. Bergé-Lavigne and Nogrix, *Rapport d'information*, no. 215, 10.

313. LPM 2003–2008, 28, 40.

314. Kromhout, "New Armée de l'Air," 43.

315. Sénateur Xavier Pintat, *Avis présenté au nom de la commission des Affaires étrangères, de la défense et des forces armées (1) sur le projet de loi de finances pour 2003, adopté par l'Assemblée Nationale,* Tome 7, *défense—air,* no. 71 (Paris: Sénat, November 2002), 49, http://www.senat.fr/rap/a02-071-7/a02 -071-71.pdf; and Craig Hoyle, "Ground Troops to Get Video via Damocles Targeting Pod," *Flight International* 175, no. 5190 (26 May–1 June 2009): 13.

316. Pintat, *Avis,* no. 71, 50; and Bergé-Lavigne and Nogrix, *Rapport d'information,* no. 215, 9–10.

317. Mercier to the author, e-mail.

318. LPM 2009–2014, rapport annexé, 7; and Lewis, "French Connection," 28.

319. Louis-Marie Clouet and Laurence Nardon, "Build European MALE: Common System Would Boost Military Markets," *Defense News,* 24 May 2010, 21; and Pierre Tran, "France May Buy Predators," *Defense News,* 24 May 2010, 1, 8.

Chapter 4

German Air Force (Luftwaffe)

On 6 May 1955, the Federal Republic of Germany was declared sovereign. Within two days, the ban on military forces was lifted, and Germany became a member of NATO. Since the country had missed a decade of vital experience in military aviation, the German Air Force had to be built up from scratch. At the beginning, the GAF hinged upon support from its allies, principally the United States, for training and equipment.[1] On 6 June 1958, the first fighter-bomber wing, equipped with American F-84s, was established.[2] In the early 1960s, the introduction of a more complex combat aircraft, the F-104 Starfighter, made visible the challenges a not yet mature but fast-growing air force had to overcome. Primarily due to a lack of experienced personnel, a significant number of fatal incidents occurred.[3]

As a frontline state, German security was dependent upon its allies. The GAF became deeply integrated into NATO and assigned all units into the NATO command structure, contributing the largest share to the integrated air defence within Central Europe.[4] The GAF had no national command and control facilities at its

Courtesy Eurofighter Jagdflugzeug GmbH, www.eurofighter.com; photo by JG (German Fighter Wing) 74

A Eurofighter Typhoon from GAF Fighter Wing 74 taking off on a QRA scramble at Neuburg AFB in southern Germany (January 2008)

disposal for the conduct of major air operations.[5] The 1985 *Defence White Book* unambiguously stated that the German armed forces could accomplish their tasks solely in the context of NATO.[6]

Up to the unification of Germany, the bulk of the German armed forces was solely geared towards territorial defence, and out-of-area operations were primarily relegated to humanitarian missions, with GAF transport aircraft shouldering the main burden.[7] Along with the integration of the remnants of the East German armed forces, this led to significant challenges in the post–Cold War era, when out-of-area operations became one of the predominant tasks of European armed forces. Only gradually did German decision makers respond to the demands of the post–Cold War era.

How Has the German Air Force Adapted to the Uncertainties Created by Shifting Defence and Alliance Policies?

This section first analyses German post–Cold War defence policy and the GAF's response to it. It then analyses Germany's alliance context and its influence on the GAF.

Defence Policy

Throughout the 1990s, German defence policy was basically articulated by two key documents, the 1992 *Defence Policy Guidelines* and the 1994 *Defence White Book*. The period was characterised by the adherence to two traditional principles: territorial defence and universal conscription. Yet, Germany's allies would have expected it to contribute to peace and stability in accordance with its political and economic weight.

In 1999 the newly elected left-wing government initiated a military reform.[8] This resulted in a progressive report by an independent commission, known as the Weizsäcker Commission.[9] Defence Minister Rudolf Scharping, who regarded the Weizsäcker Commission report as too progressive, finally came up with his own, more moderate reform proposal, which was approved by the government on 14 June 2000.[10] Only in the wake of 11 September, and due to the continuing imbalance between available means and actual operational requirements, did the newly elected defence minister, Peter Struck, deem it indispensable to undertake significant adjustments of the military reform. This process started in late 2002 and resulted in the 2003 *Defence Policy Guidelines*.[11] The guidelines can be understood as a fundamental change in paradigms towards a more proactive defence policy, where out-of-area operations are the core business of Germany's armed forces. As such, Struck's adjustments were in fact a reform in their own right. With the conservatives winning the elections of 2005, Franz Josef Jung, the new defence minister, released the *White Paper 2006: German Security Policy and the Future of the Bundeswehr*. It not only reiterated the importance of deployed operations but also emphasised the territorial defence of Germany against external threats as the German armed forces' raison d'être and core function, thus partially reversing Struck's reform.[12] In October 2009, Karl-Theodor zu Guttenberg became the new German defence minister. Despite being a member of a conservative party, as was his predecessor, Guttenberg intends to embark upon the most far-reaching reform of the post–Cold War era. While the exact corollaries of this reform cannot yet be grasped (up to this writing), they will most likely result in smaller armed forces better suited for deployed operations. As such, potential force reductions are alluded to in this chapter.

Threat and Risk Perception. The 1992 *Defence Policy Guidelines*, published on 26 November, identified recent changes on the European continent as irreversible. The reunification of Germany, the dissolution of the Warsaw Pact, and the democratisation in Eastern Europe laid a solid basis for a process of which Germany was the main beneficiary. The country was no longer in the direct vicinity of an opposing military alliance but surrounded by allies and partners. The guidelines argued that, for the foreseeable future, Russia had neither the rationale nor the economic and military potential to wage a large-scale offensive against NATO. Provided there was the capacity for a flexible reconstitution of Western armed forces, a full-scale invasion was not conceivable without at least one year of warning time. While Central Europe could benefit from immense improvements, internal conflicts and collapsing states as well as the proliferation of WMDs at the European periphery were perceived to be serious challenges. The increased interdependence in the international system meant that any form of destabilization—military or non-military—was considered to have the potential of producing ripple effects that affected the West.[13]

In 1994 the *Defence White Book* largely reiterated the views of the preceding *Defence Policy Guidelines*, and Russia's military potential still remained a significant factor in German threat perception. Warning time in case of a massive Russian military buildup, however, was no longer specified.[14] Only in 2000 did the Weizsäcker Commission report unambiguously state that, regarding a major military threat against NATO, a warning time of at least eight to 10 years would give the West sufficient time for military reconstitution.[15]

Against the backdrop of 11 September, the 2003 *Defence Policy Guidelines* particularly identified religiously motivated extremism and fanaticism in combination with international terrorism as threats to modern civilization. Concerning conventional military threats, the *Defence Policy Guidelines* takes the view that, though they could be ruled out for the foreseeable future, they could not in the long term.[16] While three years later the *White Paper 2006* identifies international terrorism as "the most immediate danger," it also refers to a "soft" security spectrum including challenges such as fragile statehood in developing countries, migration, or pandemics. The document further highlights issues that are intrinsically related to Germany's economic prosperity, such as energy and resource security or secure transportation routes and communication. Moreover, the white paper singles out illegal arms trade, regional conflicts, weapons proliferation, and military buildup in many parts of the world as destabilising factors. It concludes that "these new types of risks cannot . . . be countered by solely or predominantly using military means."[17]

Tasks of the Armed Forces. During the Cold War era, the West German armed forces were almost exclusively geared up for deterrence and defence against a potential full-scale aggression from the East. In 1992 the *Defence Policy Guidelines* added international crises and conflict management as the second main function.[18] These two main functions were broken down into five tasks:

- protecting Germany and its citizens against political blackmail and external danger,

- contributing to the military stability and integration of Europe,

- defending Germany and its allies,

- contributing to world peace and international security in accordance with the UN Charter, and

- assisting during catastrophes and emergencies as well as supporting humanitarian actions.[19]

In spite of this moderate shift towards out-of-area operations, defence continued to be the raison d'être of the armed forces and of universal conscription.[20] The 1994 *Defence White Book* reemphasised the importance of defence, arguing that, while this had become the least likely scenario, it nevertheless remained the most perilous.[21]

Territorial defence as the main function of German defence policy persisted into the new century. In his official report of June 2000, Defence Minister Scharping emphasised that territorial defence remained the constitutional bedrock for the German armed forces and that it primarily determined the size and structure of the armed forces. The missions remained basically the same as the ones stated in the *Defence Policy Guidelines* of 1992, and the geographical boundaries for crises management operations were basically confined to Europe and its periphery.[22]

Only in 2002–3 did German decision makers adopt a fresh approach to defence. Struck, who became defence minister in July 2002, took the view that defence could no longer be narrowed down to geographical boundaries but should contribute to safeguarding security wherever it is in jeopardy. Hence, the *Defence Policy Guidelines* 2003 states that conflict prevention, crisis management, and postconflict rehabilitation are an integral part of modern defence.[23] To gear up the German armed forces as an instrument for a comprehensive and proactive defence policy, four tasks are identified. The Bundeswehr

- safeguards the capacity for action in the field of foreign policy,

- contributes to stability on a European and global scale,

- ensures national security and defence and helps defend allies, [and]

- supports multinational cooperation and integration.[24]

With the *Defence Policy Guidelines* 2003, an important step was taken. Traditional defence against a conventional attack was superseded by international conflict prevention and crisis management as the determining factors of German defence policy.[25] Accordingly, the armed forces should be capable of operating anywhere in the world across the entire mission spectrum.[26] The *Defence Policy Guidelines* 2003 can hence be understood as a fundamental change in paradigms.

While *White Paper 2006* in principle reiterates the missions as outlined in *Defence Policy Guidelines* 2003, albeit with a slightly different wording, it unambiguously identifies the defence of Germany against external threats as the German armed forces' raison d'être and core function: "The central task of the Bundeswehr continues to be national and collective defence in the classical sense."[27] This partial reversal of the 2003 change in paradigms can be attributed to the elections of 2005, which were won by the conservatives. Nevertheless, *White Paper 2006* did not completely reverse the previous reform, stating that "for the foreseeable future, the most likely tasks will be the prevention of international conflicts and crisis management, to include the fight against international terrorism. They will determine the structure of . . . the Bundeswehr."[28]

Courtesy Pressestelle Mazar-i-Sharif

Recce Wing 51 reconnaissance Tornados on the flight line in Mazar-i-Sharif, Afghanistan

Towards Expeditionary Warfare and Deployable Armed Forces. Germany's reunification, later reviewed in more detail, was preceded by the so-called 2 plus 4 negotiations, which included East and West Germany, the United States, the Soviet Union, Great Britain, and France. The resulting 2 plus 4 Treaty, paving the way for reunification, foresaw amongst others the limitation of the German armed forces to 370,000 personnel.[29] Despite the resulting force reduction, both the 1992 *Defence Policy Guidelines* and the 1994 *Defence White Book* emphasised that conscription would be retained in a unified Germany and that only a minor fraction of the armed forces was earmarked for out-of-area operations with a UN mandate.[30] Germany's main contribution to NATO was rather seen in its capacity to mobilise large quantities of troops in case of collective defence. This capacity was further believed to give Germany significant leverage within the alliance.[31] Crisis reaction forces represented only a complementary element and were deliberately kept at a low level to prevent too interventionist a defence policy.[32] International obligations, however, conflicted with this self-imposed policy of restraint. Particularly, German participation in the Balkan operations during the second half of the 1990s revealed significant shortfalls and serious limitations in Germany's power projection capacities.[33]

Increasingly obsolete force structures caused the 2000 Weizsäcker Commission report to ultimately regard the German armed forces as too large and wrongly structured. The ratio between the overall size of the armed forces and readily deployable troops was considered to be completely inadequate. The commission suggested that the German armed forces should be transformed into an intervention force for international crisis management.[34]

This reform proposal, however, was deemed too progressive by Scharping, who took the view that territorial defence still determined the size and structure of the German armed forces.[35] He finally decided that they were to be reduced but that a significant number of conscripts would be retained. His reform aimed at generating 80,000 directly deployable personnel plus an additional 70,000, with a reduced degree of deployability.[36]

An increasing demand for military deployments because of 11 September, however, revealed that the structures still proved to be inadequate for prolonged and sustained out-of-area operations.[37] Clearly, a recalibration of the reform became necessary. To better respond to the more probable types of operations—conflict prevention and crisis management—it was planned in 2004 to realign the German armed forces into three force categories. These categories encompass response forces (35,000), stabilisation forces (70,000), and support forces (147,500), bringing the total up to 252,500 personnel. Response forces were conceived for joint high-intensity and evacuation operations, the stabilisation forces for joint operations of low and medium intensity during an extended period of time, and the support forces for comprehensive joint support. Response forces were especially set up to comply with international requirements such as German contributions to the NATO Response Force or the EU battle groups.[38] Despite this new emphasis upon power projection, universal conscription in an adapted form was retained, as it was considered an indispensable requirement for operational readiness, effectiveness, economic efficiency, and reconstitution for the purpose of national defence against conventional attack.[39]

The *White Paper 2006* with its more conservative focus reemphasised the value of universal conscription, offering a strong bond between society and the military. Accordingly, it announced that the number of conscripts would be increased by approximately 13,000 during 2006 and 2007, particularly to guarantee conscription equity. It can be concluded that this increase did not aim at strengthening the German armed forces' power projection capacities. In 2006 only 20 per cent of deployed personnel were extended-service conscripts.[40] The renewed emphasis upon conscription was in line with the renewed value of "national and collective defence in the classical sense."[41] Yet as of mid-2010, suspending conscription has been seriously discussed, and on 22 November 2010, Defence Minister Guttenberg formally announced the government's plan to suspend conscription effective 1 July 2011.[42]

Expeditionary warfare requires not only deployable troops but also military planning and command facilities. During the Cold War, the German armed forces had no autonomous planning capacities at the operational level and were designed solely to perform their missions within NATO's integrated military structure.[43] The lack of a national planning and command structure significantly inhibited German participation in out-of-area operations during the first half of the 1990s.[44] Consequently, the 1992 *Defence Policy Guidelines* and the 1994 *Defence White Book* called for an autonomous planning capacity.[45] Only cautiously, however, were limited national command and control structures established. In January 1995, a joint command centre was set up, the tasks of which were limited to relatively minor UN missions. As a premium was put upon the primacy of politics, the new joint command centre was primarily designed as an aid for the defence minister, with the role of the Joint Chief of Staff (Generalinspek-teur der Bundeswehr) relegated to one of an intermediary.[46]

A more robust joint command architecture was only established in autumn 2001.[47] The mission of the Operations Command in Potsdam is national planning and conduct of out-of-area operations on the operational level.[48] The Operations Command can also function as an operations headquarters in the context of EU operations, whereby a nucleus of German staff is augmented nationally and multinationally.[49] Besides this joint command and planning cell, the Berlin Decree of 21 January 2005 elevated the position of the Joint Chief of Staff as compared to the single service chiefs. Regardless of single service boundaries, the Joint Chief of Staff bears the responsibility for force structuring and for the planning and conduct of operations. The single service chiefs are responsible for the provision of combat ready units.[50]

Defence Policy and Its Impact upon the German Air Force. To respond to the challenges of the post–Cold War era, the GAF has been undergoing three major reforms. These reforms have been named after the new structures the GAF has been implementing. Air Force Structures 4, 5, and 6 have led to major force reductions, summarized in table 2. Reforms underlying these reductions are discussed next.

Table 2. GAF reductions during the post–Cold War era

	Prior to 1990 (West Germany)	Air Force Structure 6 (Planned)
Military personnel	110,000	34,500 + 28,000[a]
Civilian personnel	21,500	˜ 6,500
Combat aircraft	755[b]	262
Flying wings	18	7
GBAD systems	176	24
SAM groups/battalions	15	6
Tactical air command and control units	14	3

Adapted from Hans-Werner Jarosch, *Immer im Einsatz: 50 Jahre Luftwaffe* (Hamburg, Germany: Verlag E. S. Mittler & Sohn, 2005), 174.

[a]Air Force Structure 6 envisages 18,000 GAF personnel being allocated to the newly established Joint Service Support, with a further 10,000 engaged in training and education within the GAF, the German armed forces in general, as well as in civilian facilities. "Further Development of the Luftwaffe—Luftwaffe Structure 6," in *CPM Forum: Luftwaffe 2005—The German Air Force Today and Tomorrow*, managing director Wolfgang Flume and project director Jürgen Hensel (Sankt Augustin, Germany: CPM Communication Presse Marketing GmbH, 2005), 47.

[b]Includes Navy Tornado aircraft.

Air Force Structure 4 (see fig. 1) was implemented between 1990 and 1994.[51] One of its major goals was overall force reduction in order to abide by the limitations set by the 2 plus 4 Treaty.[52] It envisaged three command layers between the Air Staff and the operational wings/units. Directly subordinate to the Air Staff were the Air Force Office, in charge of training and administration; the German Air Force Support Command, responsible for combat service support and materiel; and the German Air Force Command. Subordinate to the GAF Command were the Air Transport Command and two regional air force commands, the Air Force Commands North and South. Two air divisions (AD) were subordinate to each of the regional air force commands.[53] Unlike previous ADs, which had been functionally configured, the ADs were henceforth geographically configured, each containing a large cross section of GAF capabilities.[54]

Total combat strength of Air Force Structure 4 contained four air defence wings, five ground-attack and one reconnaissance wing (RW), and six SAM wings.[55] This corresponded to a reduction of five flying wings and a significant reduction of SAM wings since 1989, when the West German Air Force had 11 ground-attack, two reconnaissance, and two fighter wings (FW) in its inventory.[56]

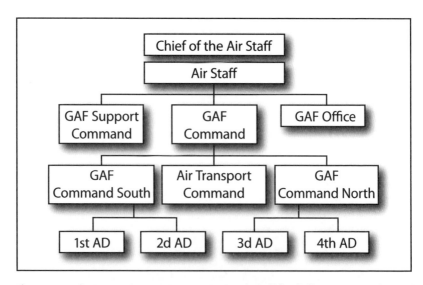

Figure 1. Air Force Structure 4 (GAF) simplified diagram. (*Adapted from* Hans-Werner Jarosch, ed., *Immer im Einsatz: 50 Jahre Luftwaffe* [Hamburg, Germany: Verlag E. S. Mittler & Sohn, 2005], 129.)

Moreover, the nuclear missile squadrons were disbanded, and the airborne nuclear component was reduced.[57]

Alongside these reductions, the air defence role received new emphasis. With German reunification and the reestablishment of full sovereignty, the GAF was tasked with the duty of air policing, previously confined to the United States and Great Britain.[58] As a result, all F-4F Phantom and MiG-29 aircraft were assigned to the air defence and air policing role.[59]

Courtesy German Air Force; photo by Toni Dahmen

F-4F Phantom II. To remain relevant pending delivery of the Eurofighter, the F-4F fleet had to undergo significant upgrades. By December 1996, 65 F-4F Phantom IIs had been delivered. The most essential components of the retrofit included the integration of the APG-65 radar and the AIM-120 AMRAAM.

The next major reform step, Air Force Structure 5 (see fig. 2), was initiated against the backdrop of the 1999–2000 defence reform initiative on 1 October 2001.[60] Air Force Structure 5 foresaw the transfer of 17,000 GAF personnel to a newly established Joint Support Service.[61] This represented a considerable thrust towards jointness as it forced all of the services to cooperate in the areas of C2, training, maintenance, and logistics, thereby reducing redundancies.[62] The single-service GAF Support Command was no longer needed. Furthermore, the C2 structure was compressed by disbanding both Regional Command North and Regional Command South.[63] Total combat strength of Air Force Structure 5 was eight operational flying wings and four SAM wings. The flying units were divided into four ground-attack wings, one reconnaissance wing, and three air defence wings.[64]

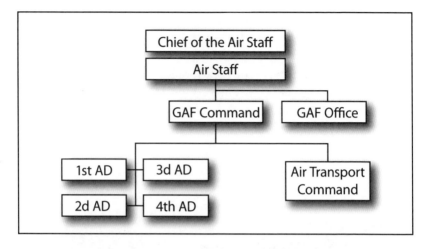

Figure 2. Air Force Structure 5 (GAF) simplified diagram. (*Adapted from* "The German Air Force Structure 5: Principles, Structures, Capabilities," in *CPM Forum: German Air Force—Structure and Organisation, Equipment and Logistics, Programmes and Perspectives*, ed. Wolfgang Flume, in cooperation with the GAF staff, MOD, Bonn, and editorial coordinator Lt Col (GS) Rainer Zaude [Sankt Augustin, Germany: CPM Communication Presse Marketing GmbH, 2003], 20.)

The new structure initiated the most substantial improvement for C2 in the history of the GAF. During the Cold War era, the GAF had no means at its disposal to conduct air combat operations above the wing level and fully hinged upon NATO C2 structures. In the post–Cold War era, with an increasing German commitment to out-of-area operations, a national capacity for the planning, conduct, and command of air combat operations became necessary. In late 2001, the GAF Air Operations Command (Kommando Operative Führung Luftstreitkräfte) was established at the former location of Air Force Command North. For the first time in its history, the postwar GAF was given an autonomous capacity for operational planning of air operations. The GAF Air Operations Command had been primarily designed as a national nucleus with 59 posts for a combined joint force air component command headquarters in the context of NATO or EU operations.[65]

Besides out-of-area operations, 11 September had the biggest impact upon the C2 structure of the GAF. As a response to renegade planes, an interministerial National Air Defence Operations

Centre (Nationales Lage- und Führungszentrum—Sicherheit im Luftraum) was established in July 2003 for ensuring airspace security. The centre coordinates interception missions between the GAF, the MOD, and civilian ministries, which allows for time-critical decision making.[66] The centre is collocated with the GAF Air Operations Command, and it is directly responsible to the chief of the Air Staff (Inspekteur der Luftwaffe) in questions of national airspace security.[67]

Defence budget cuts rendered a timely and entire adoption of Air Force Structure 5 infeasible. By late 2002, the necessity to achieve a balance between available resources and actual operational requirements had become paramount.[68] This, in turn, required further substantial reductions. To achieve savings with immediate effect, Navy Tornado combat aircraft and their maritime tasks were to be transferred to the GAF, while the equivalent of two GAF Tornado wings were simultaneously reduced. Additionally, the disbanding of Hawk and Roland SAM systems—originally planned for 2008 and 2012, respectively—was announced.[69] These quantitative adjustments ushered in Air Force Structure 6 (see fig. 3), the adoption of which commenced in 2005–6.[70]

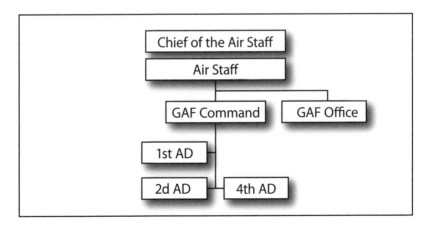

Figure 3. Air Force Structure 6 (GAF) simplified diagram. (*Adapted from* "Further Development of the Luftwaffe—Luftwaffe Structure 6," in *CPM Forum: Luftwaffe 2005—The German Air Force Today and Tomorrow,* managing director Wolfgang Flume and project director Jürgen Hensel [Sankt Augustin, Germany: CPM Communication Presse Marketing GmbH, 2005], 47.)

Air Force Structure 6 originally envisioned reducing the number of ADs from four to three and disbanding the German Air Transport Command, with the EATC assuming its responsibilities.[71] With the adoption of Air Force Structure 6, the GAF was planned to have seven flying units at its disposal.[72] The Eurofighter was to equip four operational units and a training unit, replacing the F4-F Phantom fleet as well as a large part of the Tornado fleet.[73] Only two Tornado units were planned to stay operational.[74] Once equipped with the Eurofighter, the multirole-capable 31st Fighter-Bomber Wing (FBW) and the 73d FBW, a training unit, were to be capable of contributing to national air policing alongside FWs 71 and 74.[75] Despite this move towards a more flexible employment of air power, the GAF seemed to continue emphasising specialized wings. Yet against the backdrop of an unbenign economic climate as of mid-2010, "Tranche 3B" of the Eurofighter, comprising 37 aircraft, was not likely to be acquired, which would result in one Eurofighter wing less.[76] On the positive side, all of the remaining frontline wings were planned to be used more flexibly in both the air-to-air and air-to-ground roles.[77]

Moreover, after the removal of Hawk and Roland units, three SAM wings equipped with 24 Patriot systems form a SAM nucleus in the context of Air Force Structure 6.[78] Despite the partial reversal of the 2003 reform, *White Paper 2006* affirms the foundations of Air Force Structure 6.[79]

Table 3. Flying wings within Air Force Structure 6 (GAF)

Unit	2006	Future
51st RW	Tornado	Tornado
32d FBW	Tornado ECR, SEAD	Tornado ECR, SEAD
33d FBW	Tornado (LGB, Taurus)	Eurofighter (LGB, Taurus)
31st FBW	Tornado (LGB)	Eurofighter (LGB, Taurus)
71st FW	F-4F Phantom	Eurofighter
74th FW	F-4F Phantom	Eurofighter
73d FW	F-4F Phantom	Eurofighter (training unit)

Source: Lt Col (GS) Nicolas Radke, GAF, MOD, Bonn, Germany, to the author, e-mail, 28 November 2006.

To comply with the political goal of enhancing military power projection capacities, the GAF is amongst others acquiring new capabilities such as AAR, strategic air mobility, and a deployable airspace surveillance capability.[80] Particularly, the tactical air control units, till 2005 only concerned with home and alliance defence, are to be modernised and developed into a state-of-the-art tactical air command and control service with a mobile component.[81] Since the GAF's projection capability significantly hinges upon materiel in order to cover the new mission spectrum, full adoption of Air Force Structure 6 will only be completed with the introduction of planned systems.[82]

Professional personnel are a further key criterion for expeditionary armed forces. Despite general conscription and the resulting difficulties in deploying troops, the problem does not seem as serious for the GAF as for the German Army. In the time span from 1990 to 2002, the percentage of professionals has increased from 65.5 per cent to 76 per cent.[83]

German Reunification and Its Impact upon the German Air Force. On 3 October 1990, the two Germanies were reunified. The West German Air Force was subsequently tasked to disband the East German Air Force and to integrate the remaining parts, which were unified in a single AD. The establishment of the 5th AD allowed for a smooth integration. Against the backdrop of Air Force Structure 4, the original 3d AD in West Germany was disbanded, and the 5th AD in the East was renamed 3d AD in early 1994. The headquarters of the newly formed 3d AD was relocated to Gatow, a former RAF air base in West Berlin. The 5th AD was initially equipped with systems of the East German Air Force, including MiG-29s, SAM systems, and helicopters of Soviet provenance. Its structure encompassed two GBAD wings, a supply regiment, a helicopter unit, and a MiG-29 wing, later to be augmented by West German F-4 Phantoms. With the transition to Air Force Structure 4, the formerly East German SAM systems were replaced by Hawk and Patriot systems, and the Recce Wing 51 was transferred to the newly established 3d AD.[84] On 1 January 1995, the operational units in former East Germany were allocated to NATO and became thereafter part of NATO's integrated air defence.[85]

The disbandment of the East German Air Force required the scrapping of the main bulk of its weapons systems, as they could hardly be used in a Western defence concept.[86] At the end of the Cold War, the East German armed forces had an overall strength of 767 aircraft from all services, both fixed and rotary wing.[87]

Amongst the systems that continued to be operated, 24 MiG-29s and three Airbus A310s were the most prominent ones.[88] The latter actually represented a milestone step for the GAF and formed the nucleus for the GAF's current Airbus fleet.[89] Throughout the 1990s, the MiG-29s were sought-after sparring partners for allied air forces.[90]

In early 1990, the East German military hoped that there would be separate armed forces on the former East German territory, which would not be integrated into an alliance system. High-ranking East German officers thus supported the so-called theory of two armed forces in a reunified Germany.[91] However, shortly prior to the reunification of Germany, on 28 September 1990, the remaining generals and admirals of the East German armed forces were made redundant by the last government of the German Democratic Republic.[92] Being too closely entangled with the former socialist regime, they were not offered a second chance.[93] The directives by West German defence minister Gerhard Stoltenberg were unambiguous. The armed forces of a reunified Germany were to be structured according to the example of the West German armed forces, both in form and attitude.[94]

On 3 October 1990, the reunified German armed forces reached a strength of 620,000 personnel. According to the results of the 2 plus 4 negotiations, this number had to be reduced to 370,000 by 1994, with originally 82,400 airmen planned. In light of these reductions, former East German personnel were integrated according to requirements.[95] In 1989 the peacetime strength of the East German Air Force amounted to 34,600. In September 1990, the strength had shrunk to approximately 18,000. These reductions were due to a number of reasons. For instance, conscripts no longer turned up for national service, or others did not want to collaborate with the former enemy. In the wake of reunification, numbers continued to shrink steadily. Many left the armed forces on their own initiative, some were not deemed suitable for Western standards, and others had to leave because of their involvement in the former East German internal security apparatus.[96] All airmen that aspired to be re-employed by the GAF had to undergo a screening that lasted up to two years.[97] A total of 2,500 (mostly officers) filed an application for full employment in the GAF. Of these, 1,000 were employed as officers and 500 as non-commissioned officers.[98]

Particular problems were the relatively large number of officers and the lower qualifications. Compared with West German standards, for instance, many East German officers accomplished tasks that did not require a commission.[99] The different career structures caused many reemployed officers to be degraded in rank.[100] By removing the highest ranking officers and thoroughly screening the remaining officers, a discontinuity was achieved in East Germany, and any vestiges of the former regime were eradicated.

In terms of required resources, the integration of the remaining parts of the East German Air Force cannot be accurately quantified as there were many overlying factors, such as the so-called peace dividend or overall reunification costs that led to significant reductions in defence resources at the time. As a result, the acquisition of new material had to be significantly delayed or numbers reduced.[101] These developments are examined in the section on procurement (maintaining a relevant air force).

Emphasis upon Air Power. According to a 1987 RAND study, the services' shares of West Germany's procurement budget had been divided according to a rather rigid split of 50:30:20 for the Army, Air Force, and Navy, respectively.[102] No exact figures for single-service budget allocations are available for the post–Cold War era, as the defence budget is centrally administered by the department of finance of the MOD without explicit reference to the services. According to a 1999 assessment by the chief of the Air Staff, roughly 29 per cent of the overall defence budget could be considered as the share for the Air Force, which is basically in line with Cold War budget allocations.[103] Though this represents a share which is slightly less than a third, it has to be noted that, as in other nations, German air power is not confined to the Air Force alone. Both the Army and the Navy have been operating a significant number of air assets.

In the early 1990s, prior to the unification of the German armed forces, the West German Army operated 750 helicopters, including 108 heavy CH-53G transport helicopters, and a significant amount of MANPADS and air-defence guns, including over 430 self-propelled, radar-guided Gepard systems as well as 143 Roland SAM systems. German naval aviation consisted of 104 Tornado combat aircraft; around 40 fixed-wing aircraft for SIGINT, maritime reconnaissance, search and rescue (SAR), and liaison purposes; and 19 Sea Lynx and 22 Sea King helicopters.[104] In 2002 these numbers had been reduced but nevertheless remained significant. Army aviation still encompassed 566 helicopters. Along with a considerable number of air-defence guns and MANPADS, 143 Roland SAM systems and 380 Gepard systems were still in the service of the German Army. The Navy operated 43 helicopters and 66 Tornado aircraft, amongst others.[105]

It can be concluded that the defence budget allocation to air power in general has been significantly more than the one-third assigned to the GAF. In the period from 2007 onwards, this share is growing, as the Eurofighter and A400M programmes are expected to consume a significant percentage of the overall materiel acquisition funds.[106] The Eurofighter has become the most expensive programme of the German armed forces.[107] It is evident that German decision makers perceive air power to be an effective military instrument and that the even share placed upon air power reflects German emphasis upon jointness.

Alliance Context

Germany's postwar armed forces were not conceived as an instrument of autonomous statecraft but as forces hinging upon allied command structures. During the Cold War, German armed forces—with the exception of a few territorial Army units—could only accomplish their mission within the NATO chain of command. Air defence forces were already during peacetime assigned to the NATO commander, Allied Air Forces Central Europe.[108] Germany thus deliberately transferred sovereignty on the use of military force to NATO. With the Cold War over and the country reunited, Germany did not depart from this defence policy orientation. It not only insisted on integration into the Western Alliance system but also added a new emphasis upon European integration.[109] These political paradigms were underlined by the 1992 *Defence Policy Guidelines* and by the 1994 *Defence White Book*. The former identified European integration, including a European defence identity, as well as the partnership with the United States and the significant American military presence in Europe as vital German interests.[110] The latter stated that the challenges of the new era could only be tackled by a transnational and cooperative approach and that Germany would only act in cooperation with allies and partners.[111]

Given this emphasis upon multilateralism, German decision makers have put a premium upon synergies between international institutions, particularly between NATO and an emerging European defence pillar within the EU.[112] In the early 1990s, Germany, together with France, was the main promoter of European integration and the establishment of a Common Foreign and Security Policy. The strengthening of the security and defence political dimension of the EU, however, conflicted with Germany's interest to preserve a strong transatlantic alliance and to secure involvement by the United States in Europe. As a consequence, German politicians have continued to emphasise that they regard genuine European military capabilities as a complement to NATO and not as a substitute and that strengthened European defence structures would offer the prospect of a truly transatlantic partnership.[113] This fundamental view also remained unscathed by tensions in the course of the Iraq crisis. The 2003 *Defence Policy Guidelines* unambiguously stated that NATO membership was deemed the cornerstone of German security and that the United States would remain indispensable to European security.[114] Hence, Germany's opposition to the war in Iraq could not be seen as a signal of Berlin's adherence to a Franco-German project of creating a counterweight to American hegemony in the field of defence policy.[115] On the contrary, the Federal Republic has perceived itself as one of the staunchest advocates of American involvement in European security. However, German perception of American-German relations has been continuously shifting "from a relationship based on acceptance of American leadership towards one of collaboration amongst equal partners."[116]

Though Germany has been a strong proponent of European defence initiatives and provided France with important leverage on crucial issues, Berlin's unequivocal support for NATO has often led to disappointment in Paris. A major foreign policy dilemma has been how to comply simultaneously with French and American demands. In critical decisions, Germany has tended to tilt in favour of NATO and thus of the United States' position. In the mid-1990s, Germany, for instance, did not support the French proposal for a European commander heading AFSOUTH in Naples.[117] In this sense, the 2003 *Defence Policy Guidelines* restated that the ESDP continued to be complementary with—and was not a supplant for—NATO and the transatlantic partnership, a view that was not compatible with France's position.[118] The *White Paper 2006* was even more explicit: "The transatlantic partnership remains the bedrock of common security for Germany and Europe. It is the backbone of the North Atlantic Alliance, which in turn is the cornerstone of German security and defence policy."[119] As such, it implicitly tended to relegate the ESDP to operations requiring joint civilian and military efforts, a position that is not fully reconcilable with France's position.[120]

By reiterating essential aspects of the 1992 *Defence Policy Guidelines* and the 1994 *Defence White Book*, the 2003 *Defence Policy Guidelines* and *White Paper 2006* gave German alliance policy continuity. The former confirmed that, with the exception of evacuation and rescue missions, Germany would only act multinationally and that national defence policy remained determined by European integration and the transatlantic partnership.[121] Three years later, *White Paper 2006* stated that "Germany is committed to active multilateralism. . . . Germany therefore safeguards its security interests primarily in international and supranational institutions."[122]

By deferring sovereignty to supranational institutions, Germany has sought throughout the post–Cold War era to remain a relevant international actor in defence policy and to obtain maximum leverage.

Alliance Policy and Its Impact upon the German Air Force

In line with Germany's alliance policy, the GAF has remained closely integrated into supranational military structures throughout the post–Cold War era. While it has always put a premium upon integration into NATO and a close relationship with the USAF, particularly in the field of training, the GAF also became a main driver for integrating European air power. Hence, German multilateralism on a strategic level has been carried right down to the military operational level.

ESDP Context. The 2003 *Defence Policy Guidelines* stated that assets reported to NATO and the EU would be available to both

Courtesy GAF; photo by Stefan Gygas

Eurofighter Typhoon, the GAF's latest combat aircraft

organisations.[123] Similarly to France, Germany was supposed to provide approximately 20 per cent of the air assets for potential EU operations, amounting to approximately 80 combat aircraft with corresponding support aircraft.[124] Yet, against the backdrop of logistics shortfalls and a very limited number of tanker aircraft, the number of aircraft theoretically put at the disposal of the EU was too ambitious, particularly when deploying to distant theatres.

As was explained earlier, the GAF, together with the FAF, was crucial in developing the EU Air RRC. Pending implementation of the EU RRAI, lead nations of a particular EU battle group were responsible for the various force enablers. For 2007, for instance, the GAF envisaged supporting the potential deployment of a German-led battle group with four Tornados for precision ground attack, four reconnaissance Tornados, two C-160 Transalls, four UH-1D helicopters for intratheatre airlift, and two A310s—one a multirole transport (MRT) and the other an MRTT—for medical evacuation and air-to-air refuelling.[125]

Along with these initiatives, Germany has been involved in bi- and multilateral projects that have the potential for a European dimension. As such, EU Headline Goal 2010 explicitly mentions a proposal for a Franco-German air transport.[126] These initiatives are reviewed below.

NATO Context. During the Cold War, the GAF was closely integrated into the Western Alliance. Air defence forces were subordinated to the NATO command and control structure, and offensive combat aircraft wings would have been subordinated in case of defence.[127] This emphasis upon NATO continued into the post–Cold War era. German fighter-bombers have further been assigned to the task of delivering American nuclear weapons in case of alliance defence. In compliance with the 2 plus 4 Treaty, however, no aircraft earmarked for the nuclear role have been stationed in eastern Germany.[128]

Furthermore, the GAF's command structure has remained closely linked to that of NATO. In the 1990s, harmonisation between the structures was achieved particularly through collocation of the Air Force Commands North and South with NATO CAOCs. German general officers who headed these regional commands simultaneously became commanders of the respective NATO CAOC.[129] Through these CAOCs—at first interim CAOCs due to technical deficiencies—allied air forces were in 1994 for the first time directly led by German generals.[130] In addition, the German general officer in charge of Air Force Command North also headed the multinational Reaction Forces Air Staff (RFAS).[131] This staff was established in 1995 for the planning and coordination of allied out-of-area air operations in support of Supreme Headquarters Allied Powers, Europe (SHAPE).[132] The NATO Airborne Early Warning and Control Force also provided the GAF with a high-ranking NATO command post. On a rotational basis, it has been commanded either by a USAF or a GAF major general.[133]

As was noted, with the transition to Air Force Structure 5, the GAF Air Operations Command was established. This command regularly assigns five to 10 German planning officers to the NATO Response Force CJFAC headquarters. For NRF 11/2 and 12/1 during the second half of 2011 and the first half of 2012, respectively, Germany and France are intended to be in charge of the air component. As such, the GAF Air Operations Command is supposed to provide the core of the CJFAC headquarters—amounting to 150 personnel—and to be in command of the NRF 11/2 air component.[134]

The GAF intends to make a permanent contribution to NRF rotation cycles. For the time being, bottlenecks in logistics, however, do not allow for two simultaneous major contributions from both the Army and the Air Force. Through the provision of scarce assets such as TBMD units, dedicated SEAD Tornados, and air-launched cruise missiles, the GAF aspires to gain visibility.[135]

Moreover, the setting up of NATO's Joint Air Power Competence Centre was a GAF initiative. It was the German Air Staff that identified the need for a common centre of excellence to provide subject matter expertise on allied air power concepts and doctrine.[136] On 16 December 2002, these views were officially outlined. Henceforth, Germany acted as the lead nation in setting up the JAPCC, which became operational in early 2005 and basically took over the infrastructure of the RFAS that was disbanded on 31 December 2004.[137]

Bilateral and Multilateral Organisational Relationships. In 1995–96, a Franco-German airlift exchange system was created.[138] Such initiatives had already been motivated by the 1994 *Defence White Book* that emphasised international cooperation in logistics and training. Cooperation was expected both to reduce costs and facilitate common standards and interoperability.[139] In the late 1990s, German defence minister Scharping particularly supported the setting up of a common European airlift command to partially remedy European shortcomings highlighted in the context of NATO's Defence Capabilities Initiative and the European Headline Goal.[140] At the Franco-German summit in November 1999, France and Germany declared their intention to transform their cooperation in the field of military airlift into the European Air Transport Command. The setting up of the EATC has been pursued in an evolutionary way so far. An EAG airlift study laid the foundation for the European Airlift Coordination Cell in Eindhoven, Netherlands, which was further developed into the European Airlift Centre.[141] Unlike other European states, for which transferring national sovereignty into a multilateral framework is a more sensitive issue, Germany would have been prepared to transfer national authority in the field of airlift immediately and to set up directly an independent EATC.[142]

At the Tervuren Summit on 29 April 2003, Germany, France, Belgium, and Luxembourg confirmed the EATC initiative.[143] Finally, in May 2007, an agreement between Germany, France, Belgium, and the Netherlands was signed to establish the EATC, and in February 2010, the defence ministers of the four countries declared that the multinational command, headed by a GAF major general, begin operations in Eindhoven in the second half of 2010.[144] Accordingly, the national German Air Transport Command was disbanded in July 2010.[145] Germany and France, the two EATC initiative nations, put the majority of their airlift assets under the authority of the EATC. Exceptions are AAR, special ops, CSAR, VIP transport, and medevac. Planning and execution is delegated to the EATC. In relation to the Movement Coordination Centre Europe, the EATC is at the same level as a national air transport command.[146] When the EAC was disbanded on 1 July 2007 and merged with the Sealift Coordination Centre into the MCCE, some EAC personnel were transferred into the EATC implementation team.[147]

Germany was also a key player in the establishment of the Strategic Airlift Interim Solution, available to both NATO and the EU.[148] Since early 2006, two Antonov An-124-100 heavy transport aircraft are permanently stationed at the German airport Leipzig-Halle to fill the European strategic airlift gap pending arrival of the A400M aircraft.[149]

Another traditional field for cooperation has been training. Since its early days, the GAF has had a tradition of conducting pilot training in an international environment through bilateral and multilateral agreements. In the mid-1950s, the GAF basically had to be built up from scratch. As only a minority of the pilots were WWII veterans, many pilots were trained in the United States and in the United Kingdom.[150] In the post–Cold War era, basic jet training of GAF pilots continued to take place in a multinational environment at Sheppard AFB, Texas. Moreover, German weapon system officers are trained together with their US Navy counterparts at the naval air station in Pensacola, Florida, and German SAM crews alongside their US Army counterparts at Fort Bliss, Texas.[151] For low-level flight training, cooperation was also sought with Canada. From 1980 up to 2005, Tornado, Phantom, and Transall crews were training in Goose Bay, Canada. With decreasing importance of low-level flights in modern air operations, however, the GAF did not extend its agreement with the Canadians.[152]

Early on, the GAF also built up close partnerships in Europe, particularly with the Italian Air Force. Since 1959 both air forces have used the Decimomannu Air Base in Sardinia for aerial combat training.[153] A significant step in European training cooperation was the establishment of the Tri-national Tornado Training Establishment at Cottesmore, UK, in 1979. This cooperation was particularly aimed at achieving standardisation and interoperability amongst European Tornado operators. Divergence in national midlife updates, however, made the operation of a common fleet of training aircraft impracticable in the late 1990s.[154]

After the trilateral arrangement for Tornado training ceased in March 1999, the GAF transferred basic weapon systems training and advanced training on the Tornado to Holloman AFB, New Mexico.[155] Until December 2004, when the F-4F training squadron was disbanded, the training of German Phantom crews was collocated at Holloman AFB.[156] With these steps, the GAF centralised the entire basic training of German F-4F and Tornado crews in the United States.[157]

Almost simultaneously with this increasing involvement in the United States, the GAF became a member of the EAG on 12 July 1999, a step that has fostered European standardisation and interoperability.[158]

Cooperation on an Operational Level. Besides combined air operations, examined in the following section, the GAF has deepened its international expertise through participation in integrated exercises, particularly the Flag exercises in the United States and in Canada. GAF participation at Red Flag started in 1989. In the ensuing years, the GAF, with its MiG-29s, became a sought-after training partner.[159]

In 1995 the GAF itself established an annual integrated and combined air exercise in Germany, called ELITE (Electronic Warfare Live Training Exercise). Not only flying units benefit from this integrated exercise but also German SAM units. Apart from Exercise ELITE, German SAM units have regularly participated in the annual exercise Roving Sands in Texas and New Mexico and in the Dutch exercise series Joint Project Optic Windmill.[160]

How Has the German Air Force Responded to the Challenges of Real Operations?

During the Cold War, the German armed forces were involved in a number of humanitarian operations to alleviate the effects of natural disasters. Planning and conduct of these operations were assigned to a particular service, mostly the Air Force.[161] Germany also supported the United Nations financially and logistically. In 1973, for instance, GAF units airlifted Nepalese troops to Lebanon. Operations of a predominantly military nature, however, were confined to NATO territory. Despite self-imposed restrictions on the use of military force, international expectations for a more solid German participation in international burden sharing had grown since the Federal Republic's accession to the UN in 1973.[162] In 1987, for the first time since the end of WWII, Germany was asked by its allies to deploy naval units to the Persian Gulf to protect international shipping. The Federal Republic, however, was opposed to out-of-area actions for basically four reasons. First, the German constitution was deemed not to provide a foundation for this kind of operation. Second, there was widespread concern that West German interventionism might cut off all contacts with Eastern Europe. Third, it was UN Security Council policy not to dispatch troops of divided countries such as Germany and Korea, and, finally, Germany's historical legacy exerted enormous restraint internally upon any form of military action.[163] Thus, a policy of gradual steps was pursued in the post–Cold War era, and the GAF became only incrementally involved in combat operations.

German Contributions to Western Air Campaigns

In the midst of the reunification process, German decision makers found themselves confronted with the occupation of Kuwait and its broader international consequences. For constitutional reasons, and more importantly because a German military deployment to the Gulf could have provoked the Soviet Union to put German reunification on hold, military participation in the American-led coalition was ruled out. Furthermore, more than two-thirds of the population was against any military involvement. Germany's contribution to the Gulf War finally turned out to be a significant financial support.[164] Whilst the Federal Republic did not get directly involved in Operation Desert Storm, it was asked by NATO to support Turkey against potential Iraqi incursions. Only after some hesitations did Germany finally agree to dispatch 18 Alpha Jets. The United States and Turkey were not pleased by

Germany's lukewarm response to the request. Later on, Germany reinforced its deployment by sending 11 air defence batteries.[165]

Germany's and the GAF's reactions to Desert Storm can be characterised by discomfort, both politically and militarily. Uneasiness with offensive air power doctrine in deployed operations led to a heated political debate regarding the proper interpretation of the German constitution, as is examined in the following pages. Militarily, uneasiness with offensive air power doctrine—as demonstrated during Desert Storm—led to a gap in doctrinal thought throughout the 1990s. The GAF published its first air power doctrine on 22 March 1991, immediately after Desert Storm. Yet it was still firmly embedded in a Cold War setting, and no reference was made to the revolutionary developments of Desert Storm. Unlike in the case of the RAF, no successful attempt at revising the doctrine was made to reflect the lessons of Desert Storm. This doctrinal gap was mirrored in a lack of a broader doctrinal debate in the form of publications, as is analysed further below. In terms of equipment, particularly in the area of precision-guided air-to-ground munitions, the GAF started to implement the lessons of Desert Storm only belatedly. At the time of Allied Force in 1999, there were still no LGB-capable German Tornados available. This can be related to two further issues: a strained budget situation in the wake of German reunification and the main defensive mission of the German armed forces. As regards the former, the 1994 *Defence White Book* announced that any major equipment programmes for the main defence forces would be deferred to the period after 2000. As regards the latter, Russia's military potential remained a significant factor in Germany's threat perception throughout the 1990s. Accordingly, territorial defence continued to be the constitutional bedrock for the German armed forces and determined their size and structure. This, in turn, led to a continuation of Cold War concepts. For instance, unlike the RAF, which disposed of its submunitions dispenser system partly as a consequence of Desert Storm, the GAF retained its equivalent system as the main armament of its Tornado fleet throughout the 1990s, as is analysed further below.

From October 1992, German AWACS personnel contributed to the airspace surveillance of the no-fly zone over Bosnia.[166] Whilst this represented the only German contribution to Operation Deny

Flight, it nevertheless aroused staunch political opposition.[167] The junior government party as well as the major opposition party appealed to the Federal Constitutional Court.[168] There was disagreement principally on how to interpret Article 24, paragraph 2, of the constitution. The specific article stated that the German Federation could participate in collective security systems for the preservation of peace.[169] Up to that point, this article had been considered to limit combat operations to collective territorial alliance defence. On 8 April 1994, the court declared that German participation in Deny Flight did not violate the constitution.[170] Hence, with the reestablishment of full German sovereignty and in light of NATO's first deployed combat operations, a flexible interpretation of the article was deemed appropriate. The main answer of the court as of 12 July 1994 stated that the Federal Republic could assume full responsibility as a member of a collective security system, including armed out-of-area operations. Yet, the court added that for such operations, parliamentary approval was a prerequisite.[171] This development allowed the GAF to be employed in militarily robust operations beyond alliance territory.

While the verdict of the Federal Constitutional Court did not explicitly consider a UN mandate a precondition, it was commonly assumed that a UN mandate would lay the foundation of any German military intervention, as had been the case so far. Despite this new interpretation of the German constitution, the foreign and defence ministers made it clear at the time that Germany would continue to adhere to a policy of restraint and that missions outside of Europe would remain an exception.[172]

In the course of Operation Deny Flight, particularly in late 1994, pressure upon German decision makers to make a more robust contribution grew. NATO authorities regarded German Tornado ECRs, especially conceived for the SEAD role, as valuable assets for ongoing operations. Despite the recent verdict by the constitutional court, German decision makers proved reluctant to respond immediately to any external requests and preferred to pursue the accustomed policy of self-imposed restraint.[173]

Finally on 30 June 1995, the German parliament voted in favour of a more robust military commitment in the context of the Bosnian civil war. The core of the German contribution encompassed eight Tornado ECR and six reconnaissance Tornado aircraft.[174] A deployment of combat troops was ruled out on the grounds of Germany's historical legacy. Moreover, nationally imposed rules of engagement provided a very narrow margin for German aircraft to operate in. Their employment was restricted to supporting defensive actions of the multinational RRF, which had only recently been inserted to protect UNPROFOR's potential withdrawal. As a consequence, German Tornado aircraft were not entitled to directly participate in UN mandated air campaigns such as Deny Flight.[175]

At the request of the RRF, GAF Tornados flew their first reconnaissance mission against the backdrop of Operation Deliberate Force on 1 September 1995. Approval by the German MOD was a prerequisite.[176] While the alliance could not directly draw upon German SEAD aircraft for Operation Deliberate Force, it was argued that the mere presence of German Tornado ECR aircraft inhibited hostile SAM activity.[177]

Up to this moment, German Tornado crews had not been exposed to real operations. Yet, this lack of experience was effectively compensated for by GAF participation in integrated exercises such as Red Flag or Maple Flag. The deployed Tornados were also specifically optimised for operations over the Balkans, which—despite German political caveats—provided the GAF with leverage amongst its allies.[178] Moreover, officers of the former East German Air Force provided a unique insight into Soviet procedures that the Bosnian Serb SAM crews employed.[179]

Subsequent to the air campaign, the German Tornado contingent was put at the disposal of the Implementation Force (IFOR) with the particular purpose of surveying SAM activity as well as the disentanglement of hostile troops. Alongside the Tornado aircraft, the GAF provided 12 C-160 Transalls.[180]

Events in the Balkans, particularly the massacre of Srebrenica, proved to have a considerable impact upon German public opinion and the political decision-making process in the medium term. The slogan "never war again from German soil" was replaced by the slogan "never Auschwitz again, never Srebrenica again." Accordingly, public support for German participation in IFOR was robust. However, this shift in public opinion did not yet embrace classical combat operations.[181]

At the end of the 1990s, the Kosovo crisis confronted the Federal Republic with the dilemma of contributing to an allied air campaign against the backdrop of a humanitarian disaster, but without a UN mandate. German decision makers eventually argued in favour of participating militarily.[182] The German contingent contained 10 Tornado ECRs and four reconnaissance Tornados after the decision on 16 April 1999 to withdraw two reconnaissance Tornados and to supplant them with two Tornado ECRs.[183] German combat aircraft accounted for 1.37 per cent of the allied aircraft fleet, which in the course of the campaign was increased to 1,022 assets, or 1.33 per cent of allied sorties.[184] Nevertheless, German Tornados released a significant number of HARM missiles—236 out of a total of 743 expended throughout the entire campaign.[185]

Though the military significance of the German contribution was limited, the political significance was more important. For the first time, the GAF participated as an "equal partner" in a peace-enforcement operation.[186] With regard to German public opinion, support amounted to slightly less than two-thirds and remained stable throughout the air campaign. Any potential employment of ground forces in a high-intensity scenario was, however, clearly declined.[187]

Courtesy GAF; photo by Herbert Albring

Recce Wing 51 reconnaissance Tornado in Mazar-i-Sharif, Afghanistan

Germany's participation in Allied Force, which pursued clear-cut humanitarian goals, did not mean that German politicians would in the future easily participate in any allied air campaign. As such, the GAF's contribution to Enduring Freedom was relegated to NATO AWACS aircraft deployments to the United States and to military airlift in support of deployed naval, Army, and special forces units as well as of allies.[188] Though the GAF played a vital role in the running of Kabul International Airport in Afghanistan, as described in the following section, the German Parliament decided to dispatch reconnaissance Tornados to Afghanistan in support of the ISAF only in early 2007.[189] Three newly acquired Heron-type MALE RPAs reinforced the six Tornados in early 2010.[190] These Tornado aircraft were deployed back to Germany in late 2010.[191]

Against the backdrop of Operation Iraqi Freedom, German crews manning NATO AWACS aircraft patrolling Turkish airspace in early 2003 again aroused sensitive political debates. In May 2008, the Federal Constitutional Court concluded that those missions would have actually required parliamentary consent, as they did not represent routine missions.[192]

Though Germany pursued an evolving path throughout the 1990s, which finally culminated in the GAF's participation in Allied Force, the main emphasis of Germany's military contribution has remained upon stabilization operations.[193] Despite Germany's strong commitment to the NRF and the EU battle groups in recent years, it is unlikely that German decision makers will commit combat forces if a humanitarian purpose is not obvious. Moreover, sensitive political debates and the need for parliamentary consent might inhibit German participation in multinational rapid reaction formations. The prerequisite of support by a broad parliamentary majority for armed operations was formalised in March 2005, when the so-called Parliamentary Participation Act came into effect.[194]

Humanitarian Operations and Troop Deployments

In the post–Cold War era, airlift missions continued to be the GAF's main contributions to out-of-area operations. In the immediate aftermath of Desert Storm, the German armed forces embarked with approximately 2,000 troops upon their most comprehensive humanitarian operation thus far in the Iranian-Turkish border area.[195] The operation included the establishment of an intratheatre air bridge which was run by GAF and German Army aviation units.[196] Almost simultaneously, two GAF C-160 Transall and three Army aviation CH-53 helicopters were deployed to Bahrain and Iraq to support the UNSCOM in Iraq.[197]

The year after, it was decided that GAF transport aircraft were to participate in humanitarian missions both in Bosnia and Somalia.[198] In August 1992, two C-160 Transalls were dispatched to Kenya and participated in an airlift effort to relieve Somalia's capital of Mogadishu. With the arrival of a German logistics battalion in mid-1993 in Somalia, the GAF transferred its base from Kenya to Djibouti.[199]

Simultaneously, from 4 July 1992, two GAF Transalls participated in the international air bridge to Sarajevo.[200] For the first time since WWII, German transport aircraft were exposed to a constant GBAD threat, giving this mission an entirely new quality. Aircraft had to be fitted immediately with a self-protection suite and additional armour.[201] Nevertheless, these measures could not prevent a German Transall from being damaged by GBAD in February 1993.[202] Despite the fact that German aircraft were shot at several more times, the GAF was, alongside the USAF, the RAF, the FAF, and the Canadian Air Force, the only air force that constantly contributed to the air bridge to Sarajevo.[203] Furthermore, GAF Transalls air-dropped relief supplies in Eastern Bosnia together with American C-130 Hercules and French C-160 Transalls from 28 March 1993 to 19 August 1994. The missions had to be flown at night and entailed a significant exposure to Bosnian Serb GBAD.[204]

GAF transport aircraft and helicopters were also involved in various other humanitarian operations such as in Iran (1990), Rwanda (1994), and Mozambique (2000), which had been seriously devastated by floods.[205] From October 1999 to February 2000, two German Transalls, configured as medevac aircraft, made it as far as Darwin in northern Australia when Germany received a request from the UN to evacuate wounded from crisis-torn East Timor.[206] Medevac as such has become one of the GAF's specialities.[207]

GAF transport aircraft proved indispensable not only for humanitarian missions but also for the deployment and resupply of

German contingents. Airlift shuttles between Germany and the Balkans and, since 2002, between Germany and Afghanistan have benefited both the German armed forces as well as Germany's allies. Furthermore, in Operation Artemis in mid-2003, German aircraft transported 300 tons of materiel to Uganda, and in late 2004, the GAF deployed troops of the African Union to Sudan.[208]

Courtesy GAF; photo by Helge Treybig

GAF Airbus A310 supporting the African Union's mission to Darfur, Sudan

The Federal Republic's participation in ISAF operations put one of the highest demands upon GAF airlift capacities, revealing significant shortfalls in the GAF's airlift capacities. Ukrainian and Russian transport aircraft had to be chartered in the initial phase (as described in the section on air mobility). Nevertheless, the GAF finally managed to establish an air bridge to Kabul in early 2002. It ran from Cologne through Termez Airport in Uzbekistan to Kabul. Whereas for the first leg Airbus A310s were used, the second leg was served by C-160 Transall aircraft, which were equipped with self-protection suites. At Termez Airport in Uzbekistan, a German base was established that hosted up to seven C-160s, one of them configured as a medevac aircraft. This route became an important lifeline not only for the German but also for the entire ISAF contingent. Furthermore, from February 2003 till

June 2004, the GAF was responsible for the running of Kabul International Airport. What originally was supposed to be a six-month assignment ended as a 15-month assignment, as no other nation was prepared to take on the succession.[209] As of 2010, GAF C-160 Transalls were providing significant airlift support for the ISAF, operating out of Mazar-i-Sharif.[210]

In 2006 the German armed forces played a significant role in Operation EUFOR in the Democratic Republic of Congo. After Operation Artemis in 2003, EUFOR was the second autonomous military operation of the European Union in Africa, which aimed at providing security during the elections in conjunction with other civilian crisis management cells of the EU. With a German general in charge of the conduct of the operation and the Operations Command in Potsdam serving as the operations headquarters, the GAF and the German Army deployed airlift assets, including heavy CH-53 helicopters and C-160 Transalls.[211]

Air Policing and Subsidiary Tasks

Since 19 November 1990, with the reestablishment of full German sovereignty, GAF fighter aircraft and tactical air control units have been assigned with air policing, requiring constantly available resources. To draw upon NATO's C2 network and sensors, air policing has been conducted in the framework of the alliance's integrated air defence. Engagement of hostile aerial vehicles, however, requires consent from German authorities.[212] Prior to 11 September, air policing procedures were strictly confined to military targets. In the immediate aftermath of the terror attacks, the chief of the Air Staff intensified airspace surveillance and enhanced the readiness status for air defence fighters.[213] In contrast to military targets, the interception of civilian aircraft has to be conducted solely with national assets and under national command.[214] Despite this restriction, German fighter aircraft can be supported by NATO AWACS during major events such as the 2006 football world championship.[215] Though the new importance of air policing was especially underlined by the 2003 *Defence Policy Guidelines*, the potential downing of a civilian aircraft led to intense disputes.[216] On 15 February 2006, the Federal Constitutional Court issued a verdict stating that the downing of civilian airliners is not reconcilable with the constitution.[217] This again shows the complexity of civil-military relations in Germany.

On behalf of NATO, the GAF has also regularly contributed to air policing over the Baltic. For this particular purpose, German F-4F Phantoms were for the first time dispatched to Latvia from 30 June to 30 September 2005. Since Lithuania, Estonia, and Latvia do not have adequate aircraft for this type of mission at their disposal, other NATO nations alternately provide a detachment of four fighter aircraft for a period of approximately three months.[218] In September 2009, the GAF deployed its Eurofighters to the Baltic, representing the first operational deployment to involve this aircraft type. Since the start of the alliance mission in April 2004 up to September 2009, 14 NATO member states provided air policing duties. As of 2009, this multinational effort was expected to last until around 2018.[219] Besides the Baltic, the GAF started to contribute to NATO air policing missions over Iceland in 2010.[220]

Furthermore, a number of subsidiary tasks in support of civil authorities fall within the GAF's responsibility. The 1994 *Defence White Book* explicitly states that the GAF gathered and forwarded

Courtesy GAF; photo by Toni Dahmen

During floods in Germany, GAF units provided assistance and flew rescue missions. In particular, the GAF A310 medevac fleet evacuated patients from hospitals.

findings about environmental pollution and assisted the civilian SAR service.[221] Particularly during the 1997 and 2002 floods in Germany, GAF helicopters provided assistance and flew rescue missions.[222] In the course of the 2002 floods, Tornados were also tasked with reconnaissance missions, and the GAF medevac fleet evacuated patients from hospitals.[223] Moreover, Airbus medevac aircraft have evacuated wounded German and European citizens on numerous occasions from abroad.[224]

How Has the German Air Force Responded to the New Intellectualism in Air Power Thinking and Doctrine?

During the second half of the Cold War, German airmen were particularly concerned with the GAF's role within the Western Alliance and especially with the strategy of flexible response. With the end of the Cold War, there was no longer a need to ponder these issues. Instead, German air power thinking had to be directed towards the new challenges of the post–Cold War era. Particularly, the recent defence reforms have given German air power thinking a new impetus.

Air Power Doctrine

The establishment of the Air Power Centre (Luftmachtzentrum or LMZ) represents the GAF's most significant step in tackling conceptual and doctrinal aspects in the post–Cold War era. The centre itself is a product of the adoption of Air Force Structure 6.[225] As such, it can also be regarded as an attempt to respond to the requirements of military transformation. The Air Power Centre is responsible for the development of national air power doctrine, German contributions to allied air power doctrine, and concept development and experimentation.[226] Moreover, as previously mentioned, Germany was also the lead nation in setting up the JAPCC, which became operational in early 2005 in Germany.

During the Cold War, the GAF did not have national basic air power doctrine manuals. According to a RAND study of 1986, such documents were believed to violate the "spirit and purpose" of the GAF's integration into NATO.[227] Consequently, attempts to publish a national strategic air power doctrine failed during the 1970s and 1980s, and NATO air power doctrine documents remained the bedrock of German air power education.[228] This did not mean, however, that the GAF did not operate according to distinct national doctrinal features. For instance, while the 1984 USAF basic doctrine stated that the principal objective of each service is to win the battle in its specific environment, the German chief of the Air Staff argued in the late 1970s that there is "no independent battle on land, in the air, or on the sea."[229]

Towards the late 1980s, however, supporters of a national basic air power doctrine grew in number.[230] In March 1991, the GAF finally published its first Air Force service instruction, LDv 100/1, which corresponds to a basic doctrine document. The introduction states that LDv 100/1 complements NATO doctrine documents by elaborating on the fundamental nature of air power.[231] The document itself was doctrinally still anchored in the later stages of the Cold War. While the role of aerial forces for alliance defence, particularly in the context of flexible response, is dealt with in depth, air power in support of peace support operations is not treated.[232] Regarding the conduct of the air war, the document reveals some specific German features. It argues, for instance, that the primary goal is to gain a favourable air situation and not necessarily air superiority.[233]

In the ensuing years, the staff college in Hamburg made two attempts to revise LDv 100/1. Yet, these attempts did not lead to revised editions. This was partly due to the fact that NATO doctrine documents were considered to provide a solid basis for the conduct of national air operations.[234] Finally in late 2005, the chief of the Air Staff ordered the 1991 edition to be revised.[235] The ongoing reform of the German armed forces highlighted the need to doctrinally anchor the GAF.[236] National, NATO, and EU documents provided the framework for the revision. Guidelines for the GAF's C2 structure and air policing were principally drawn from national documents. For planning processes, however, the instruction strictly adhered to NATO procedures since the GAF had deliberately foregone implementing a national process. The revised LDv 100/1 was primarily designed for staff officer courses.[237] While it was issued in April 2009, it was not released to the wider public.[238]

Teaching of Air Power Thought

The Führungsakademie (FüAk), their Armed Forces Command and Staff College, in Hamburg is the highest military educational institution of the German armed forces. Its main responsibility is the training and education of German staff officers. In the wake of the reconstitution of the West German armed forces, the FüAk was inaugurated in 1957.[239] It runs a basic staff officers' course as well as other courses, but the flagship of the FüAk is the national general/admiral staff course.[240] On average, 17 per cent of the officer corps of the three service branches is selected for attending this two-year course. Education and training embrace basically five areas of study—the military within the state and society, the conduct and employment of armed forces, the German armed forces, capabilities and structures of armed forces, and personal leadership.[241]

The educational emphasis is unambiguously placed upon the conduct and employment of armed forces, with slightly less than half of the course devoted to this subject matter, which also encompasses the "conduct of aerial forces."[242] The latter is—with the exception of joint planning exercises—exclusively taught to Air Force officers and lays the foundation for air power related input into both national and multinational joint planning exercises. Allied Joint Publication (AJP)-5, *Allied Joint Doctrine for Operational Planning*; NATO's *Guidelines for Operational Planning*; AJP-01, *Allied Joint Doctrine*; and NATO AJP-3.3, *Joint Air and Space Operations Doctrine* form the bedrock for air power education and guarantee commonality in thinking with NATO and other European allies. The evolution of air power education at the FüAk has therefore become closely interrelated with the revision of NATO planning processes and doctrine.[243]

The primary goal of air power education is to make future general staff officers understand the planning responsibility of an air component commander. Consistent with this goal, course participants are first familiarised with operational planning principles and then participate in two major planning exercises at the end of the module on conduct of aerial forces. While the course emphasises operational planning processes and, to a lesser degree, the aerospace industry, literature by air power thinkers and historians is in general not studied.[244] This omission was confirmed by an RAF officer attending the national general staff course at the FüAk in the first half of the 1990s. Nevertheless, this RAF officer argued that the great advantage of the FüAk was that original, even maverick, thought was not obstructed. For instance, he was free to choose his own air power subject for study in-depth—in his case the "fallacy of the strategic air campaign," for which he received the Clausewitz Medal.[245] In the course from September 2006 to September 2008, a week's seminar on air campaigns was introduced. Students primarily studied six case studies—ranging from the Cuban missile crisis, the Vietnam War, the Soviet-Afghan War, and the Falklands War to Desert Storm and Iraqi Freedom. While this seems to be an improvement, there was—in the view of an international course participant—still no adequate treatment of the theoretical aspects of air power. American air power theorists such as John Warden or Robert Pape were used as a point of reference, yet on a very superficial level.[246]

Nevertheless, depending upon personalities, the FüAk curriculum is flexible enough to allocate time for air power history and theory. Col John Olsen, a well-known European air power theorist from the Royal Norwegian Air Force, experienced a four-week air power theory and history seminar series when attending the FüAk from August 2003 to September 2005. His tutor had a keen interest in air power doctrine, theory, and history and asked Olsen to co-organise the seminar series. Various schools of thought ranging from Douhet, Marshal of the RAF Hugh Trenchard, and Mitchell to Warden and Pape were discussed, and air campaigns were accordingly examined. Yet, Olsen added a caveat, arguing that his course might rather represent an exception.[247] Hence, it can be concluded that while personal attempts succeed in boosting air power history and theory in the air power studies curriculum, they have so far failed to be properly institutionalised.

The overall rather operational and technical approach to air power is also reflected in the general staff course dissertations and in other course writings. In 2005–6, dissertation topics ranged from European missile defence and intelligence gathering in the theatre of operations to multinational logistics in out-of-area operations and military use of the national airspace. Against this backdrop, topics such as effects-based operations or the challenges of modern conflicts are rather rare.[248]

Education and training at the FüAk have been influenced by the major reforms of the German armed forces since the end of the 1990s. For instance, the primary educational goal of making students understand the planning responsibilities of an air component commander goes hand in hand with the German armed forces' thrust towards more autonomous planning responsibility, which became particularly apparent with the establishment of the GAF Air Operations Command in late 2001. Furthermore, in line with the reform of the German armed forces and the particular emphasis upon jointness, the FüAk presented in 2004 a concept for the reform of the staff officer courses.[249] Consequently, education and training at the FüAk have to a very high degree become joint regarding both organisation and content.[250] Along with this reform, teaching modules—which can be attended by staff officers during the remainder of their professional careers—have been introduced.[251] Air power modules have so far focused on the GAF's ongoing reforms.[252]

In line with Germany's deep integration into NATO and the emerging ESDP, about a quarter of the participants on the national general/admiral staff course come from NATO and/or EU countries. Besides the national course, there also exists an international general/admiral staff course for participants from non-NATO/EU countries, which lasts one year. Furthermore, an American exchange officer from the USAF is integrated in the Air Force branch of the teaching staff, and four to eight GAF officers are annually sent abroad to staff colleges in France, the United Kingdom, and the United States.[253] As in the case of France, this thrust towards multinational education has been complemented since 2001 by the Euro exercise CJEX.[254] Conducted in close cooperation with the staff colleges of the United Kingdom, France, Italy, and Spain, this exercise represents a high spot at the FüAk.[255]

While air education at the FüAk thoroughly prepares officers for staff work, the teaching of the underlying theory and history of air power is hardly dealt with and certainly not properly institutionalised. Moreover, as of 2006, no link existed between the military historical research office in Potsdam and the FüAk with regard to teaching.[256] Hence, one can make the argument that

Germany's historical legacy has led to an ahistorical approach towards air power teaching. This view was cautiously shared by a retired high-ranking GAF general officer who particularly highlighted the German public's sensitivity with regard to the country's history. Moreover, the ahistorical approach can be regarded a product of *Innere Führung*, or moral military leadership—a cornerstone of the German officer's self-image. While the concept emphasises tradition and focuses on the Prussian Army reforms of the early nineteenth century, the military resistance to the National Socialist regime, and the history of the postwar German armed forces, the Bundeswehr, it implicitly blanks out whole segments that are particularly pertinent to the evolution of air power theory.[257]

Courtesy GAF; photo by Ingo Bicker

F-4F Phantom II from Fighter Wing 71 Richthofen getting readied for takeoff

Dissemination of Air Power Thought

Since the mid-1980s, the FüAk has hosted annual air power seminars. The circle of participants, including the chief of the Air Staff, has been restricted to a small number of GAF officers and officers from other service branches. Topics ranged from doctrinal issues such as air attack, through the consequences of the Helsinki Headline Goals for the GAF, to logistical support for out-of-area operations.[258] In 2000 the air power seminar was held together with the first EURAC air power colloquium. The seminar particularly highlighted recent developments of the ESDP and NATO and their impact upon European air power.[259]

These days, Army and Navy officers are routinely present at the air power seminars, reflecting the German armed forces' thrust towards jointness.[260] A significant change also took place regarding content. Whereas until 2000, the primary goal of the air power seminars was to educate lieutenant colonels and colonels on a specific air power topic and to further air power thinking, the air power seminars have since become product oriented. The new focus is to make a concrete contribution to the GAF's conceptual development that the Air Staff can draw upon.[261] From 2002 to 2006, topics included network-centric warfare, conduct of joint operations, RPAs, and the GAF's position in relation to space.[262]

Regarding the dissemination of German air power thought in journals and books, there used to be a GAF yearbook between 1964 and 1981. Topics ranged from weapon systems and force structuring to air power in light of NATO's deterrence strategy and doctrinal issues, such as the meaning of depth and time for aerial forces.[263] Furthermore, though the GAF has never published a journal similar to the *Royal Air Force Air Power Review* or *Penser les ailes françaises*, GAF officers used to pronounce their views during the 1980s in defence journals such as *Wehrtechnik* (*Defence Technique*), *Europäische Wehrkunde* (*European Defence Information*), or, on a more international level, *NATO's Sixteen Nations*. The articles were commonly written by high-ranking GAF general officers as well as mid-career officers working at the MOD. Topics included force planning, mission priorities, air defence, missile defence, and joint operations.[264]

In the post–Cold War era, GAF officers posted at the MOD primarily voiced their views in the journal *Europäische Wehrkunde*, which was renamed *Europäische Sicherheit* (*European Security*) in January 1991. The emphasis of the various contributions has been clearly put upon force structuring, procurement, airlift, or air defence, and accounts of Western air campaigns in the post–Cold War era have been rare, and very often with a predominantly technical focus, as was the case in the wake of Operation Desert Storm.[265] Only after Operation Allied Force can the first true account of a modern air campaign be found in *Europäische*

Sicherheit.[266] This is in line with Germany's first full integration into a combat operation.

Almost in the same vein as the old yearbook, the GAF has had an annual journal published since 2002, the so-called *CPM Forum*. The publication appears alternately in English and German and basically provides a comprehensive survey of the adoption of new Air Force structures as well as the GAF's new capabilities profile. But more doctrinal issues have also been dealt with recently, such as the GAF's perspective upon network-centric operations.[267]

Regarding books on the postwar GAF, the military historical research office in Potsdam published in 2006 *Die Luftwaffe 1950 bis 1970* (*The Air Force from 1950 to 1970*), scrutinising organisation, technical aspects, and the integration into NATO.[268] Up to the time of writing, this book represents the only major military historical research project on the postwar GAF. In general, there is no research that deals with the relationship between German military history and modern GAF doctrine, as was already argued by an American study of 1987.[269] A retired high-ranking GAF general officer underlined that a premium had always been put upon discontinuing Wehrmacht (refers to the German armed forces of WWII) traditions. As such, German WWII history is to be avoided in military education as well as in air power doctrine.

How Have German Defence Planners Attempted to Maintain a Relevant Air Force in Light of Escalating Costs and Advanced Technologies?

Against the backdrop of massive unification costs, the 1992 *Defence Policy Guidelines* stated that any security concept had to take into account the strained budgetary situation.[270] Moreover, there was a growing expectation of significant peace dividends.[271] In light of a continuing tense budget situation, the 1994 *Defence White Book* went so far as to argue that the equipment of the main defence forces, then representing the bulk of the German armed forces, need only be modernised after 2000.[272]

In this context, international cooperation in major defence programmes was identified as a way to reduce costs.[273] During the Cold War, Germany had already pursued a path towards industrial cooperation. The most prominent example is the trilateral British-Italian-German Tornado aircraft programme. Though, as outlined previously, there was evidence that European cooperative ventures cost up to a third more than hypothetical purchases from US firms, the absence of a European aerospace industry might have exposed European air forces to US monopoly prices.

Moreover, from a German point of view, there is more to industrial cooperation than just cost savings, as it has contributed significantly to closer alliance and European relations and therefore has enhanced European security.[274] Accordingly, all of the GAF's current major procurement programmes are perceived to strengthen European as well as transatlantic relationships.[275] Hence, NATO and EU initiatives such as NATO's Defence Capabilities Initiative or the European Headline Goal have been considered the GAF's principal guidelines for identifying areas of action and removing military deficits.[276] In line with Germany's 2002–3 defence reform, force structuring has no longer been guided by worst-case scenarios but by most likely scenarios and Germany's international commitments.[277]

Combat Aircraft

With the end of the Cold War and against the backdrop of Air Force Structure 4, entire aircraft systems were phased out in order to reduce costs. While the light ground-attack aircraft Alpha Jet as well as the reconnaissance aircraft RF-4E Phantom were decommissioned, the fighter aircraft F-4F Phantom and the Tornado were kept in service.[278] Henceforth, the Tornado IDS (interdiction-strike) fleet was to shoulder both aerial reconnaissance and air attack.[279]

With German participation in real operations over Bosnia in 1995, ad hoc upgrades for the Tornado aircraft involved became necessary. The earmarked aircraft were equipped with enhanced communication devices and GPS. Moreover, the engines were modified in order to increase thrust.[280] In the late 1990s, the entire Tornado fleet was about to undergo various life extension and upgrade programmes, including improvement of avionics as well as the integration of a laser-designator pod for the employment of LGBs.[281]

With the upgrade programmes still under way, however, Operation Allied Force was too "early" for the GAF. While ECR and reconnaissance Tornados were ready for employment, the main bulk of the German Tornado fleet was not, as it could not yet deliver PGMs.[282] Even in the case of the Tornado ECR, ad hoc improvements, such as the integration of night goggles, proved necessary.[283]

Against the backdrop of Air Force Structure 6, it was decided to reduce the Tornado fleet to 85 aircraft by 2013.[284] While the remaining aircraft were planned to undergo further upgrades, the retired airframes have become a cost-effective source of spare parts.[285]

In the air defence role, German reunification offered the opportunity of integrating an F-4F Phantom and an East German MiG-29 squadron into one wing. Combined, both aircraft types could complement each other.[286] To remain relevant pending delivery of the Eurofighter, however, the F-4F had to undergo significant upgrades. By December 1996, 65 F-4F Phantom IIs had been delivered. The most essential components of the retrofit included the integration of the APG-65 radar and the integration of the AIM-120 AMRAAM missile.[287]

The MiG-29 was also considered an interim air defence aircraft. Though it had excellent flying performance, particularly regarding climb rate and manoeuvrability, it also had significant shortfalls in avionics and communications. Moreover, the engines consumed too much fuel for Western standards, significantly reducing range, and the airframe had a very limited lifespan.[288] For these reasons, the 1994 *Defence White Book* states that, beginning in 2002, the MiG-29 and the F-4F Phantom fleet would be replaced by a modern fighter aircraft.[289] In 2003–4, the German MiG-29 aircraft were finally transferred to the Polish Air Force.[290]

Through German participation in the quadrilateral Eurofighter project, a European design has been envisaged for the replacement of the F-4F fleet. Since the technical specifications had been laid down during the Cold War with the ultimate aim of tackling new capabilities of the Warsaw Pact air forces, the project faced significant problems during the early 1990s.[291] The Eurofighter (original designation Fighter Aircraft 90) was conceived in 1985 to be optimised for air superiority with a limited ground attack capability. At the time, a 1995 in-service date was agreed upon, and Germany and the UK planned to acquire 250 aircraft each.[292]

With the end of the Cold War, the project became highly contentious, particularly in Germany, where unification costs started to have an impact upon the defence budget.[293] Germany seriously threatened to bail out of the common project in the early 1990s, and alternative options to the Eurofighter were being discussed.[294] Amongst the discussed alternatives were either American designs, such as the F-16, the F-18, the F-15, and the F-22, or the Soviet MiG-29. Bailing out would, however, have caused a considerable loss of trust among Germany's European allies.[295] Furthermore, some were concerned that if an American fighter aircraft were procured, Germany would become fully dependent upon the United States for any future system upgrades.[296] Moreover, it was recognised that bailing out of the project would represent a major blow to the German aerospace industry and would imperil a significant number of jobs as well as the technological base.[297]

In the early 1990s, future GAF Eurofighters were conceived solely for the air-to-air role, and Germany indicated that it would commit itself to a specific number of aircraft only at later stages in the programme, and then closer to 140 than 250, as originally intended.[298] Technical difficulties and the German reassessment finally led to significant delays in the project.[299] In April 2004, nine years after the original in-service date, the first Eurofighters were delivered to a German fighter wing.[300] Against the backdrop of Air Force Structure 6, the GAF planned to procure a total of 180 Eurofighters equipped with an advanced self-protection suite, the defensive aids subsystem, and the Link-16 terminal MIDS.[301] Yet as of mid-2010, this number was likely to be reduced to 143 aircraft.[302]

For aerial combat, the GAF has been introducing new weapons. Besides the AIM-120B AMRAAM for medium ranges, the Eurofighter is being equipped with the IRIS-T for short ranges. The AMRAAM will be supplanted by the Meteor missile from 2014 onwards.[303] Both the Meteor and the IRIS-T are co-development projects of several European nations. The lead nation in the development of the latter, Germany plays an important role in the replacement of the aging AIM-9L Sidewinder missile.[304] In the medium term, the GAF's air-to-air armament will thus be entirely of European origin.

In the field of air-to-ground armament, the GAF started to implement the lessons of modern warfare relatively late. Throughout the 1990s, the GAF retained the submunitions dispenser system MW-1, designed for low-level attacks against area targets, as the main armament of its Tornado IDS fleet.[305] Precision-guided standoff weapons and LGBs were only introduced in the wake of the major air campaigns over the Balkans. Defence Minister Scharping's key document of 2000 explicitly argues that, while main weapon systems were to be reduced, standoff and precision strike capabilities had to be improved.[306] In 2001 the GAF acquired a precision strike capability by means of the GBU-24 LGB.[307] In 2005 a further milestone was reached when the first of 600 Taurus cruise missiles were delivered and integrated on the Tornado attack aircraft.[308] Moreover, the need for weather-independent, short-range PGMs became apparent. The German Eurofighter fleet is to be equipped with the all-weather-capable GBU-48 PGM, combining a semiactive laser with a combined inertial measuring unit/GPS receiver, from 2012 onwards. As a standoff weapon, Taurus cruise missiles are planned to be integrated on German Eurofighters at a later stage.[309]

Throughout the post–Cold War era, the GAF has retained specific niche capabilities. This has been particularly apparent in the domain of SEAD. Growing concerns over improvements in Warsaw Pact air defences in the later stages of the Cold War rendered a robust SEAD capability indispensable. Accordingly, a batch of 35 specialised Tornado ECR aircraft carrying the American HARM missile was procured.[310] Due to changing requirements in the post–Cold War era and the need to avoid collateral damage, the United States, together with the European HARM users Germany and Italy, conducted a HARM upgrade programme.[311] Along with this programme, Germany also embarked upon a national and a bilateral programme together with France to develop a new anti-radiation missile in the late 1990s. France, however, bailed out of the bilateral programme, and the national programme called Arminger was abandoned, too.[312]

Air Mobility

The end of the Cold War revealed significant shortfalls in the GAF's airlift capacity. In early 1991, for instance, the GAF was not

Courtesy Eurofighter Jagdflugzeug GmbH, www.eurofighter.com

German Eurofighter in flight test with an A310 MRTT as part of the clearance process for air-to-air refuelling (October 2008)

in a position to deploy German SAM batteries to Erhac, Turkey.[313] Instead, American C-5 Galaxies had to execute this task.[314] The GAF's air transport fleet proved insufficient not only for cargo capacities but also for self-protection systems, which became particularly essential in an era of zero-casualty tolerance. When the GAF started to participate in the airlift effort to Sarajevo, protection of the C-160 Transall transport aircraft was limited to armour plates beneath the cockpit. In an unbureaucratic fashion, financial means were made available to remedy these significant shortfalls. Soon, German Transall aircraft flying to Sarajevo were fitted with radar warning receivers, chaff, flares, and other self-protection devices.[315] While the C-160 Transall transport aircraft, acquired from 1968 onwards, continued to be the German workhorse for medium ranges, the capacity for long-range transportation was increased through the acquisition of seven A310 MRT aircraft in the 1990s, the first three of which were former East German aircraft. The C-160 fleet was planned to be supplanted by the Future Large Aircraft (FLA), later known as the Airbus A400M.[316] In 2000 Germany signed a preliminary contract for the acquisition of 73 aircraft, the largest share among the European partners.[317]

Significant improvements in the GAF's power projection had not been reached by the turn of the century. Consequently, German equipment for operations in Afghanistan had to be deployed by An-124 heavy-lifters from the Ukraine and Russia.[318] Moreover,

GAF Transall transport aircraft were initially only able to dispatch German troops to Turkey. For the final leg, Turkey to Kabul, they were flown in Dutch C-130 Hercules aircraft, as not all German Transall transport aircraft had adequate self-protection suites.[319]

Though Defence Minister Scharping's key document of 2000 explicitly stated that improvements in Germany's power projection capability would have first priority, the financing of the German A400M became a major bone of contention in the German parliament. This, in turn, caused significant delays for the entire international airlift programme and irritation amongst Germany's allies, France and Great Britain, which wanted the A400M to be delivered as soon as possible.[320] In addition to the delays, the German A400M order was reduced to 60 aircraft.[321] Mired in the adverse economic conditions of 2010–11, the aircraft order was again further reduced to 53.[322] Yet the GAF might field only 40 aircraft, with the remainder being sold.

Throughout the 1990s, the only aircraft in the GAF's inventory to be used in the air-to-air refuelling role was the Tornado by means of a buddy-buddy refuelling suite. The air-refuelling capacity was, however, very limited. Hence, employment of Tornados as tanker aircraft has been primarily relegated to training, which has been essential for German participation in combined air operations.[323] Only within the context of Air Force Structure 6 did the GAF acquire a strategic AAR capability by the conversion of four Airbus A310 MRTs into A310 MRTTs.[324] In addition, a number of A400M aircraft are planned to be equipped for air-to-air refuelling of both combat aircraft and helicopters.[325] These planned assets represent a significant improvement, though still falling short of British or French AAR levels.

A further specialised element of air mobility is medevac, which has become one of the GAF's specialities. Since 1994 the GAF has continuously built up its medevac capability, starting with a medevac installation kit for its C-160 Transall aircraft and ending with two dedicated A310 MRT medevac Airbuses. On an international level, this capability is state of the art.[326] The GAF's medevac capacities are further increased by the acquisition of two new Airbus A319 CJ and two secondhand Airbus A340 VIP aircraft from early 2010 onwards, which can be converted into medevac platforms.[327]

Courtesy Airbus Military; photo by L. Olivas

A400M on its first flight. The GAF is likely to field between 40 and 53 A400M transport aircraft.

C4ISTAR

In the field of C4ISTAR, the GAF has been particularly able to benefit from multilateral cooperation. In the later stages of the Cold War, the GAF became one of the most prominent contributors to NATO's Airborne Early Warning and Control Force, based on German territory. Germany has shouldered a third of the programme, and the country is also likely to bear a significant share of the NATO Alliance Ground Surveillance Core project.[328] Through NATO AGS Core, multilateral cooperation is transferred to the field of large-area ground surveillance. Early on, it was clear to German decision makers that the existing capability gap in this area could only be closed through acting together with partner nations.[329]

In the field of satellite reconnaissance as well, Germany has been taking significant steps in recent years. While during the 1990s, German participation in the French-led Helios II project failed due to financial bottlenecks in the defence budget, the project of establishing a national strategic reconnaissance capability has been pursued vigorously since the beginning of the defence reform in 1999–2000.[330] In late 2006, the first of five synthetic-aperture radar Lupe satellites was launched.[331] In addition to this national capability, an agreement to secure a satellite data exchange between Germany and France was implemented in early 2008.[332] In accordance with this new emphasis upon strategic reconnaissance assets, Germany seeks to make a meaningful contribution to the EU and NATO in early warning of crises.[333]

Throughout the first two decades of the post–Cold War era, the GAF's principal airborne reconnaissance asset was the Tornado aircraft with its reconnaissance pod integrating electro-optical and infrared sensors. In the second half of 2009, the RecceLite reconnaissance pod, providing real-time intelligence by means of data link, became operational on German Tornados in Afghanistan. Simultaneously, one of two Tornado reconnaissance squadrons was disbanded and replaced by an RPA squadron initially operating the Israeli MALE RPA system Heron. In the field of HALE RPA systems, the GAF is about to acquire the Euro Hawk for SIGINT, based upon the Global Hawk platform, with Germany developing the SIGINT avionics. Development began in 2005, and the first test flight was conducted in California on 29 June 2010.[334]

With regard to C2, the GAF has always been a strong proponent of centralised control over NATO aerial forces. To overcome the divisions between the different command posts for air defence and offensive air operations within NATO, the GAF by the 1980s already supported the development of the allied ACCS. It provides a single C2 structure for composite air operations combining defensive, offensive, and supporting air warfare elements of all NATO nations in Central Europe.[335] After an extended development phase, ACCS is being introduced as an integral part of Air Force Structure 6.[336] This thrust towards network-centric warfare is further enhanced through the introduction of MIDS Link-16 terminals on all major weapon systems, including the remaining Tornado aircraft.[337]

Moreover, parallel to the adoption of Air Force Structure 5 and 6, the tactical air control service has been fundamentally modernised and cut by half. Despite this reduction, a qualitative capability improvement is expected, and in line with Germany's thrust towards expeditionary warfare, a deployable air surveillance capability is being established.[338]

Ground Based Air Defence

In the post–Cold War era, the GAF adopted the concept of extended air defence, which aims at countering not only aircraft and helicopters but also theatre ballistic and cruise missiles.[339] The concept does not, however, cover strategic ballistic missile defence.[340] To meet the goals of extended air defence, German Patriot systems underwent various upgrade programmes, particularly with regard to sensor performance and C2. The PAC-3 missile was to be introduced with half of the remaining German Patriot units from 2007 onwards, giving German Patriot systems a TBMD capability against theatre ballistic missiles with a range of up to 1,000 km.[341]

With the Patriot the only viable system for the full spectrum of EAD, Hawk and Roland SAM systems were disbanded by 2005, earlier than expected, to achieve rapid cost-saving effects.[342] Currently and in the medium term, 24 Patriot units represent the GAF's only GBAD assets in the context of EAD.[343] These drastic reductions occurred against the background of the Struck reform, which saw a further adaptation of Air Force Structure 5 and the elaboration of Air Force Structure 6 as indispensable in light of a strained defence budget.

The 2003 *Defence Policy Guidelines* explicitly underlines the importance of EAD in general and TBMD in particular. It emphasises the protection of deployed forces against rockets and missiles.[344] There is a resultant shift from territorial defence to expeditionary warfare in the field of GBAD.

In the medium term, the acquisition of 12 Medium Extended Air Defence System (MEADS) units was planned. Yet the necessity of this highly mobile system, configured for the PAC-3 missile and a surface-launched version of the IRIS-T, was being reassessed in the first half of 2010, particularly against the backdrop of depleted public budgets in general and of delays and cost increases to the trinational US-German-Italian project.[345]

Conclusion

In the post–Cold War era, the GAF has evolved from a force primarily geared up to forward defence in NATO's central region and restrained by Germany's particular historical legacy to a force

Courtesy GAF; photo by Oliver Fischer

GAF Patriot unit during Exercise ELITE in Germany

that is preparing itself to commensurately take on its share in the international arena.

At the outset of the post–Cold War era, German reunification required the GAF to swiftly take on the responsibility for air policing across the entire reunified territory, a task that had so far been accomplished by the USAF and RAF in Western Germany. Moreover, integration of former East German Air Force personnel presented a particular challenge that was successfully met. For the remainder of the 1990s, however, German decision makers only reluctantly transformed the German armed forces and adhered to Cold War principles. A reorientation started only in 1999–2000. Particularly, Struck's adjustments of the ongoing reform during 2002–3 represented the most decisive milestone step in Germany's military transformation since the end of the Cold War. No longer did the most perilous operations determine the structure of the German Armed Forces, but the most likely operations. As a consequence, out-of-area operations became the defining criterion. As such, Struck's adjustments were de facto a reform in their own right. Only three years later, however, the *White Paper 2006* partially reversed the progressive keynote of the *Defence Policy Guidelines* 2003 by reemphasising national and collective defence in the classical sense. Nevertheless, the *White Paper 2006* affirmed the GAF's ongoing transformation.

In line with the realignment of Germany's defence policy and due to financial constraints, the GAF has undergone a continuous force transformation and significant force reductions. For instance, flying wings were reduced from 18 to seven and might drop even lower. Yet at the same time, new capabilities and assets have been or are planned to be acquired such as an autonomous air operations planning capacity, HALE and MALE RPA systems, a precision strike capability, strategic air mobility, AAR, deployable air command and control units, or a TBMD capability. With regard to intertheatre airlift, the GAF is about to emerge as an important European player, as Germany placed a significant order for A400M transport aircraft. Moreover, it has retained specific niche capabilities such as SEAD, which provide leverage against the backdrop of alliance operations, as was the case in the course of Operation Allied Force. The GAF released slightly less than a third of all HARM missiles employed.

In the field of alliance policy, it has always been a core interest of German policy to balance American and European—primarily French—interests. Consistent with this policy, the GAF has played a bridging function between the FAF and the American armed forces, be it through training establishments in the United States or through ventures such as setting up the European Air Transport Command with France. Germany's deep embedding into alliance structures also required the GAF to make meaningful contributions to cooperative ventures in the field of rapid force deployments such as the NRF or the EU Air RRC. Yet, these will require a further balancing between the shaft and the spear. Moreover, from a political point of view, a dilemma between Germany's strong emphasis upon the transatlantic partnership as the bedrock for German security and the country's reluctance to employ military force across the spectrum of military force, a prerequisite when closely operating with American forces, is likely to persist. It is to be seen whether the GAF's new operational potential will be translated into effective operational output.

In line with Germany's alliance policy, defence programmes cannot solely be judged in terms of value for money. From a German point of view, a multilateral defence programme is not only a means to an end but also an end in itself, as it deepens alliance cohesion as well as European relations. Consistently, Germany has taken a leading role in European and transatlantic defence programmes such as the NAEW&CF, the A400M, or the Eurofighter. All current major defence programmes of the GAF are multilateral. As such, they decisively shape Europe's defence landscape and further European integration as well as transatlantic cohesion. Given the requirements of deployed operations and the allies' expectation for increased German responsibility in international security matters, the GAF has gradually evolved into an increasingly balanced air force. Yet, it has to be underlined that the acquisition of a precision strike capability or AAR certainly represents an improvement, albeit from a very low starting point, and brings the GAF finally to a reach and strike standard that has, with the exception of air-launched cruise missiles, long been held by the RAF or the FAF. The GAF's AAR capacity still falls short of British or French capacity levels. This shortfall is compounded by the German armed forces' general shortcomings in the area of logistics, as is highlighted by Germany's NRF contributions. Due to logistics bottlenecks, the Army and Air Force can only make major contributions consecutively. Given these logistics and AAR shortcomings, theoretical plans to contribute up to 80 combat aircraft for potential EU operations appear too ambitious. Moreover, the GAF lacks the experience of deploying large numbers of combat aircraft to distant theatres. For Operation Allied Force, for instance, the GAF made available only 14 Tornados which could operate from well-prepared NATO air bases in Italy. Yet, a significant step in the GAF's power projection capability can be expected with the introduction of the A400M transport aircraft.

Germany and the GAF reacted to the offensive air power doctrine as demonstrated in Desert Storm with uneasiness. Only gradually did the GAF become involved in deployed operations. Though it has—due to political and constitutional restraints—contributed in a limited way to Western air campaigns, the GAF nevertheless proved to be a professional force that can seamlessly plug into highly complex air operations. In this regard, GAF participation in major integrated air exercises such as Red Flag or Maple Flag proved pivotal and made up for lack of experience. While the GAF has been gradually geared up for expeditionary warfare, operations at the upper level of the spectrum of force are likely to remain an exception. A clear-cut humanitarian purpose will certainly remain a crucial prerequisite. Nevertheless, with the earmarking of combat aircraft to the NRF and to German-led EU battle groups, missions across the spectrum of force might become a reality at any time. Reluctance to contribute to offensive missions, however, did not imply that the GAF was not ready to expose its personnel to increased levels of risk, as was revealed by Germany's contribution to airlift operations in Bosnia.

Germany's unease about deployed offensive operations might have led to a lack of top-down support for doctrine development and might have caused a gap in doctrinal thought throughout the 1990s. Only Struck's reform of 2002–3 helped to put an end to doctrinal stagnation. Against the backdrop of Air Force Structure 6, the Air Power Centre was established in order to grasp the latest developments doctrinally. In the same vein, the revision of the Air Force Service Instruction LDv 100/1, a basic air power doctrine document, was ordered in late 2005. Yet, air power thinking is still cultivated in a very top-down manner. For the top echelons of the GAF, there is an annual air power seminar series. However, forums for broader audiences to further a genuine air power debate to shape the Air Force's future—conferences or a specific air power journal—do not yet exist. This top-down manner is also reflected by the fact that the revised LDv 100/1, published in April 2009, is not available to the wider public. In the specific German case, air power thinking is moreover inhibited by the country's historical legacy. Historical aspects with regard to both military education and air power doctrine are hardly dealt with. Officers attending the staff college in Hamburg acquire intensive training in staff work for combined air operations. Yet, they are not systematically exposed to air power theory and history. On the positive side, a joint approach to teaching has been adopted, reflecting the integrated nature of deployed operations.

Notes

1. Thompson, *Political and Military Components*, 200–202.
2. Ibid., 205.
3. Ibid., 209–12.

4. Hans-Werner Jarosch, ed., *Immer im Einsatz: 50 Jahre Luftwaffe* (Hamburg, Germany: Verlag E. S. Mittler & Sohn, 2005), 170.

5. "Das Kommando Operative Führung Luftstreitkräfte," in *CPM Forum: Luftwaffe 2004*, managing directors/ed. Harald Flex and Wolfgang Flume, in cooperation with the GAF staff, MOD, Bonn, and editorial coordinator Lt Col (GS) Rainer Zaude (Sankt Augustin, Germany: CPM Communication Presse Marketing GmbH, 2004), 56.

6. Bundesministerium der Verteidigung (Federal Ministry of Defence, Germany), *Weißbuch 1985 zur Lage und Entwicklung der Bundeswehr* [White book 1985: the state and development of the Bundeswehr] (Bonn, Germany: MOD, 1985), § 160.

7. MOD, Germany, *Weißbuch 1994 zur Sicherheit der Bundesrepublik Deutschland und zur Lage und Zukunft der Bundeswehr* [White book 1994: the security of the Federal Republic of Germany and the state and future of the Bundeswehr] (Bonn, Germany: MOD, 1994), § 604; and Meiers, *Zu neuen Ufern?*, 267.

8. MOD, Germany, *Verteidigungspolitische Richtlinien für den Geschäftsbereich des Bundesministers für Verteidigung: Erläuternder Begleittext* (Berlin: MOD, 21 May 2003), chap. 2.

9. Richard von Weizsäcker, director, *Gemeinsame Sicherheit und Zukunft der Bundeswehr: Bericht der Kommission an die Bundesregierung* (Berlin: Kommission Gemeinsame Sicherheit und Zukunft der Bundeswehr, May 2000), 3.

10. Rudolf Scharping, *Die Bundeswehr sicher ins 21. Jahrhundert: Eckpfeiler für eine Erneuerung von Grund auf* (Bonn/Berlin: MOD, 2000); and Tom Dyson, "German Military Reform 1998–2004: Leadership and the Triumph of Domestic Constraint over International Opportunity," *European Security* 14, no. 3 (September 2005): 367–68.

11. MOD, Germany, *Verteidigungspolitische Richtlinien*, 2003, chap. 2.

12. MOD, Germany, *White Paper 2006: German Security Policy and the Future of the Bundeswehr* (Berlin: Federal Ministry of Defence, August 2006), 9.

13. MOD, Germany, *Verteidigungspolitische Richtlinien für den Geschäftsbereich des Bundesministers der Verteidigung* (Bonn, Germany: MOD, 26 November 1992), § 9, 19, 20–21, 23, 25.

14. MOD, Germany, *Weißbuch 1994*, § 230, 233–34.

15. Von Weizsäcker, *Gemeinsame Sicherheit*, § 15.

16. Peter Struck, *Defence Policy Guidelines for the Area of Responsibility of the Federal Minister of Defence* (Berlin: MOD, 21 May 2003), § 19, 31, 80.

17. MOD, Germany, *German Security Policy*, 19–20.

18. MOD, Germany, *Verteidigungspolitische Richtlinien*, 1992, § 37.

19. Ibid., § 44.

20. Ibid., § 38.

21. MOD, Germany, *Weißbuch 1994*, § 508.

22. Scharping, *Die Bundeswehr sicher ins 21 Jahrhundert*, § 3, 15, 20, 22.

23. Struck, *Defence Policy Guidelines*, § 5.

24. Ibid., § 71.

25. See also MOD, Germany, *Verteidigungspolitische Richtlinien*, 2003, chap. 11.

26. Struck, *Defence Policy Guidelines*, 2003, § 57.

27. MOD, Germany, *German Security Policy*, 9.

28. Ibid., 9.

29. Bundeszentrale für politische Bildung, "Verhandlungen mit den Vier Mächten: Zwei-plus-Vier-Verhandlungen," *Deutsche Teilung—Deutsche Einheit*, http://www.bpb.de/themen/9VUO61,0,0,Verhandlungen_mit_den_Vier_M%E4chten.html.

30. MOD, Germany, *Verteidigungspolitische Richtlinien*, 1992, § 38, § 47; MOD, *Weißbuch 1994*, § 517; and Meiers, *Zu neuen Ufern?*, 320.

31. MOD, Germany, *Weißbuch 1994*, § 533.

32. Meiers, *Zu neuen Ufern?*, 326–27.

33. Ibid., 312.

34. Von Weizsäcker, *Gemeinsame Sicherheit*, § 4, 6.

35. Meiers, *Zu neuen Ufern?*, 199.

36. Scharping, *Die Bundeswehr sicher ins 21. Jahrhundert*, § 54, 56.

37. Meiers, *Zu neuen Ufern?*, 349.

38. MOD, Germany, *Outline of the Bundeswehr Concept* (Berlin: MOD, 10 August 2004), 23–25.

39. Struck, *Defence Policy Guidelines*, § 3, 62.

40. MOD, Germany, *German Security Policy*, 61.

41. Ibid., 9.

42. Henning Bartels, "Kommentar: Die Reform der Bundeswehr und das Sparen," *Europäische Sicherheit*, no. 7 (July 2010): 3; and "Bundeswehrtagung in Dresden: Signal der Veränderung," Bundesministerium der Verteidigung (MOD) website, 22 November 2010, http://www.bmvg.de.

43. MOD, Germany, *Weißbuch 1985*, § 160.

44. Meiers, *Zu neuen Ufern?*, 322.

45. MOD, Germany, *Verteidigungspolitische Richtlinien*, 1992, § 48; and MOD, Germany, *Weißbuch 1994*, § 604.

46. Meiers, *Zu neuen Ufern?*, 323–25.

47. Jarosch, *Immer im Einsatz*, 154.

48. Susanne Lichte, "Einsatzführung—Alles in einer Hand," Bundeswehr, http://www.bundeswehr.de, 11 September 2006.

49. Bundeswehr, "Das Operative Hauptquartier der Europäischen Union in der Henning-von-Tresckow-Kaserne in Geltow," accessed 28 September 2006, http://www.einsatz.bundeswehr.de.

50. MOD, Germany, "Berliner Erlass: Weisung zur militärischen Spitzengliederung vom 21. Januar 2005," http://www.bmvg.de.

51. Jarosch, *Immer im Einsatz*, 128–29.

52. MOD, Germany, *Weißbuch 1994*, 118.

53. Ibid., 118.

54. Lt Gen Axel Kleppien, GAF, retired (commander, 5th Air Division, 1 October 1991–mid-1993), telephone interview by the author, 16 July 2008.

55. MOD, Germany, *Weißbuch 1994*, 118.

56. International Institute for Strategic Studies, *The Military Balance 1989–90* (London: Brassey's for the IISS, 1989), 64.

57. Jarosch, *Immer im Einsatz*, 130; and MOD, Germany, *Weißbuch 1994*, § 532.

58. Lt Gen Jörg Kuebart, "Perspektiven und Parameter der Luftwaffenstruktur 4: Umbau nicht zum Null-Tarif," *Europäische Sicherheit*, no. 9 (September 1991): 504; and Jarosch, *Immer im Einsatz*, 136.

59. MOD, Germany, *Weißbuch 1994*, § 636.

60. "The German Air Force Structure 5: Principles, Structures, Capabilities," in *CPM Forum: German Air Force—Structure and Organisation, Equipment and Logistics, Programmes and Perspectives*, ed. Wolfgang Flume, in cooperation with the GAF staff, MOD, Bonn, and ed. coordinator Lt Col (GS) Rainer Zaude (Sankt Augustin, Germany: CPM Communication Presse Marketing GmbH, 2003), 20.

61. Ibid., 19.

62. Scharping, *Die Bundeswehr sicher ins 21. Jahrhundert*, § 71.

63. "German Air Force Structure 5," 20.

64. Jarosch, *Immer im Einsatz*, 156.

65. "Das Kommando Operative Führung Luftstreitkräfte," 56–57.

66. Jarosch, *Immer im Einsatz*, 158.

67. Lt Col (GS) Guido Leiwig, GAF (MOD, Bonn, Germany), interview by the author, 26 April 2006; and "Further Development of the Luftwaffe—Luftwaffe Structure 6," in *CPM Forum: Luftwaffe 2005—The German Air Force Today and Tomorrow*, managing director Wolfgang Flume and project director Jürgen Hensel (Sankt Augustin, Germany: CPM Communication Presse Marketing GmbH, 2005), 46.

68. MOD, Germany, *Verteidigungspolitische Richtlinien*, 2003, chap. 2.

69. Jarosch, *Immer im Einsatz*, 165.

70. Lt Col (GS) Nicolas Radke, GAF, MOD, Bonn, Germany, to the author, e-mail, 28 November 2006.

71. MOD, Germany, *Outline of the Bundeswehr Concept*, 32.

72. "Luftwaffenstruktur 6," *Wehrtechnischer Report: Fähigkeiten und Ausrüstung der Luftwaffe*, no. 3 (Bonn: Report Verlag GmbH, 2006), 14.

73. MOD, Germany, *Outline of the Bundeswehr Concept*, 32.

74. "Luftwaffenstruktur 6," 14.

75. Lt Col (GS) Eberhard Freiherr von Wintzingerode-Knorr, "Weiterentwicklung des 'Jägers' Eurofighter für den Luft-/Boden-Einsatz," *Europäische Sicherheit*, no. 11 (November 2006): 43.

76. "Rote Liste für den Baron," *sueddeutsche.de*, accessed 7 July 2010, http://www.sueddeutsche.de/politik/sparkurs-im-verteidigungsministerium-rote-liste-fuer-den-baron-1.971018; and Thomas Newdick, "A Change in the Air," *Jane's Defence Weekly* 47, no. 22 (2 June 2010): 30.

77. Lt Col (GS) Frank Gräfe, "Die Mehrrollenfähigkeit des Waffensystems EUROFIGHTER," *Europäische Sicherheit*, no. 4 (April 2010): 55.

78. "Further Development of the Luftwaffe—Luftwaffe Structure 6," 45–46.

79. Federal Ministry of Defence, *German Security Policy*, 92–94.

80. MOD, Germany, *Outline of the Bundeswehr Concept*, 32; and Luftwaffe, "Luftwaffenstruktur 6—Die Luftwaffe im Wandel," accessed 2 August 2006, http://www.luftwaffe.de.

81. "Further Development of the Luftwaffe—Luftwaffe Structure 6," 46–47.

82. "Luftwaffenstruktur 6," 14.

83. International Institute for Strategic Studies, *Military Balance 1990–91*, 68; and International Institute for Strategic Studies, *Military Balance 2002–03*, 43.

84. Kleppien, interview.

85. Lt Gen Walter Jertz, GAF, retired, to the author, e-mail, 26 June 2008. In 1995 Jertz was the GAF contingent commander for air operations over the Balkans. He left his final posting as commandant, German Air Force Command, in June 2006.

86. Ibid.; and Jarosch, *Immer im Einsatz*, 112.

87. "Verwertung des Wehrmaterials," *Information für die Truppe: Armee der Einheit*, no. 5 (May 1998): 46.

88. Heinz von Knobloch, *Bundesluftwaffe intern: Aufbau, Wandel, Einsätze* (Stuttgart, Germany: Motorbuch Verlag, 2008), 257.

89. Lt Gen Hans-Werner Jarosch, GAF, retired, telephone interview by the author, 16 July 2008; and Jarosch, *Immer im Einsatz*, 113. The Airbuses might have had a dual-use function, militarily for the East German Air Force and in a civilian role for the East German airline *Interflug*. They do not appear in the orders of battle as provided by *The Military Balance*. See International Institute for Strategic Studies, *Military Balance 1990–91*, 49.

90. Jertz to the author, e-mail, 26 June 2008.

91. Rüdiger Wenzke, "Die Nationale Volksarmee 1989/90," *Information für die Truppe: Armee der Einheit*, no. 5 (May 1998): 15.

92. Ove Ovens, *Die Nationale Volksarmee der DDR zwischen "Wende" und Auflösung: Der Untergang der NVA im Lichte des Zusammenbruchs der DDR* (Regensburg, Germany: Universität Regensburg, 2003), 417.

93. Jarosch, *Immer im Einsatz*, 115.

94. Ibid., 110–11.

95. Von Knobloch, *Bundesluftwaffe intern*, 255; and Jarosch, interview.

96. Lt Gen Axel Kleppien, GAF, retired, to the author, e-mail, 17 July 2008; and Kleppien, interview.

97. Jertz to the author, e-mail, 26 June 2008; and Norbert Straka, "Die Teilung überwinden: Erfahrungen eines Offiziers mit der deutschen Einheit," in Jarosch, *Immer im Einsatz*, 247.

98. Kleppien to the author, e-mail; and Kleppien, interview.

99. Jarosch, *Immer im Einsatz*, 122.

100. Kleppien, interview; and Jarosch, interview.

101. Jarosch, interview.

102. Thompson, *Political and Military Components*, 40–41.

103. Lt Gen Rolf Portz, Inspector, GAF, "The Future of Air Power (IV): Germany," *Military Technology*, no. 10 (1999): 14.

104. International Institute for Strategic Studies, *Military Balance 1990–91*, 67–68.

105. International Institute for Strategic Studies, *Military Balance 2002–03*, 43.

106. Sascha Lange, *Neue Bundeswehr auf altem Sockel: Wege aus dem Dilemma: SWP-Studie* (Berlin: Stiftung Wissenschaft und Politik, 2005): 14–15.

107. Ibid., 7.

108. MOD, Germany, *Weißbuch 1985*, § 160.

109. Hanns W. Maull, "Germany and the Use of Force: Still a 'Civilian Power'?," *Survival* 42, no. 2 (Summer 2000): 69.

110. MOD, Germany, *Verteidigungspolitische Richtlinien*, 1992, § 8.

111. MOD, Germany, *Weißbuch 1994*, § 255, 319.

112. MOD, Germany, *Verteidigungspolitische Richtlinien*, 1992, § 29–36.

113. Meiers, *Zu neuen Ufern?*, 125, 136–37, 167–69.

114. Struck, *Defence Policy Guidelines*, § 32, 40, 46.

115. Franz-Josef Meiers, "Germany's Defence Choices," *Survival* 47, no. 1 (Spring 2005): 153.

116. Hubert Zimmermann, "Security Exporters: Germany, the United States, and Transatlantic Cooperation," in *The Atlantic Alliance under Stress: US-European Relations after Iraq*, ed. David M. Andrews (Cambridge, UK: Cambridge University Press, 2005), 129, 150–51.

117. Meiers, *Zu neuen Ufern?*, 233, 246.

118. Struck, *Defence Policy Guidelines*, § 33.

119. MOD, Germany, *German Security Policy*, 24.

120. Ibid., 41.

121. Struck, *Defence Policy Guidelines*, § 11, 41.

122. MOD, Germany, *German Security Policy*, 22.

123. Struck, *Defence Policy Guidelines*, § 52.

124. Radke to the author, e-mail; and Wolsztynski, "La contribution de l'Armée de l'Air," 11.

125. Trautermann to the author, e-mail.

126. European Union, *Headline Goal 2010*, 3.

127. MOD, Germany, *Weißbuch 1985*, § 443.

128. MOD, Germany, *Weißbuch 1994*, § 626.

129. Ibid., § 632.

130. Jarosch, *Immer im Einsatz*, 129–30.

131. MOD, Germany, *Weißbuch 1994*, § 633.

132. Lt Col (GS) Jens Asmussen, GAF, MOD, Bonn, Germany, to the author, e-mail, 12 January 2007.

133. NATO, "NATO Airborne Early Warning and Control Force," 1.

134. Schmidt, interview.

135. Ibid.

136. Lt Gen Hans-Joachim Schubert (GAF, executive director, JAPCC, Kalkar, Germany), interview by the author, Üdem, Germany, 5 May 2006.

137. JAPCC, "History"; and Asmussen to the author, e-mail.

138. Gregory, *French Defence Policy*, 138.

139. MOD, Germany, *Weißbuch 1994*, § 541.

140. Lebert, "Einrichtung eines Europäischen Lufttransportkommandos," 20.

141. Havenith, interview; and Lt Col (GS) Armin Havenith, GAF, MOD, Bonn, Germany, to the author, e-mail, 19 June 2006.

142. Lebert, "Einrichtung eines Europäischen Lufttransportkommandos," 20.

143. Havenith, interview.

144. Jean-Laurent Nijean, "L'Armée de l'Air et l'Europe: gros plan—au coeur de l'Europe de la défense," *Air actualités*, no. 605 (October 2007): 38; and Luftwaffe, "Startschuss für das Europäische Lufttransportkommando," 25 February 2010, http://www.luftwaffe.de.

145. Luftwaffe, "Abschied vom Lufttransportkommando in Münster," 6 July 2010, http://www.luftwaffe.de.

146. Havenith, interview; and Havenith to the author, e-mail.

147. Lt Col (GS) Carsten Rietsch, GAF, assistant air attaché, German Embassy in London, to the author, e-mail, 5 August 2009.

148. Bundeswehr, "16 Nationen, eine Transportlösung," Leipzig-Halle, 23 March 2006, accessed 9 August 2006, http://www.bundeswehr.de.

149. BMVg, "Verlässlicher Zugriff."

150. Jarosch, *Immer im Einsatz*, 14.

151. Bernd Vetter and Frank Vetter, *Luftwaffe im 21. Jahrhundert* (Stuttgart, Germany: Motorbuch Verlag, 2005), 112, 118, 133.

152. Ibid., 141–42.

153. Ibid., 146.

154. Mason, *Air Power*, 249–50.

155. Luftwaffe, "Das Fliegerische Ausbildungszentrum der Luftwaffe," accessed 18 September 2006, http://www.luftwaffe.de.

156. Vetter and Vetter, *Luftwaffe im 21. Jahrhundert*, 122–23.

157. Jarosch, *Immer im Einsatz*, 153.

158. European Air Group, "History of the EAG."

159. Jarosch, *Immer im Einsatz*, 149.

160. Ibid., 149–51.

161. Dieter Stockfisch, "Das Führungszentrum der Bundeswehr," in *Von Kambodscha bis Kosovo: Auslandeinsätze der Bundeswehr*, ed. Peter Goebel (Frankfurt am Main, Germany: Report Verlag, 2000), 183.

162. Klaus Dau, "Auslandeinsätze zwischen Politik und Verfassungsrecht," in Goebel, *Von Kambodscha bis Kosovo*, 22–23.

163. André Herbrink, *Bundeswehr Missions Out-of-Area*, NUPI Report no. 227 (Oslo, Norway: The Norwegian Institute of International Affairs, 1997), 13.

164. Ibid., 15.

165. Lawrence Freedman and Efraim Karsh, *The Gulf Conflict* (London: Faber and Faber, 1994), 355.

166. MOD, Germany, *Weißbuch 1994*, 71.

167. Jarosch, *Immer im Einsatz*, 139.

168. Meiers, *Zu neuen Ufern?*, 272–73.

169. *Grundgesetz für die Bundesrepublik Deutschland* (Basic Law for the Federal Republic of Germany [German constitution]), 23 May 1949, art. 24, para. 2, http://www.gesetze-im-internet.de/bundesrecht/gg/gesamt.pdf.

170. Jertz, *Im Dienste des Friedens*, 14.

171. Bundesverfassungsgericht (Federal Constitutional Court of Germany), *AWACS 1:* BverfG 2 BvE 3/92 u.a. (Karlsruhe, Germany: 12 July 1994), http://www.jur-abc.de/cms/index.php?id=606.

172. Ibid.; and Meiers, *Zu neuen Ufern?*, 275–77.

173. Ibid., 281.

174. Walter Jertz, "Einsätze der Luftwaffe über Bosnien," in Goebel, *Von Kambodscha bis Kosovo*, 140.

175. Meiers, *Zu neuen Ufern?*, 283–84.

176. Jertz, "Einsätze der Luftwaffe über Bosnien," 146–47.

177. Meiers, *Zu neuen Ufern?*, 287. See also Jarosch, *Immer im Einsatz*, 140.

178. Lt Gen Walter Jertz, GAF, retired, to the author, e-mail, 23 June 2007.

179. Ibid., 26 June 2008.

180. Jertz, "Einsätze der Luftwaffe über Bosnien," 152–53.

181. Meiers, *Zu neuen Ufern?*, 291–92.

182. Ibid., 295–96.

183. Jarosch, *Immer im Einsatz*, 141; and Jertz, *Im Dienste des Friedens*, 146.

184. Meiers, *Zu neuen Ufern?*, 301.

185. Jarosch, *Immer im Einsatz*, 145; and Lambeth, *NATO's Air War for Kosovo*, 110.

186. Meiers, *Zu neuen Ufern?*, 302.

187. Ibid., 308.

188. Jarosch, *Immer im Einsatz*, 159–60.

189. Robert Wiedmann and Herbert Albring, "'Professionelle Gelassenheit' bei den Immelmännern vor ihrem Auslandeinsatz," 14 March 2007, Luftwaffe, http://www.luftwaffe.de.

190. Nicholas Fiorenza, "First German Heron 1 Arrives in Afghanistan," *Jane's Defence Weekly* 47, no. 11 (17 March 2010): 32.

191. Presse- und Informationsstab BMVg, "Abzug der Aufklärungs-Tornado aus Afghanistan," Bundesministerium der Verteidigung, 23 September 2010, www.bundeswehr.de.

192. Bundesverfassungsgericht, "Einsatz deutscher Soldaten in AWACS-Flugzeugen über der Türkei bedurfte der Zustimmung des Bundestags," press release no. 52/2008, judgement of 7 May 2008—2 BvE 1/03, Karlsruhe, Germany, http://www.bundesverfassungsgericht.de/pressemitteilungen/bvg08-052.html.

193. Lange, *Neue Bundeswehr auf altem Sockel*, 9.

194. MOD, Germany, *German Security Policy*, 56.

195. MOD, Germany, *Weißbuch 1994*, 72.

196. Jarosch, *Immer im Einsatz*, 137.

197. Meiers, *Zu neuen Ufern?*, 254.

198. Peter Goebel, "Von der Betroffenheit zur Selbstverständlichkeit," in Goebel, *Von Kambodscha bis Kosovo*, 15.

199. Jarosch, *Immer im Einsatz*, 137.

200. MOD, Germany, *Weißbuch 1994*, 70.

201. Jarosch, *Immer im Einsatz*, 139.

202. Roger Evers, "Transportflieger in humanitärem Auftrag," in Goebel, *Von Kambodscha bis Kosovo*, 94.

203. Jarosch, *Immer im Einsatz*, 139; and Evers, "Transportflieger in humanitärem Auftrag," 91–92.

204. Evers, "Transportflieger in humanitärem Auftrag," 86, 100.

205. Reinhart Hoppe, "First In and Last Out: Lufttransport im weltweiten Einsatz," in Jarosch, *Immer im Einsatz*, 214.

206. Ibid., 213.

207. Jarosch, *Immer im Einsatz*, 146.

208. Ibid., 145–46.

209. Ibid., 160–61.

210. Lt Gen Aarne Kreuzinger-Janik, chief of the Air Staff, "Die Luftwaffe im Gesamtsystem Bundeswehr," *Europäische Sicherheit*, no. 6 (June 2010): 24.

211. Nijean, "L'Armée de l'Air et l'Europe: opération EUFOR RDC," 34–35.

212. Jarosch, *Immer im Einsatz*, 136.

213. Ibid., 157.

214. Schmidt, interview.

215. Leiwig, interview.

216. See Struck, *Defence Policy Guidelines*, § 86; and "Kein Abschuss entführter Flugzeuge in Deutschland: Urteil des Verfassungsgerichts," *NZZ Online*, 15 February 2006, http://www.nzz.ch/2006/02/15/al/newzzEJPFU4C8-12.html.

217. Bundesverfassungsgericht, "Abschussermächtigung im Luftsicherheitsgesetz nichtig," press release no. 11/2006, judgement of 15 February 2006—1 BvR 357/05, Karlsruhe, Germany, http://www.bundesverfassungsgericht.de/pressemitteilungen/bvg06-011.html.

218. Jan Kuebart, "Air Policing Baltikum," *Europäische Sicherheit*, no. 12 (December 2005): 32.

219. "German Eurofighters Arrive for Baltic Air Policing Debut," *Flight International* 176, no. 5205 (8–14 September 2009): 18; see also Hoyle, "Baltic Exchange," 33.

220. Kreuzinger-Janik, "Die Luftwaffe im Gesamtsystem Bundeswehr," 23.

221. MOD, Germany, *Weißbuch 1994*, § 627.

222. Hoppe, "First In and Last Out," 214.

223. Jarosch, *Immer im Einsatz*, 145; and Hoppe, "First In and Last Out," 214–15.

224. Ibid., 215–16.

225. Col (GS) Reinhard Vogt, GAF, LMZ, Cologne Wahn, Germany, to the author, e-mail, 26 October 2006.

226. Col (GS) Reinhard Vogt (GAF, LMZ, Cologne Wahn, Germany), interview by the author, 28 April 2006.

227. Thompson, *Political and Military Components*, 9.

228. Vogt to the author, e-mail.

229. Thompson, *Political and Military Components*, 76–77.

230. Vogt to the author, e-mail.

231. MOD, Germany, *Luftwaffendienstvorschrift: Führung und Einsatz von Luftstreitkräften*, LDv 100/1 VS-NfD (Bonn, Germany: MOD, March 1991), introduction.

232. Ibid., § 501–15.

233. Ibid., § 407.

234. Lt Col Manfred Brose, GAF, FüAk, Hamburg, Germany, to the author, e-mail, 9 November 2006.

235. Vogt, interview.

236. Brose to the author, e-mail.

237. Lt Col (GS) Arnt Kuebart, GAF, LMZ, Cologne Wahn, Germany, to the author, e-mail, 20 October 2006.

238. Rietsch to the author, e-mail, 28 August 2009.

239. Lt Col Manfred Brose (GAF, FüAk, Hamburg, Germany), interview by the author, 3 May 2006.

240. Führungsakademie der Bundeswehr, "Lehrgänge an der Führungsakademie der Bundeswehr," 22 July 2008, http://www.fueakbw.de/index.php?Show Parent=345&show_lang=de; and Führungsakademie der Bundeswehr, "Nationaler Lehrgang Generalstabs-/Admiralstabsdienst," 6 April 2009, http://www .fueakbw.de/index.php?ShowParent=244&show_lang=de.

241. Brose, interview.

242. FüAk Kommandeur, *Lehrplan für den Lehrgang Generalstabsdienst/ Admiralstabsdienst National* (Hamburg, Germany: FüAk, 20 July 2005), 11.

243. Brose, interview.

244. Col (GS) Thomas Lorber, GAF (FüAk, Hamburg, Germany), interview by the author, 3 May 2006.

245. Simon Pearson, to the author, e-mail, 19 January 2007. Pearson served three years as an RAF officer on exchange with the Luftwaffe, JaboG 49, at Fürstenfeldbruck, Germany, followed by two years at the FüAk in Hamburg.

246. International course member attending the FüAk, Hamburg, Germany, to the author, e-mail, 11 September 2007.

247. Lt Col John Olsen, dean of the Norwegian Education Command and head of the Department for Strategic Studies at the Norwegian Command and Staff College, Oslo, Norway, to the author, e-mail, 6 September 2007.

248. List of the 2005–6 *Lehrgangsarbeiten* provided by Lt Col Manfred Brose, FüAk, Hamburg, Germany, 3 May 2006.

249. Führungsakademie der Bundeswehr, "Konzeption Stabsoffiziersaubildung," 10 November 2008, http://www.fueakbw.de/index.php?ShowParent=2678& show_lang=de.

250. Lorber, interview.

251. Führungsakademie der Bundeswehr, "Module an der Führungsakademie der Bundeswehr," 24 November 2009, http://www.fueakbw.de/index.php?Show Parent=446&show_lang=de.

252. Führungsakademie der Bundeswehr, "Seminar Luftwaffe in der Transformation," accessed 20 October 2006, http://www.fueakbw.de; and Führungs-

akademie der Bundeswehr, "Seminar Perspektiven von Luftmacht im 21. Jahrhundert," 14 October 2009, http://www.fueakbw.de/index.php?showParent=2614&show_lang=de.

253. Brose, interview.

254. CID, "Enseignement Opérationnel."

255. FüAk Kommandeur, *Lehrplan*, 16.

256. Lt Col (Dr.) Wolfgang Schmidt, GAF, Military-Historical Research Office, Potsdam, Germany, to the author, e-mail, 13 June 2006.

257. MOD, Germany, *German Security Policy*, 59.

258. Col (GS) Burkhard Potozky and Lt Col (GS) Stefan Klenz, GAF (MOD, Bonn, Germany), interview by the author, 26 April 2006.

259. Lt Col (GS) Stefan Klenz, GAF, MOD, Bonn, Germany, to the author, letter, 6 July 2006.

260. Potozky and Klenz, interview.

261. Klenz, letter.

262. Potozky and Klenz, interview.

263. Examples include Frank Dorn, "Bessere Handhabung durch bessere Technik—am Beispiel des Waffensystems Tornado," in *Jahrbuch der Luftwaffe*, no. 14 (Munich, Germany: Bernard und Graefe, 1981), 68–78; Lt Gen Eberhard Eimler, "Die Alliierten Luftstreitkräfte Europa-Mitte," in *Jahrbuch der Luftwaffe*, no. 14, 13–22; Lt Col Manfred Erl, "Die Bedeutung der Faktoren Raum und Zeit für Luftstreitkräfte," in *Jahrbuch der Luftwaffe*, no. 14, 80–85; Lt Col Hans Felde, "Der taktische Wert des Tornado," in *Jahrbuch der Luftwaffe*, no. 14, 60–66; Lt Col Hans-Werner Jarosch, "Tiefe des Raumes: ein Faktor für Land- und Luftstreitkräfte von Warschauer Pakt und NATO," in *Jahrbuch der Luftwaffe*, no. 14, 51–58; Maj Gen Jörg Kuebart, "Die Luftverteidigung Mitteleuropas unter Beteiligung der deutschen Luftwaffe," in *Jahrbuch der Luftwaffe*, no. 14, 24–31; Lt Gen Gerhard Limberg, "Herausforderung: Die Kampfkraft der Luftwaffe muss erhöht werden," in *Jahrbuch der Luftwaffe*, no. 12 (Bonn-Duisdorf, Germany: Wehr und Wissen, 1975), 15; Lt Gen Friedrich Obleser, Inspector GAF, "Die Rolle von Luftstreitkräften in Mitteleuropa und der Beitrag der Luftwaffe," in *Jahrbuch der Luftwaffe*, no. 14, 8–12; Lt Gen Guenter Rall, Inspector GAF, "Die Luftwaffe im NATO Buendnis: Die Entwicklung der Luftwaffe in Abhängigkeit von den militär-strategischen Konzeptionen der NATO," in *Jahrbuch der Luftwaffe*, no. 10 (Darmstadt, Germany: Wehr und Wissen, 1973), 14–21; Col Heinrich Starke, "Die neuen Waffensysteme der Luftwaffe," in *Jahrbuch der Luftwaffe*, no. 12, 118–24; Lt Gen Johannes Steinhoff, chief, Air Staff, "Wir planen die Luftwaffe der 80er Jahre," in *Jahrbuch der Luftwaffe*, no. 7 (1970), 8–11; and Bernard Wolf and Konrad Wolf, "Trends bei der Entwicklung von Kampfflugzeugen," in *Jahrbuch der Luftwaffe*, no. 11 (Koblenz, Germany: Wehr und Wissen, 1974), 25–33.

264. Examples include Maj Gen Jörg Bahnemann, "Air Defense in Central Europe," in *NATO's Sixteen Nations*, no. 7 (1985): 40–46; Lt Gen Eberhard Eimler, "Die Luftverteidigung im NATO-Bereich," in *Europäische Wehrkunde*, no. 9 (1983): 410–20; Lt Gen Eberhard Eimler, "Besonders wichtig ist die Verlage-

rung des Planungsschwerpunktes zur Luftverteidigung," in *Wehrtechnik*, no. 12 (December 1983): 24–29; Wolfgang Flume, "Luftverteidigung heute und morgen: neue Bedrohung durch taktische ballistische Raketen."

265. Examples of this emphasis include Lt Gen Gerhard Back, Inspector GAF, "Reform der Luftwaffe—Konzentration auf unmittelbare Einsatzaufgaben," *Europäische Sicherheit*, no. 6 (June 2001): 14–19; Brig Gen Dirk Böcker, "Operative Fähigkeiten der Luftwaffe in der Zukunft," *Europäische Sicherheit*, no. 12 (December 1994): 600–604; Dieter Büscher, "Die Zukunft im Lufttransport," *Europäische Sicherheit*, no. 3 (March 1996): 19–22; Eberhard Eckert, "'Gerätevorhaben der Luftwaffe,'" in *Europäische Sicherheit*, no. 3 (March 1995): 26–30; Lebert, "Einrichtung eines Europäischen Lufttransportkommandos," 20–24; Lt Gen Bernhard Mende, Inspector GAF, "Die Luftwaffe auf dem Weg in die Zukunft," *Europäische Sicherheit*, no. 5 (May 1996): 12–18; Ferdinand Mertes, "Wahrung der Unversehrtheit des Luftraumes: Luftverteidigung unter neuen Rahmenbedingungen," *Europäische Sicherheit*, no. 5 (May 1992): 273–80; Holger H. Mey and Karl P. Sasse, "Erweiterte Luftverteidigung zwischen europäischer und transatlantischer Kooperation," *Europäische Sicherheit*, no. 9 (September 1994): 466–71; Lt Col Reinhard Mittler, "Luftangriff im Aufgabenspektrum der Luftwaffe," *Europäische Sicherheit*, no. 12 (December 2001): 31–39; Christian Philipp, "Abstandswaffen für die Luftwaffe," *Europäische Sicherheit*, no. 6 (June 1995): 33–34; Pitsch, "Modulare Abstandswaffe Taurus," 39–43; Lt Gen Rolf Portz, Inspector GAF, "Bedeutung von Luftmacht bei der Krisenbewältigung," *Europäische Sicherheit*, no. 5 (May 1998): 10–15; Lt Col Hans-Joachim Ratzlaff, "Lufttransport-Erfahrungen aus Krisenreaktionseinsätzen," *Europäische Sicherheit*, no. 1 (January 2000): 35–38; Col Gero Schachthöfer and Lt Col Stefan Preschke, "Sicherheit im Luftraum," *Europäische Sicherheit*, no. 9 (September 2003): 15–18; Scheibl and Schad, "Combat Search and Rescue," 42–45; Ludwig Schiller, "Future Large Aircraft: Ein 'Wooden Wonder' oder Realität," *Europäische Sicherheit*, no. 11 (November 1995): 28–32; Stütz, "Luftbetankung," 17–20; Lt Col Klaus Vogelsang, "Lufttransportkommando—Grenzen der Leistungsfähigkeit," *Europäische Sicherheit*, no. 12 (December 1996): 28–32; Lt Col Wilhelm Vitzthum, "Lufttransport im 21. Jahrhundert," *Europäische Sicherheit*, no. 7 (July 1997): 16–20; and Niklas von Witzendorff, "Luftverteidigung—wohin?," *Europäische Sicherheit*, no. 5 (May 1995): 9–10.

266. Lt Gen Jürgen Höche, "Erfahrungen aus den Luftoperationen der NATO im Kosovo," *Europäische Sicherheit*, no. 2 (February 2000): 24–27.

267. See *CPM Forum* issues 2002–2005.

268. Luftwaffe, "Neue Einblicke," Potsdam, Germany, 23 March 2006, accessed 17 October 2006, http://www.luftwaffe.de.

269. Lt Col (Dr.) Wolfgang Schmidt, GAF, Military-Historical Research Office, Potsdam, Germany, to the author, e-mail, 12 June 2006; and Thompson, *Political and Military Components*, 94.

270. MOD, Germany, *Verteidiguntgspolitische Richtlinien*, 1992, § 4.

271. Ibid., § 43.

272. MOD, Germany, *Weißbuch 1994*, § 571–72.

273. MOD, Germany, *Verteidiguntgspolitische Richtlinien*, 1992, § 43; and Schön and Schetilin, "Material- und Ausrüstungsplanung der Luftwaffe," 48.

274. Thompson, *Political and Military Components*, 22–23.

275. Maj (GS) Markus Schetilin, "Material- und Ausrüstungsplanung der Luftwaffe," *Europäische Sicherheit*, no. 10 (October 2005): 34.

276. "German Air Force Structure 5," 22.

277. Schön and Schetilin, "Material- und Ausrüstungsplanung der Luftwaffe," 38–39; and Groß, "Revision der Reform," 118.

278. Jarosch, *Immer im Einsatz*, 131.

279. Lt Col Walter Fleischer, "Schwert und Schild der Luftwaffe," *Europäische Sicherheit*, no. 5 (May 1991): 293.

280. Jarosch, *Immer im Einsatz*, 139–40; and Jertz, "Einsätze der Luftwaffe über Bosnien," 142.

281. Volker Schmitt, "Modern Aircraft and Helicopters for the German Armed Forces," *Military Technology* 23, no. 2 (February 1999): 60.

282. Meiers, *Zu neuen Ufern?*, 302.

283. "Wirken gegen Ziele am Boden," *Wehrtechnischer Report: Fähigkeiten und Ausrüstung der Luftwaffe*, no. 3 (2006): 50.

284. Col (GS) Hans-Dieter Schön, GAF (MOD, Bonn, Germany), interview by the author, 27 April 2006.

285. "Wirken gegen Ziele am Boden," 49–50.

286. Jarosch, *Immer im Einsatz*, 148.

287. "European Military Aircraft Programmes Revisited," 8.

288. Franz Mendel, "Jagdflugzeug 90: Eine Dauerdiskussion ohne Ende," *Europäische Sicherheit*, no. 11 (November 1991): 652; and "MiG-29 bleibt," *Europäische Sicherheit*, no. 8 (August 1991): 437.

289. MOD, Germany, *Weißbuch 1994*, § 582.

290. Stefan Petersen, "Die 'Rote Diva' fliegt nach Polen," *Aktuell: Zeitung für die Bundeswehr*, no. 14 (5 April 2004): 9.

291. Willett, Gummet, and Clarke, *Eurofighter 2000*, 1.

292. Ibid., 4–5.

293. Ibid., 6; and Fides Krause-Brewer, "Jäger 90: Die Kosten des Ausstiegs—vier bis fünf Milliarden allein für die Partner?," *Europäische Wehrkunde*, no. 9 (September 1990): 544.

294. Willett, Gummet, and Clarke, *Eurofighter 2000*, 9.

295. Krause-Brewer, "Jäger 90," 546; and Mendel, "Jagdflugzeug 90," 650.

296. Krause-Brewer, "Jäger 90," 546.

297. Mendel, "Jagdflugzeug 90," 652.

298. Krause-Brewer, "Jäger 90," 546; and Willett, Gummet, and Clarke, *Eurofighter 2000*, 7.

299. Ibid., 39.

300. Schetilin, "Material- und Ausrüstungsplanung der Luftwaffe," 24; and Schön and Schetilin, "Material- und Ausrüstungsplanung der Luftwaffe," 40.

301. Von Wintzingerode-Knorr, "Weiterentwicklung des 'Jägers' Eurofighter," 42–43.

302. Newdick, "Change in the Air," 30.

303. Gräfe, "Mehrrollenfähigkeit des Waffensystems EUROFIGHTER," 50–51.

304. "Wirken gegen Ziele in der Luft," 44.

305. Lt Col Wilfrid Schmidt, "Bewaffnung für die Kampfflugzeuge der Luftwaffe: Sachstand-Konzepte-Planungen," *Europäische Sicherheit*, no. 10 (October 1998): 38.

306. Scharping, *Die Bundeswehr sicher ins 21. Jahrhundert*, § 49.

307. Lt Gen Walter Jertz, "Unser Schwerpunkt ist der Einsatz: Das Luftwaffenführungskommando auf dem Weg in die Zukunft," *Strategie & Technik*, March 2006, 23.

308. Ibid., 22.

309. Gräfe, " Mehrrollenfähigkeit des Waffensystems EUROFIGHTER," 52–54.

310. Thompson, *Political and Military Components*, 88.

311. Alfons Erwes, "Zukünftige SEAD-Flugkörper," *Europäische Sicherheit*, no. 7 (July 2001): 25.

312. Lehmann, "Further Development of Luftwaffe's Equipment," 53–54; and Erwes, "Zukünftige SEAD-Flugkörper," 26.

313. Goebel, "Von der Betroffenheit zur Selbstverständlichkeit," 14.

314. Dr. Hermann Hagena, brigadier general, GAF, retired, "FLA, Antonow 70 oder C-130J: wird nur einer gewinnen?," *Europäische Sicherheit*, no. 4 (April 1999): 37.

315. Evers, "Transportflieger in humanitärem Auftrag," 90–91.

316. Lehmann, "Further Development of Luftwaffe's Equipment," 54–55.

317. Agüera, "Zwischen Hoffnung und Verzweiflung," 47.

318. Hagena, "Charter oder Leasing?," 35.

319. Agüera, "Zwischen Hoffnung und Verzweiflung," 47.

320. Scharping, *Die Bundeswehr sicher ins 21. Jahrhundert*, § 48; and Agüera, "Zwischen Hoffnung und Verzweiflung," 48–49.

321. Schetilin, "Material- und Ausrüstungsplanung der Luftwaffe," 26.

322. "Frankreich attackiert Deutschland wegen A400M," *Handelsblatt*, 31 January 2011, http://www.handelsblatt.com/unternehmen/industrie/airbus-frankreich-attackiert-deutschland-wegen-a400m;2743458.

323. Lt Col Peter Stütz, "Luftbetankung," *Europäische Sicherheit*, no. 7 (July 1999): 18.

324. International Institute for Strategic Studies, *The Military Balance 2010* (Abingdon, Oxfordshire: published by Routledge for the IISS, February 2010), 136.

325. Schön and Schetilin, "Material- und Ausrüstungsplanung der Luftwaffe," 40.

326. Hoppe, "First In and Last Out," 215; and Schön, interview.

327. Luftwaffe, "Luftwaffe modernisiert Regierungsflotte mit A319 CJ," 25 March 2010, http://www.luftwaffe.de.

328. Lange, *Neue Bundeswehr auf altem Sockel*, 10.

329. Lehmann, "Further Development of Luftwaffe's Equipment," 52.

330. Meiers, *Zu neuen Ufern?*, 156; Scharping, *Die Bundeswehr sicher ins 21. Jahrhundert*, § 48; and Struck, *Defence Policy Guidelines*, § 92.

331. Schön, interview.

332. *LPM 2009–2014*, rapport annexé, 6.

333. "Nachrichtengewinnung und Aufklärung," *Wehrtechnischer Report: Fähigkeiten und Ausrüstung der Luftwaffe*, no. 3 (2006): 33.

334. "Auflösung der 2. Aufklärungsstaffel," *Strategie & Technik*, October 2009, 49; Schetilin, "Material- und Ausrüstungsplanung der Luftwaffe," 30; Luftwaffe, "Recce Lite: Ein großer Fortschritt," 10 November 2009, http://www.luftwaffe.de; and Luftwaffe, "Erfolgreicher Erstflug des Euro Hawk symbolisiert Zukunft der Luftwaffe," 2 July 2010, http://www.luftwaffe.de.

335. Thompson, *Political and Military Components*, 115–16.

336. Jertz, "Unser Schwerpunkt ist der Einsatz," 22.

337. Schön, interview.

338. "German Air Force Structure 5," 22; and MOD, Germany, *Outline of the Bundeswehr Concept*, 32.

339. Lehmann, "Further Development of Luftwaffe's Equipment," 52.

340. Von Hoyer-Boot, "Erweiterte Luftverteidigung," 506.

341. Lt Col Ralf Gosch, "Luftverteidigung—Veränderungen in der neuen Struktur," *Europäische Sicherheit*, no. 3 (March 2005): 48; and Schön, interview.

342. Jarosch, *Immer im Einsatz*, 165.

343. Gosch, "Luftverteidigung," 46.

344. Struck, *Defence Policy Guidelines*, § 92.

345. "Germany Faces Pressure to Re-evaluate MEADS," *Jane's Defence Weekly* 47, no. 11 (17 March 2010): 5.

Chapter 5

Royal Netherlands Air Force
(Koninklijke Luchtmacht)

The Royal Netherlands Air Force (RNLAF), or Koninklijke Luchtmacht, has its roots back in 1913, when it consisted of a single aeroplane and a unit of merely six men. After World War II, in 1953, the RNLAF was formed as an independent service.[1] Fifty-seven years later, it had become a professional service branch employing approximately 9,500 personnel.[2] Particularly in the post–Cold War era, the RNLAF has gained a reputation of a highly efficient and capable Air Force out of proportion to its relatively small size, providing the Netherlands government with an effective tool for combined out-of-area operations.

When Gen Charles A. Horner, the air component commander of Operation Desert Storm, was asked to comment upon the RNLAF, he particularly highlighted the Air Force's outstanding qualities regarding professionalism, equipment, training, and attitude. Moreover, he took the view that in the course of Allied Force, the RNLAF proved to be highly interoperable and able to cover a broad panoply of air power missions, including offensive air combat missions. As such, the air component commander could rely upon the RNLAF to execute difficult missions, with other NATO

Courtesy Netherlands Institute for Military History; photo by Frank Visser

RNLAF F-16 at Kandahar Airfield in November 2008

allies relegated to more supporting roles due to deficiencies in equipment, training, or attitude.[3]

By examining how the RNLAF has responded to the air power challenges of the post–Cold War era, the following examination explores how a relatively small nation, whose defence spending is below the European NATO average, can have an effective Air Force. The RNLAF has also been one of the main drivers of European air power integration while retaining its strong transatlantic link.

How Has the Royal Netherlands Air Force Adapted to the Uncertainties Created by Shifting Defence and Alliance Policies?

This section first analyses the Netherlands' post–Cold War defence policy and its influence on the RNLAF's evolution. The Netherlands' alliance policies are described next, along with a view towards how they have affected its Air Force.

Defence Policy

Between 1990 and 2003, four major defence reports were published—the *Defence White Paper 1991*, the *Defence Priorities Review* 1993, the *Defence White Paper 2000*, and the *Prinsjesdag Letter* 2003. The Netherlands' proactive defence policy exerted far-reaching impulses for Dutch air power and put the RNLAF relatively swiftly on an expeditionary footing. As such, the flexible and responsive Dutch defence policy turned the RNLAF into a relevant foreign policy tool.

Threat and Risk Perception. The Netherlands' defence posture has been closely related to its threat and risk perception, as military reforms followed the most recent security assessments. The *Defence White Paper 1991* clearly recognised that European security had markedly improved and emphasised crises and armed conflicts outside Europe. While the Soviet Union was still considered a major military power, a strategic surprise attack was ruled out as a contingency. Nevertheless, the white paper suggested that NATO ought to maintain an equal and balanced military capability to meet that of the Soviet Union.[4]

Partly as a reaction to the dissolution of the Soviet Union in late 1991, the *Defence Priorities Review* was published in 1993. It considered the armed forces of the states of the former Soviet Union no longer a major factor in the Netherlands' security equation.[5] Moreover, the *Defence Priorities Review* defined the new security environment in political, social, economic, ecological, and humanitarian terms instead of a merely military dimension. Countries in the developing world that were experiencing internal and external conflicts shifted into the focus of Dutch defence policy.[6]

More than half a decade later, the *Defence White Paper 2000* clearly recognised the unpredictability of future crises as the biggest problems that planners are confronted with: "where, when, with whom, under what circumstances and for what task the units of our armed forces are deployed in the future is uncertain."[7] To meet these future challenges, the *Defence White Paper 2000* particularly called for further improvements in combat readiness and a significant enhancement of the deployability of the Netherlands armed forces.[8]

The 2003 *Prinsjesdag Letter* partly reiterated the views of the 1993 *Defence Priorities Review* and the 2000 *Defence White Paper*, stating that the Netherlands need no longer be concerned about a large-scale attack with conventional weapons on alliance territory. The situation in the Balkans had stabilised, though it needed constant attention, and the expansion of NATO and the EU were fostering security in and around Europe.[9] As a consequence of the 11 September attacks, however, the *Prinsjesdag Letter* argued that the world as a whole had not become a safer place. The contemporary external security situation was linked to the future internal security of the Netherlands. Therefore, it was considered essential that Western values be protected and promoted together with the Netherlands' allies and partners.[10]

Tasks of the Armed Forces. According to this shift in the threat and risk perception from a potential major onslaught on NATO alliance territory to asymmetric threats, the missions of the Netherlands armed forces have changed. Whereas the *Defence White Paper 1991* still perceived the Soviet Union as a potential threat, it already laid the foundation for the future of expeditionary warfare. Hence, it stressed the importance of mobility, flexibility, rapid deployment, interoperability, and the ability to carry out operations outside the NATO treaty area.[11] Though the Netherlands govern-ment emphasised the importance of non-military means for con-flict prevention, it clearly stated that "if necessary . . . the Nether-lands is prepared to use military means to uphold the international rule of law, to withstand aggression and to safeguard fundamental economic interests [outside the NATO treaty area]."[12]

Two years later, the *Defence Priorities Review*, aside from the defence of national and allied territory, particularly emphasised out-of-area crisis management operations.[13] Seven years later, the *Defence White Paper 2000* reiterated these broad guidelines of Dutch defence policy and outlined the core tasks of the armed forces as follows:

- protecting the integrity of national and Allied territory, in-cluding the Netherlands Antilles and Aruba;

- advancing the international rule of law and stability; [and]

- assisting the civil authorities in the context of law enforce-ment, disaster relief, and humanitarian aid, both nationally and internationally.[14]

Published in 2003, the *Prinsjesdag Letter* was in many ways a response to the events starting with the terrorist attacks on 11 September. It reconfirmed the importance of expeditionary war-fare, as internal security can be threatened by conditions far away from the national territory.[15] It also concluded that the armed forces must be able to make a credible contribution to national and international peace and security, even if this might involve the risk of casualties.[16]

Towards Expeditionary Warfare and Professional Armed Forces. In the period from 1990 to 2003, the Netherlands armed forces transitioned from being primarily relegated to territorial defence to an expeditionary defence posture. While, at the outset of the post–Cold War era, a premium was put upon peacekeeping operations at the lower end of the spectrum of military force, a gradual shift towards more robust operations occurred.[17] The 2003 *Prinsjesdag Letter* states that operations based on chapter 7 of the UN Charter—which do not require the consent of the warring parties—had become the preferred mode of operating. It further underlines that such crisis management operations were conducted under the aegis of NATO, the EU, or an ad hoc coalition, with the UN's role limited to issuing the mandate.[18]

Because of these developments, a premium was put upon the ability to operate at the high end of the spectrum of force, even in traditional peacekeeping operations. Hence, it was deemed necessary to concentrate on technologically advanced military contributions. Only by so doing would the Netherlands remain interoperable with the most important allies.[19] In 2004 the Netherlands defence minister, Henk Kamp, reconfirmed the importance of expeditionary warfare: "Participation in non-Article 5 crisis response operations must now also be regarded as a contribution to the security of our territory and that of our citizens, in its broadest sense. The fact is we are going to the problems before they come to us."[20]

Courtesy Netherlands Institute for Military History; photo by Frank Visser

RNLAF AH-64D at Tarin Kowt in the province of Uruzgan in November 2008

Emphasis upon out-of-area operations across the entire spectrum of military force led to the abolition of the obligation to enlist for military service in early 1996.[21] A further major reform step occurred in 2005, when the commanders in chief of the individual services were phased out and superseded by a single joint commander, relegating the single services to "mere" force providers.[22] The reorganisation was completed in 2006. Besides the Army,

Navy, and military police, the Air Force became one of four operational commands. The Command of the Air Forces directly reports to the commander of the armed forces.[23] Given the great autonomy Dutch single-service commanders wielded throughout the Cold War, this represents a significant step towards more integrated armed forces.[24]

Defence Policy and Its Impact upon the Royal Netherlands Air Force. In 1991 the RNLAF continued the nuclear task of its F-16s since the potential military threat of the Soviet Union had not yet completely vanished.[25] Nowadays, only one F-16 squadron is assigned with a nuclear task.[26] In terms of conventional air power, great importance was attached to a good air defence capability of the F-16 fighter aircraft, as the Soviet Air Force continued to receive a new generation of fighter aircraft. The Netherlands government thereby reconfirmed its interest in an integrated air defence of Europe. Air defence as a purely national responsibility was deemed to lead to the disintegration of the Deep Multi-Layered NATO Integrated Air Defence System into small national pockets with a consequent reduction in warning and reaction times.[27] Both measures—the maintenance of the nuclear task of the F-16 and the emphasis on an advanced air defence capability—reflect the security assessment of the *Defence White Paper 1991*. Yet, there was also change due to the relaxation of Cold War tensions. As such, it was intended to reduce the number of operational combat aircraft available to NATO from 162 to 144. These remaining F-16s were planned to undergo a midlife update (MLU).[28] Also, a more decentralised and flatter C2 and organisational structure was assumed by the mid-1990s. One outcome is that squadron commanders became fully responsible for combat readiness of their units.[29]

The role of the F-16 fighter aircraft shifted considerably in the post–Cold War era. Whereas the *Defence White Paper 1991* put a premium upon an advanced air defence capability, the *Prinsjesdag Letter* stated in 2003 that the Netherlands military capabilities were foremost geared up to influence the situation on the ground.[30] To make a relevant contribution to deployed operations, an efficient air-to-ground capacity was considered essential by Dutch decision makers.[31] This development clearly reflects the Netherlands' shift from territorial alliance defence to out-of-area operations.

While the *Defence White Paper 1991* announced a reduction in the RNLAF's F-16 fleet, a need for improved mobility and flexibility called for an enhanced air transport capability against the backdrop of a more active UN peacekeeping role outside the NATO treaty area. Particularly through the acquisition of tanker/ transport aircraft, it was intended to establish a strategic airlift capacity.[32] Regarding intratheatre mobility, the *Defence White Paper 1991* announced the establishment of an airmobile brigade, integrating helicopters and light infantry.[33] The helicopter force was subsequently put under the command of the Air Force and underwent far-reaching developments from light reconnaissance helicopters such as the Alouette III to some of the most advanced combat and transport helicopters forming the Tactical Helicopter Group (THG). The THG was fully incorporated into Air Force structures, with the Royal Netherlands Army determining the operational requirements.[34]

Table 4 shows the development of the RNLAF's aircraft inventory from 1991 to 2005. To meet the political demands of enhanced mobility and combat readiness, the combat aircraft fleet

Courtesy Netherlands Institute for Military History; photo by H. Keeris

An RNLAF CH-47 Chinook on Exercise Snow Falcon in Norway. The THG, combining transport and Apache combat helicopters, and the Royal Netherlands Army's 11th Air Mobile Brigade are integrated into the "blue-green" Air Manoeuvre Brigade.

was reduced, upgraded, and standardised. The F-5A, procured in the second half of the 1960s, was phased out at the beginning of the post–Cold War era.[35] Reduction and standardisation also applied to the RNLAF's GBAD systems. The GBAD inventory was solely concentrated upon the Patriot system, as it provides a limited TBMD capability and fully meets the requirements of deployed operations. Yet, the necessary swift buildup in intra- and intertheatre mobility led to a heterogeneous force structure.

Table 4. RNLAF aircraft/GBAD inventory and personnel

1991	1998	2005
Combat Aircraft		
181 F-16A/Bs (plus 23 in store)	170 F-16A/Bs (plus 11 in store)	108 F-16A/B MLUs (plus 29 in store) (all aircraft converted under the European MLU programme)
Transport Aircraft		
14 Fokker F-27 tactical transport aircraft	2 C-130 Hercules, 6 Fokker F-50/F-60 tactical transport aircraft	2 C-130 Hercules, 6 Fokker F-50/F-60 tactical transport aircraft
AAR Aircraft		
None	2 KDC-10s	2 KDC-10s
Helicopters		
4 SAR helicopters	81 helicopters (including 12 AH-64A combat helicopters, 13 CH-47D Chinooks, 17 Cougars)	72 helicopters (including 30 AH-64D combat helicopters, 13 CH-47D Chinooks, 17 Cougars)
GBAD		
4 Patriot fire units, 16 Hawk fire units, 75 radar-controlled guns and 25 radars, 100 Stingers	4 Patriot fire units, 16 Hawk fire units, 75 radar-controlled guns and 25 radars, 100 Stingers	4 Patriot fire units (being upgraded to PAC-3), some Stingers
Personnel		
12,500 (plus 3,500 conscripts and 11,200 reservists for immediate recall)	11,980 (plus 10,000 reservists for immediate recall)	11,050 (plus 5,000 reservists for immediate recall)

Adapted from International Institute for Strategic Studies (IISS), *The Military Balance 1991–92* (London: Brassey's for the IISS, 1991), 66–67; IISS, *The Military Balance* 1998–99 (Oxford, England: Oxford University Press for the IISS, October 1998), 60–2; and IISS, *The Military Balance 2005–06* (Oxford, England: Oxford University Press for the IISS, October 2005), 83. Data on Patriot and Hawk GBAD fire units was provided by Col Peter Wijninga, RNLAF, commander, De Peel AFB, to the author, e-mail, 30 July 2010 (forwarded by 1Lt Ton Steers, RNLAF).

Major developments after 2003 were outlined in the *Prinsjesdag Letter*. It announced that the F-16 arsenal would be further reduced from 137 to 108, aimed at lowering operating costs and improving deployability of the remaining aircraft. Along with this reduction, one of three air bases was to be closed by 2007.[36] In contrast, air transport capacities were boosted by two additional C-130s and one DC-10.[37] Moreover, the increasing demands in intertheatre airlift capacities led to cooperative multi- and bilateral arrangements as well as shared ownership. These developments are examined further below.

Developments in the post–Cold War era also impacted the personnel structure (see table 4). However, the abolition of conscription was not a revolutionary step for the RNLAF since it always consisted mainly of professionally employed personnel.

In the 1990s, the RNLAF developed from a static air force geared up for the defence of Central Europe into an expeditionary air force deployable across the entire spectrum of military operations. While certain specialised capabilities such as dedicated SEAD assets or CSAR have not yet been acquired, the RNLAF has been transformed into a more balanced air force. Balance was achieved through a shift in resource allocation from the combat aircraft fleet and GBAD units to the buildup of the THG as well as to the acquisition of C-130 Hercules and modified DC-10 aircraft. In this transformational process, role specialisation was to be avoided. Accordingly, the *Prinsjesdag Letter* of 2003 stated, "The Netherlands armed forces concentrate on a limited number of advanced capabilities, without following a path of excessive role specialisation and without unduly limiting the choice of deployment options."[38]

Following the announcements of further budget cuts in July 2007, the RNLAF further reduced its F-16 fleet to 87 aircraft, including 72 assigned to five operational squadrons. Despite this reduction, the number of operational squadrons was kept constant. Accordingly, squadron strength was reduced from 18 to 14 or 15 aircraft.[39]

Another important step was the establishment of the Defence Helicopter Command on 4 July 2008. This new joint command is in charge of all military helicopter operations and reports to the

Command of the Air Forces and, as such, absorbed the Royal Netherlands Navy's Maritime Helicopter Group.[40]

Emphasis upon Air Power. In 2003 the Netherlands spent 1.6 per cent of its gross domestic product (GDP) on defence, below the European NATO average of two per cent.[41] Also, from a long-term perspective, the Netherlands has never allotted a relatively high percentage of its GDP to defence.[42] In the chart below, the single-service budget allocations throughout the 1990s are presented. (Tabular comparisons of the national defence expenditures are provided in chap. 7.)

The Army received by far the largest part of the annual defence budget, though a slight trend towards increased spending on the sister services became visible. This imbalanced funding has its roots in the Cold War. In an effort to stifle interservice feuds, the Netherlands introduced a division of the defence budget accord-

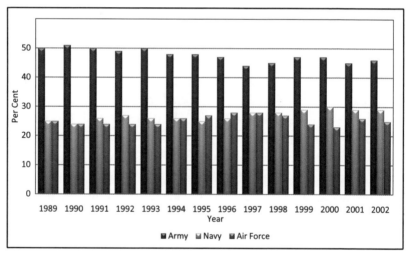

Figure 4. Budget allocation to the single services (Netherlands). (*Adapted from* J. Franssen, chairman, *Van wankel evenwicht naar versterkte defensieorganisatie: advies van de Adviescommissie Opperbevelhebberschap* [*From a precarious balance to a reinforced defence system: advice from the advisory committee supreme command*] [The Hague: Adviescommissie Opperbevelhebberschap [Advisory Committee Supreme Command], 19 April 2002], background information, part B, *De verdeling van de budgetten die in de afgelopen tien jaren aan de krijgsmachtdelen zijn toebedeeld* [*Budget allocation to the single services in the last ten years*].)

ing to the 1:2:1 formula, allocating half of the budget to the Army and approximately a quarter each to the other services.[43]

According to the broad approach to air power which this book takes, the air power assets of the Navy and Army have to be included as well. In the 1990s, the Army operated between 60 and 95 Gepard self-propelled, radar guided anti-aircraft guns, between 60 and 131 towed anti-aircraft guns, and approximately 300 Stinger man-portable air defence systems. Furthermore, the Navy operated 22 Lynx helicopters and 13 P-3C Orion anti-submarine warfare aircraft.[44] While the P-3Cs were sold to Germany and Portugal, the 21 remaining Lynx airframes, including four kept in non-flyable condition, were brought under the auspices of the Command of the Air Forces in 2008.[45] In addition to these aircraft, the Navy operated between 10 and 14 frigates, all equipped with air defence systems.[46] If these systems are included, air power expenditure is greater than the funding of the single services might suggest. However, it is still considerably lower as a proportion than that of the United States. This state of affairs prompted an RNLAF lieutenant colonel to maintain, "I see no recognition of the fact that . . . air power is at the core, at the heart of the European military capability gap."[47]

Despite the relatively low budget, the RNLAF became from the early 1990s up to 2002 the weapon of choice in operations at the upper end of the spectrum of military force. According to a Dutch commentator, the RNLAF gained quite a reputation among politicians in the wake of the air operations over the Balkans. Contrary to the Army's troublesome experience in Bosnia, Dutch air power was perceived to add to the Netherlands' prestige abroad. As a result, from 1999 to 2002, the RNLAF was indeed the preferred instrument—at least in the eyes of major factions in The Hague and in the media. An example is the Parliament's insistence that the Dutch land contribution to the operation United Nations Mission in Ethiopia and Eritrea (UNMEE) in 2000–2001 was to be accompanied by Apache combat helicopters. Yet, according to an interviewee, the Dutch perspective on air power has become less enthusiastic.[48] This might be because stabilisation operations in Iraq and Afghanistan have required relatively large ground force contingents. Against this backdrop, the Royal Netherlands Army has also gained respect for its international commitments. During

232 ROYAL NETHERLANDS AIR FORCE

2004 no less than 40 per cent of the Army's operational personnel were deployed overseas, well above the NATO average. In the future, the Dutch Parliament might also deploy the Royal Netherlands Army for high-intensity warfare. From January to June 2005, 4,000 Dutch troops—mostly combat—went on standby for NATO Response Force 4.[49]

Alliance Context

In 1985 the Dutch scholar Jan Siccama observed that "the failure of neutrality in 1940 enhanced the tendency towards alignment after World War II in the same way as the success of neutrality in 1914 had fostered the continuation of that policy during the interbellum period."[50] Since the signature of the North Atlantic Treaty on 4 April 1949, Dutch security has been viewed exclusively in transatlantic terms. The Netherlands' colonial legacy and once more prominent position in international affairs have created a self-image of an influential, albeit small, nation. Cooperation has been regarded as the main lever for gaining influence on the international stage.[51]

Since a premium was put upon a firm American commitment to European security, attempts at establishing an independent European defence structure outside NATO in the early 1990s were feared to undermine cohesion within the NATO alliance. Yet, a shift from an outspoken transatlantic orientation to a more European stance was discernable in the mid-1990s. While the first Bush administration discouraged attempts at European defence cooperation, President Clinton supported the development of the European Security and Defence Identity and encouraged European cooperation through initiatives such as the CJTF concept within NATO. The Clinton administration argued that a more integrated European defence policy would serve peace and security better than fragmented national defence initiatives. This shift in American behaviour prompted Dutch defence policy to move in a more European direction. The Dutch government has been vigorously pursuing the goal of integrating its armed forces into supranational defence structures. Accordingly, the Royal Netherlands Marine Corps has been closely integrated with units of the British Royal Marine Corps since the days of the Cold War, and in August 1995, the Netherlands integrated its Army into a Dutch-German

Corps. As for naval forces, the Dutch and Belgian navies established a combined operational staff in 1995.[52]

A Dutch scholar concluded in the mid-1990s that shifts in Dutch defence policy were the consequence of not only shrinking resources for the armed forces but also international circumstances. Dutch defence planners realised that American commitment to NATO and Europe would paradoxically be put at risk if European NATO allies refused to integrate their efforts and to provide a more robust contribution to the transatlantic defence burden.[53] Against the backdrop of transatlantic and European tensions over the Iraq crisis in 2003, the *Prinsjesdag Letter* particularly underlined that NATO continued to be the most important pillar of Dutch defence policy, as it epitomised the alliance with the United States. Hence, the ESDP was regarded as complementing NATO.[54]

Since larger unilateral military expeditions are inconceivable for the Netherlands, Dutch military units are to operate closely with allied armed forces under the auspices of international organisations or ad hoc coalitions. Already in the *Defence White Paper 1991*, it was stated that the Netherlands would only act in conjunction with other countries and that the contribution of the Netherlands would therefore always be of a complementary kind.[55] In alliance with others, preferably the United States, Dutch politicians and militaries have been willing and capable of exercising violence and of using air power as a coercive tool.[56]

Due to the Netherlands' strong commitment to NATO and European integration, a premium is put upon interoperability. The *Defence White Paper 2000* emphasises the ability of the armed forces to participate in an international context. As such, the Netherlands armed forces are considered as a system of modules, each of which can operate in groups led by NATO, the UN, the EU, or an ad hoc coalition.[57] The capability of easily fitting into multinational forces is considered to provide the Netherlands armed forces with the required flexibility for multinational missions. The Netherlands government is firm in its belief that without far-reaching international cooperation, smaller countries would be unable to maintain relevant and affordable armed forces over the long term.[58] Hence, the pooling of national resources with other countries has been seen as an attractive way of dealing with the soaring costs of defence expenditure. On top of that, bilateral and multilateral cooperation have been believed to foster mutual trust and interoperability.

Alliance Policy and Its Impact upon the Royal Netherlands Air Force

Encouraged by NATO initiatives such as the CJTF concept of 1994, the RNLAF has been at the forefront of European cooperative ventures, which it has considered as important force multipliers against the backdrop of constrained national defence resources.

ESDP and NATO Context. In the 2003 *Prinsjesdag Letter*, Dutch decision makers expressed their intention to reduce the Netherlands' contribution to out-of-area operations in the framework of NATO or the ESDP from three to two squadrons, each with 18 fighter aircraft.[59] This reduction was probably based upon a more realistic assumption of the force level that can actually be sustained in deployed operations over a longer period of time. In terms of combat aircraft, this represented almost half of the number potentially put at the disposal of the EU by Germany or France, with the GAF and the FAF originally intending to contribute up to 80 combat aircraft. Hence, the RNLAF has been striving to make a contribution which is out of proportion to the country's size. Moreover, the Netherlands has subscribed to the importance of the NRF by allocating 12 F-16s to the response force on a structural basis, as well as a significant fraction of its other main weapon systems.[60]

Bilateral and Multilateral Organisational Relationships. The RNLAF was the driving force behind the creation of the European Participating Air Forces' Expeditionary Air Wing (EPAF EAW), a multinational European F-16 wing. The EPAF EAW is rooted in the mid-1990s, when air operations over the former Yugoslavia were in full swing. The RNLAF and the Belgian Air Force signed an MOU in October 1994. It was intended that the Belgians would take on about one-third of the RNLAF's efforts in Operation Deny Flight. However, the first Belgian F-16 detachment was deployed to Italy only in October 1996—well after Operations Deny Flight and Deliberate Force.[61] The Belgians teamed up with their Dutch counterparts in the framework of the newly established Deployable Air Task Force (DATF). The DATF enabled the two air forces to work closely together in Operation Allied Force in 1999, reducing redundancies in areas such as logistics and operational planning.[62] For instance, the DATF incorporated a combined planning cell for the preparation of contingency plans.[63] Moreover, the Luxembourg Army provided a tailored ground security force for deployed operations.[64] The concept was particularly furthered because both air forces were operating F-16 fighter-bombers and because it was fully in line with NATO's combined joint task force concept, endorsed by the alliance in January 1994.

Prior to the establishment of the DATF, the RNLAF together with the Belgian, Danish, and the Norwegian Air Forces set up the European Participating Air Forces (EPAF). EPAF members were all F-16 customers, and the concept was originally conceived for procurement purposes to pool national requirements and to gain bargaining leverage (an issue that is discussed in the section on procurement). The EPAF particularly laid out the common requirement for the MLU of the European F-16 Block 15 A/B.[65]

With a Dutch, Danish, and Norwegian EPAF deployment to Central Asia in 2002, DATF concepts were transferred into an EPAF context. Together with the Danish and Norwegian Air Forces, the RNLAF participated in the air operations over Afghanistan.

Courtesy Netherlands Institute for Military History; photo by H. Keeris

A KDC-10 refuelling an F-16 in midair over Central Asia in the course of Operation Enduring Freedom

The combined EPAF deployment proved to be quite effective despite some legal and procedural obstacles. To further improve cooperation between the European F-16 users, Gen D. L. Berlijn, the RNLAF commander in chief, took the initiative to approach his Belgian, Danish, Norwegian, and Polish counterparts to ask for their views upon a possible expansion of the DATF concept, finally resulting in the EPAF EAW. The participating air forces intend to make optimum use of available and complementary assets in out-of-area operations to increase efficiency. For example, during the operations over Afghanistan, the Netherlands and Denmark provided targeting pods free of charge for common use, while Norway provided a hangar and a deployable communication module.[66]

The EPAF EAW is inherently flexible, as it allows for deployments involving two or more air forces, depending on the particular circumstances. Through this approach, national sovereignty is respected, with each participating nation defining its level of commitment.[67] Moreover, it was also intended to make the EPAF EAW available to the NRF.[68] The MOU for the EPAF EAW was finally signed by the defence ministers of Belgium, Denmark, the Netherlands, Norway, and Portugal during the NATO summit in Istanbul on 28 June 2004.[69] The essential benefit of the EPAF EAW concept is synergy. Through their combined commitment, EPAF nations as a group can deploy more robust and sustainable force packages than autonomous national efforts would allow.

While the EPAF EAW is a form of enhanced cooperation and pooling, it does not extend to role specialisation, which is seen to curtail deployment options. This thrust is completely in line with Dutch decision makers' ideas about military cooperative arrangements. While the government views cooperation as a prerequisite for smaller countries to retain relevant armed forces, it considers a certain degree of autonomy essential to remain functional.[70]

Analogous to the DATF model, the Netherlands proposed combining American, German, and Dutch Patriot guided missile units into a deployable combined unit for TBMD in deployed NATO operations.[71] In 1999 the trinational Extended Air Defence Task Force was established in Germany. Its tasks are to enhance interoperability by planning training and exercises and to provide liaison for US, Dutch, and German Patriot units.[72] Thus far, Dutch ambitions regarding the establishment of multinational units for deployed operations turned out to be too ambitious.

In addition to these cooperative ventures, the RNLAF became a full member of the EAG on 12 July 1999 to further shape European standardisation and interoperability.[73] Moreover, as the host nation of the EAC, the Netherlands signed in May 2007 the agreement to establish the EATC in Eindhoven, Netherlands.[74]

As regards pilot training, the RNLAF has been pursuing a transatlantic path. After elementary pilot training in the Netherlands, both fighter and helicopter pilots are trained in the United States. Like their German counterparts, future fighter pilots undergo basic jet training at Sheppard AFB, Texas, in a multinational environment. Subsequently, initial F-16 training takes place within the Netherlands Detachment Springfield Ohio (NDSO). The NDSO was established in April 2007, after 306 Squadron, the RNLAF's operational conversion unit, relocated to Springfield Air National Guard (ANG) Base to provide initial qualification training. The squadron is part of the Ohio ANG's 178th Fighter Wing and operates six dual-seat and eight single-seat F-16s. Prior to the establishment of the NDSO, initial qualification training was provided by the Arizona ANG's 162d Fighter Wing at Tucson, using Tucson-based Dutch F-16s from 1990 to 1995 and the wing's own F-16s thereafter. Mission qualification training takes place in the Netherlands in one of the operational squadrons.[75]

Dutch helicopter pilots attend the initial-entry rotary wing course at Fort Rucker, Alabama. While future Apache and Chinook pilots proceed to type-specific training in the United States, Cougar pilots learn to fly the UH-60 Black Hawk before converting to their destined type in the Netherlands and in simulators abroad. Initial mission qualification training of Dutch Apache pilots is provided by the Netherlands Apache Training Detachment (NATD) at Fort Hood, Texas. The detachment also conducts annual tactical training courses for operational Apache pilots. Moreover, Dutch Apache training includes a mountain flying course in Italy and weapons qualification in Germany. Analogous to the combat helicopter syllabus, the RNLAF considered, as of 2009, establishing a permanent Chinook detachment for advanced training in the United States.[76]

Cooperation on an Operational Level. Against the backdrop of Operation Allied Force, General Horner highlighted that the Dutch had been easy to integrate into an American-led multinational force with regard to equipment as well as training.[77] Operational ties with the USAF have been strong in the post–Cold War era. For instance, American exchange pilots were embedded into RNLAF F-16 detachments during Enduring Freedom. In the course of Operation Iraqi Freedom, RNLAF Patriot units stationed in Turkey were in a position to receive American intelligence on potential missile launches.[78] Hence, a high degree of mutual trust and understanding has indeed existed between the USAF and the RNLAF. Moreover, the RNLAF has only participated in air campaigns with the United States acting as lead nation, which is expected to remain the case.[79]

Due to its embedding into NATO, the RNLAF has in general acquired great experience in operating with allied air forces. It has been participating in the most demanding integrated exercises, such as Red Flag in the United States and Maple Flag in Canada. Furthermore, 323 TACTESS (Tactical Training, Evaluation, and Standardisation Squadron), the Dutch "Top Gun" squadron, organises its own annual integrated exercise. Up to eight nations regularly participate in exercise Frisian Flag over the Netherlands and the North Sea.[80] Analogous to its flying counterparts, the RNLAF Missile Group has been organising TBMD defence exercises in the Joint Project Optic Windmill series since 1996.[81] Throughout the 1990s, the RNLAF also conducted low-level flying exercises in Goose Bay, Canada. Yet, in the aftermath of Allied Force, low-level flying diminished in importance, and it was decided to terminate the low-flying training programme in Canada.[82] The definite field test for combined operations was the succession of air campaigns over the former Yugoslavia.[83]

How Has the Royal Netherlands Air Force Responded to the Challenges of Real Operations?

In the post–Cold War era, the Netherlands has been using its armed forces as an instrument for intervention abroad. Nowadays, the Netherlands armed forces are almost exclusively geared towards expeditionary warfare.[84] Dutch politicians have been willing to employ air power across the entire spectrum of force to fulfil the goals set by their defence policy. In a speech in early 2004, the Dutch defence minister underlined that allied solidarity must not only be apparent from a country's military capabilities but also from its willingness to share risks. Hence, the ability to engage at the higher end of the military spectrum of force became an important pillar of Dutch foreign and defence policy. He particularly argued that if politicians are to contribute to stability and security, they must have the courage to shoulder responsibility for operations entailing fatal casualties.[85]

In line with the Netherlands' defence policy, all of the RNLAF's main weapon systems have been engaged in deployed operations since the end of the Cold War. From 1993 to 2001, Dutch F-16s were stationed in Italy as part of the air operations over the Balkans. From late 2002 up to 2010, they were contributing to operations in Afghanistan. RNLAF AH-64 helicopters have also been employed across a broad spectrum of force in out-of-area operations. Not only the RNLAF's airborne assets but also its guided missile units have proven to be important in the post–Cold War environment. Dutch Patriot units were deployed in the context of Desert Storm in 1991 and Iraqi Freedom in 2003. On top of these combat contributions, the RNLAF has played a role in humanitarian relief operations and troop deployments with its airlift fleet.

Dutch Contributions to Western Air Campaigns

As early as 1991, the Dutch government was willing to commit F-16 fighter aircraft to combat missions. During the 1990–91 Gulf crisis, the Netherlands was actively seeking participation in the coalition against Saddam Hussein. Due to the concentration of extremely large numbers of aircraft in the Gulf region, though, all hosting options for Dutch F-16s turned out to be exhausted.[86]

The Netherlands' immediate reaction to Desert Storm was to draw lessons from the near deployment to the Gulf area. On a political level, it sought to speed up the political decision-making process for any future crisis management operations. On an operational level, squadrons were geared up for rapid deployment.[87] In the remainder of the 1990s and beyond, Dutch decision makers indeed proved capable of swiftly committing the armed forces to operations across the spectrum of military force, and the RNLAF was able to respond effectively to the challenges of real operations. Overall, Desert Storm was not so much about a paradigm change, as in the case of the FAF, but a matter of recalibrating already existing Dutch defence capabilities. It is noteworthy that in doctrine development and the acquisition of an air-to-ground precision strike capability, Operations Deny Flight and Deliberate Force proved to be more critical for the RNLAF than Desert Storm (as is analysed later). Hence, there appears to be a difference between just observing an operation and fully participating in it.

While the RNLAF could not immediately contribute to Desert Storm, it did so indirectly. After Iraq invaded Kuwait in August 1990, NATO decided to send Patriot units to Turkey for protection against Iraqi SCUD missiles. Two Dutch Patriot missile batteries were sent to Diyarbakir in eastern Turkey. This deployment was soon to be reinforced with two more Hawk missile squadrons and Stinger units. The RNLAF also deployed a Patriot missile battery to Israel.[88] Whereas Dutch fighter aircraft were not able to participate in Operation Desert Storm, the rapid deployment of Patriot and Hawk missile units was a clear sign of the Netherlands government's willingness to actively contribute to international crisis management. Moreover, Patriot deployments in the TBMD role fitted with NATO's new extended air defence and rapid deployment concepts. With the Cold War over, the RNLAF's guided missile units were withdrawn from the Deep Multi-Layered NATO Integrated Air Defence System in Germany and freed for out-of-area operations.[89]

When Operation Deny Flight was launched on 12 April 1993, Dutch F-16s and US Navy F/A-18s flew the first CAP missions over Bosnia on the same day. A mere 10 days later, RNLAF F-16s were among the first to fly night CAP missions. The initial Dutch detachment consisted of 18 F-16s, roughly 10 per cent of the overall allied commitment to Operation Deny Flight. After the UN had authorised air-to-ground strikes including CAS in mid-1993, Dutch F-16s were prepared for the execution of these missions. The RNLAF's fighter-bombers were basically configured for three roles—air defence, air-to-ground strikes, and tactical reconnaissance.[90] The first air strikes against ground targets by Dutch F-16 aircraft were conducted on 21 November 1994. The strikes were directed at the airfield at Udbina in the Kraijna region. In the planning process, RNLAF personnel played a key role, and a Dutch major acted as the overall tactical mission commander during the attack itself.[91]

In the Srebrenica massacre in July 1995, where lightly armed Dutch peacekeepers played a controversial role, the inflexible chain of command—wherein NATO and UN consent were required—delayed air strikes. By the time Dutch F-16s were authorised to attack ground targets, Srebrenica had already been overrun by Bosnian Serb troops.[92]

During Operation Deliberate Force, RNLAF F-16s were again used in a wide spectrum of missions covering air defence and air-to-ground strikes, as well as reconnaissance missions.[93] Out of a total of 1,026 weapons released, the RNLAF accounted for approximately 13 per cent, with Dutch F-16s exclusively dropping unguided bombs.[94] In an interview with the author, a high-ranking RNLAF officer explained that Dutch F-16s had been able to place unguided bombs accurately by diving below the minimum flight altitude. Yet putting a premium upon avoiding unintended harm and destruction, the American air component commander assigned targets involving less risk of collateral damage to NATO air forces without precision weapons at their disposal.[95]

The RNLAF displayed further distinct national approaches to air power in the air campaigns over Bosnia. Towards the end of the Cold War, the RNLAF introduced the so-called swing-role concept. According to this concept, aircraft and pilot can execute multiple tasks on the same sortie.[96] While other European F-16 customers were employing the aircraft in one fixed role only, the RNLAF had been exploring the swing-role concept since the mid-1980s.[97] All Dutch F-16 pilots have been trained in both the air-to-air and air-to-ground roles.[98] Because Dutch F-16s could be re-tasked while executing a mission, the CAOC at Vicenza, Italy, had some extra leeway.[99] During Deliberate Force, on several occasions RNLAF F-16s were flying air defence missions with bombs attached to their underwing pylons and, minutes later, were re-rolled for an air-to-ground mission.[100]

Dutch aircrews put a premium upon flexibility not only for the swing-role concept but also for force packaging. When it was announced in November 1994 that all missions had to be escorted by SEAD aircraft, some allies, including Dutch pilots, expressed their doubts about this far-from-flexible way of operating.[101] Instead of overreliance on SEAD and standoff jamming assets, Dutch pilots preferred to make their flight patterns as unpredictable as possible.[102]

Though the air operations over Bosnia represented the first real operations for Dutch fighter pilots, some of the most experienced aircrews were appointed leaders of large multinational formations scheduled to perform difficult missions. According to Lt Gen H. J. W. Manderfeld, RNLAF commander in chief from 1992 to1995, factors in this outstanding performance were sound training and education.[103]

Courtesy Netherlands Institute for Military History; photo by R. Frigge

A Dutch F-16 with two AMRAAMs and two LGBs leaves the busy flight line at Amendola in southern Italy during Operation Allied Force

For Operation Allied Force, a total of 20 Dutch F-16s and two KDC-10 tanker aircraft were made available to NATO. Transport aircraft were flying almost round the clock carrying the right type of munitions on schedule to Amendola Air Base in southern Italy, from where the Dutch F-16s were operating. Through these efforts, a high operational tempo could be achieved and an average degree of readiness of over 95 per cent could be sustained. Throughout Allied Force, Dutch F-16s flew approximately 700 air-to-air, 450 air-to-ground, and 50 reconnaissance and battle damage assessment sorties over the Federal Republic of Yugoslavia. The RNLAF delivered more than 850 air-to-ground weapons, including 246 LGBs and 32 Maverick missiles. Given these figures, the RNLAF played a substantial role in Allied Force and provided, according to Dutch sources, approximately 7.5 per cent of the offensive NATO sorties.[104] To put the number of expended PGMs into context, the RNLAF accounted for approximately 3.5 per cent of the total number of PGMs employed during the campaign and released slightly more than the RAF, which employed 244 LGBs and six ALARMs.[105]

On top of this, a Dutch F-16 downed one of three Serb MiG-29s destroyed by the alliance during the first night of the operation. Indicating the sensitivity of this incident, the Dutch chief of defence, Adm Lukas Kroon, expressed concerns about too much publicity to Gen Wesley Clark.[106] Moreover, the option of ground forces was never considered by the Netherlands government.[107]

Three years later during the first half of 2002, the RNLAF supported Operation Enduring Freedom by dispatching a KDC-10 tanker aircraft to Qatar's Al Udeid airfield, from where it carried out missions in close cooperation with US tanker aircraft for nearly three months.[108] RNLAF participation in operations over Afghanistan became more robust in autumn 2002, when a combined F-16 detachment—consisting of 18 F-16s from the Netherlands, Denmark, and Norway—supported by an RNLAF KDC-10 tanker aircraft deployed to Manas International Airport in Kyrgyzstan, its base for operations over Afghanistan.[109] The Dutch aircraft alone logged 804 sorties and 4,640 flying hours, regularly providing CAS to ground troops. Approximately one year later, the European F-16 detachment returned home.[110]

In 2005 an EPAF EAW F-16 detachment was directly deployed to Kabul International Airport for NATO operation ISAF. In the second half of 2005, Belgium and the Netherlands provided four aircraft each; in February 2006, Norwegian F-16s replaced the Belgian contingent. The multinational detachment regularly carried out reconnaissance and CAS missions.[111] In November 2006, Dutch F-16s relocated to Kandahar Airfield against the backdrop of NATO's expanded ISAF mission in southern Afghanistan.[112]

As in the case of Desert Storm, the RNLAF did not directly participate in the high-intensity phase of Operation Iraqi Freedom, but indirectly by deploying guided missile units to Turkey in early 2003. In February the Dutch government received an official Turkish request for three Patriot units and additional Stinger teams to be stationed in eastern Turkey.[113] Operation Display Deterrence was primarily designed to protect NATO air bases. In less than a month from the receipt of the Turkish request, these Patriot units were operational in Turkey.[114]

A Dutch Patriot unit at the Turkish air base Diyarbakir in 2003 during Operation Iraqi Freedom. This was the second deployment to the Middle East since the fall of the Berlin Wall. In 1991 during Operation Desert Storm, RNLAF Patriots were dispatched to Turkey and Israel.

Against the backdrop of the Netherlands' defence spending—below the NATO average—real operations throughout the post–Cold War era have offered a new opportunity to show alliance solidarity. Air power has been perceived by Dutch politicians as allowing participation in high-intensity conflicts without the prospect of suffering heavy casualties. Furthermore, the RNLAF's excellent performance in conflicts at the upper end of the spectrum of military force restored the image of the Netherlands armed forces in the wake of Srebrenica.[115] Yet this excellent performance—particularly during the course of Allied Force—was deliberately played down internally to avoid too militaristic of an impression, potentially irritating the Dutch constituency.[116] The issue of having relatively fragile political support for military operations came to the fore in February 2010, when the main coalition partner of the ruling party did not support a further extension of the Dutch military presence in Afghanistan. The Netherlands armed forces began to leave Afghanistan in August 2010.[117]

AH-64D Apache Combat Helicopter Deployments

The massacre of Srebrenica in 1995, where Dutch peacekeepers could not prevent genocide among the Muslim population, taught a crucial lesson to Dutch decision makers. In 2000 a Netherlands Marine unit together with four heavy Chinook transport helicopters was deployed to the Horn of Africa to support a UN mission in Ethiopia and Eritrea. This time, the Dutch parliament requested that a detachment of four Apaches be stationed in nearby Djibouti, providing the Marine unit with escalation dominance and guaranteeing a secure retreat if necessary.[118] According to Dutch sources, the deployment to Djibouti was the first combat deployment of the AH-64D version worldwide, and the RNLAF developed new concepts for the employment of the AH-64D under desert conditions.[119]

Soon afterwards, the RNLAF Apaches became a key asset for alliance operations. On 30 January 2004, the Netherlands govern-

Courtesy Netherlands Institute for Military History; photo by 302 Squadron

RNLAF AH-64D Apache combat helicopters deployed to Djibouti in early 2001.

ment agreed to a NATO request and put six Apache combat helicopters at the disposal of ISAF in Afghanistan.[120] Reportedly, these
helicopters measurably increased the ISAF's self-protection and
reconnaissance capabilities.[121] Moreover, on 21 May 2004, it was
announced that the Netherlands would send six Apache combat
helicopters to Iraq to provide the Dutch ground contingent with
enhanced situational awareness and firepower.[122] While Dutch
forces pulled out of Iraq in the spring of 2005, Dutch Apaches returned to Afghanistan in April 2006, supporting ISAF operations.[123]

Humanitarian Operations and Troop Deployments

The RNLAF has also played a role in humanitarian relief and
crisis management operations throughout the 1990s and beyond.
Though very modest at the outset of the post–Cold War era,
RNLAF airlift capacities have grown over the years from small
transport planes such as the Fokker F-27s to larger aircraft such as
the C-130 Hercules and the KDC-10 strategic airlift and air-to-air
refuelling aircraft.

After the collapse of the Berlin Wall and in the wake of Desert
Storm, humanitarian relief missions targeted Eastern Europe and
the Middle East. With the advent of the first C-130 Hercules, the
RNLAF was able to shift its attention to hot spots in Africa and
elsewhere. The genocide in Rwanda triggered a considerable number of Dutch C-130 airlift sorties.[124] Combining combat support
and humanitarian missions, the RNLAF's transport aircraft fleet
has excelled in flexibility. During Allied Force, for instance, an
RNLAF KDC-10 conducted an air-to-air refuelling mission over
the Adriatic and subsequently flew to Macedonia, evacuating
refugees to the Netherlands.[125] The RNLAF has also conducted
relief missions in the wake of natural disasters. When the city of
Bam in Iran was shattered by a heavy earthquake in late 2003, a
KDC-10 transport aircraft delivered 45 tonnes of emergency relief
supplies on behalf of the Netherlands Red Cross.[126]

Supporting operations in Afghanistan, the RNLAF maintained
an air bridge to Kabul from January to February 2002.[127] Additionally, it airlifted German troops to Afghanistan.[128] Not only fixed
wing but also rotary wing assets have contributed to stabilisation
operations in the Balkans, the Middle East, and Afghanistan.[129]

Courtesy Netherlands Institute for Military History; photo by H. Keeris

A Dutch C-130 Hercules over the impressive landscape of Central Asia

Air Policing

In the context of NATO's integrated air defence, the RNLAF permanently maintains a two-ship of F-16 aircraft on QRA. After 11 September, aircraft on high readiness were increased from two to six for approximately one month. In the aftermath, Dutch QRA aircraft became—alongside NATO duties—also available for national air policing tasks.[130]

After the Baltic states had become members of NATO in 2004, four Dutch F-16s were deployed to Lithuania in 2005 to conduct QRA and air policing missions over the Baltic. Since Lithuania, Estonia, and Latvia do not have adequate aircraft for this type of mission at their disposal, other NATO nations alternately provide a detachment of four fighter aircraft for approximately three months.[131] As of 2009, this alliance mission was expected to last until around 2018.[132]

How Has the Royal Netherlands Air Force Responded to the New Intellectualism in Air Power Thinking and Doctrine?

The RNLAF's close relationship with the USAF and its early involvement in real operations facilitated sound approaches to deal with the intellectual challenges of the post–Cold War era.

Air Power Doctrine

In the latter half of the Cold War, experiences acquired during integrated NATO exercises as well as NATO doctrine laid the foundation for the RNLAF's approach to air power. As thinking was dominated by NATO doctrine, RNLAF officers rarely took the initiative to promulgate a national air power doctrine.[133]

Only in the first half of the 1990s was a doctrine advisory group (DAG) set up with the aim of producing an indigenous air power doctrine. The DAG was a non-permanent committee consisting of experienced officers with fighter jet, GBAD, logistics, and administrative backgrounds. Furthermore, it included two defence college graduates as well as ad hoc members.[134] In the environment of operations in the Balkans and elsewhere, a conceptual grasp of air power doctrine was deemed essential to translate limited political goals into military actions.[135] In his foreword to the first formalized RNLAF air power doctrine in 1996, the commander in chief, Lt Gen B. A. C. Droste, established an explicit link between the employment of the RNLAF in deployed operations during the early 1990s and the need for adequate air power doctrine.[136]

According to Lt Gen P. J. M. Godderij, former chairman of the DAG, the Balkans operations—particularly Operation Deny Flight, representing the RNLAF's first real F-16 deployment—gave a great impetus for Dutch doctrine development. During that period, the RNLAF gained a lot of experience in operational deployments, and it was considered necessary to write the lessons learned in black and white—both from a doctrinal as well as from an organisational point of view. The goal was for Dutch officers to acquire enough air power doctrine knowledge to put them on par with their most experienced—particularly American—counterparts in deployed operations.[137] While the first RNLAF air power doctrine in 1996 embraced many traditional concepts such as the counterair or the anti-surface force campaigns, it began to reflect the altered context of the post–Cold War era.[138] It is noteworthy that the doctrine divided the principles of war—in Dutch parlance, principles of military action—into three groups: principles for operations without consent of all parties, general principles for all operations, and principles for operations conducted with the consent of all parties and humanitarian operations. Principles such as transparency, credibility, minimal use of force, impartiality, and mutual respect specifically revealed awareness of the multi-layered and complex environment of peace support operations.[139]

The DAG must have drawn to a great extent upon UK air power thought when devising the 1996 air power doctrine. Its chapter structure is almost identical to that of the first and second editions of the RAF's AP 3000 air power doctrine manuals published in 1991 and 1993, respectively. The chapter on air power in peace support and humanitarian operations, though, can be regarded as genuinely Dutch.[140]

To make the first edition of the RNLAF air power doctrine available for international scrutiny, it was translated into English in 1999. Special correspondence was established with Dr. Alan Stevens, RAAF historian, and GPCAPT Shaun Clarke, Royal New Zealand Air Force. Though the outcome of this international correspondence is not available, Dr. Stevens reportedly proved to be a valuable interlocutor for the RNLAF due to his extensive study on doctrinal issues within the RAAF, as did Group Captain Clarke due to his specific study on small air forces. The RNLAF was particularly keen to receive input from the RAAF because of the similar size of both air forces.[141]

In 2002 the RNLAF published an updated version of the RNLAF air power doctrine. In contrast to its predecessor, it did not simply copy the chapter structure of a foreign air power doctrine, and it highlighted RNLAF specifics. One section was devoted to the swing-role concept, which had allowed Dutch F-16 aircraft to carry out several tasks on the same mission, and also related the benefits of European airlift cooperation.[142] Moreover, the Netherlands government's willingness to employ the armed forces across the spectrum of military force was reflected in the appendix "Future Developments," which attempted to anticipate trends in the future operational and strategic environment such as enhanced European integration or an increase in joint and expeditionary operations.[143]

It becomes apparent that while the first edition represents an effort to come to terms with the new strategic environment and with air power doctrine as a conceptual construct, the second edition reflects a shift towards an increased self-awareness by featuring distinct national specifics. Successful participation in major Western air campaigns over the Balkans certainly contributed to this increased self-awareness. Operations in Central Asia are likely to further shape RNLAF air power doctrine.

In 2005 the joint defence staff published a national defence doctrine. In his foreword, Gen D. L. Berlijn, chief of the defence staff, underlines that the Netherlands Defence Doctrine (NDD) filled the gap between the single-service doctrines and Dutch defence policy.[144] As a consequence, future single-service doctrines were to be harmonised with the NDD.[145] While the joint defence staff, the Army, and the Navy established doctrinal departments institutionalising the doctrinal development process, the RNLAF has—as of mid-2010—not yet installed a dedicated doctrine entity or published an air power doctrine in line with the NDD.[146] This current relative lack of doctrinal activities leads to the realization that the 1990s—with its numerous Western air campaigns—provided a more fertile ground for air power doctrine in the Netherlands.

Since the RNLAF is fully aware of its limited size, one of its goals is interoperability with the larger air forces not only in terms of hardware and software but also in terms of doctrine. In the mid-1990s, RNLAF officers were already stressing the value of international cooperation in air power doctrine. At a Dutch air power colloquium in 1997, an air commodore raised the idea of establishing a European air warfare centre, bringing together multinational expertise for developing doctrine, conducting education and training, and planning and organising multinational operations.[147] The establishment of NATO's Joint Air Power Competence Centre in 2005 is fully in line with the air commodore's vision of a combined approach to developing air power doctrine. Again, the RNLAF can draw upon a multinational institution as a force multiplier to compensate for relatively limited national resources.

Teaching of Air Power Thought

In 1992 the Netherlands armed forces collocated the single-service advanced staff courses and established the Netherlands Defence College (NDC), underlining the thrust towards joint education. A limited number of midcareer officers are selected to attend the year-long advanced staff course. It originally consisted of several modules, including one on air power and strategy.[148]

Courtesy Netherlands Institute for Military History; photo by Frank Visser

RNLAF F-16s before takeoff at Kandahar Airfield in November 2008

In the later stages of the Cold War, the air power curriculum began to shift from an emphasis not only on staff work but also on military and air power history and theory. In 1987 it was more a coincidence than a well-considered RNLAF policy that led to a relationship between the Netherlands Air Force Staff College and the USAF's Air University. Nevertheless, the RNLAF leadership's support proved essential. When Lt Col Jan van Angeren received orders to become an air power strategy instructor in the advanced staff course, he suggested that—for the purpose of thorough prep-aration—he might be sent either to AU or to the Führungsakademie in Hamburg. This finally led to close contact with AU. Based upon Colonel Van Angeren's advice, the RNLAF leadership continued to send his successors to AU. Prof. Dennis Drew at SAASS proved to be indispensable in integrating the RNLAF officers.[149] Through-out the 1990s, the assistance to the RNLAF was a self-paced, year-long reading programme, personally directed by Professor Drew and conducted on an ad hoc basis.[150] This programme was for-malised in 2000 as the Air Power Strategy Instructor Course, which only RNLAF officers have attended up to the time of this writing.[151] The Dutch example is a good case in point of how a bottom-up approach—with the necessary top-down backing—can radically alter and improve the quality of teaching.

RNLAF officers returning from the United States have developed a comprehensive air power studies curriculum. While they were influenced by the US approach, they went beyond it to include British thinking on international relations and strategy as well as specific investigations of a more national and European flavour. Grant T. Hammond, director of the Center for Strategy and Technology, AU, appraises the air power studies syllabus at the NDC as rigorous, wide-ranging, and theoretically oriented to assess the current debates on what air power can and cannot do.[152]

The air power and strategy course at the NDC was designed with the belief that a thorough study of military theory and history would better prepare Air Force officers to anticipate and tackle the problems and issues of the future. It emphasises strategy, the evolution of air power, NATO air power doctrines, the operational decision-making process in designing an air campaign and conducting an air war, and the utility of air power as an instrument of foreign policy.[153] To establish a link to the more staff-work-related aspects of an air campaign, RNLAF students attended joint air campaign planning exercises at the Joint Services Command and Staff College in the UK towards the end of the course.[154]

In late 2004, a movement towards enhanced joint teaching developed at the NDC. The 11-week single-service air power and strategy course was supplanted by a 13-week joint "art of war" module. The module includes a four-week block in which students separate to study the theories and concepts of their own environments.[155]

Furthermore, the air power faculty at the NDC began to run seminars and courses with its Belgian counterparts in 2001.[156] This cooperation on an educational level perfectly complemented the operational cooperation both air forces had realized for several years by then. The international dimension is also fostered by the invitation of foreign guest speakers to the NDC. Experts such as Tony Mason, Drew, Mark Clodfelter, and James Corum, as well as military experts from Israel and other countries who had gained experience in various conflicts, presented their thoughts and theories to Dutch air power students.[157] Moreover, foreign air force officers from Germany, Belgium, and Norway have attended the advanced staff course, despite the teaching language being Dutch. In addition, the RNLAF sends one student every year to the Air War College at Maxwell Air Force Base and another to the Belgian

Staff College. Furthermore, a Dutch Air Force officer is sent to the Führungsakademie and to the JSCSC in the UK every four years.[158]

Dissemination of Air Power Thought

Though the RNLAF had already organised air power symposia during the days of the Cold War, it was particularly in the aftermath of major air operations over Bosnia that the RNLAF began to organise air power symposia.[159] Hence, they can be understood as a means to come to terms with the new security environment and air power's role in deployed operations.

Symposia and colloquia were specifically designed to benefit doctrine development not only within the RNLAF but also beyond by discussing the challenges of joint operations.[160] Particularly interesting are a symposium organised by the NDC in April 1996 and a colloquium organised by the DAG in November 1997. Both events lasted an entire week, during which the RNLAF hosted some of the most esteemed air power practitioners and theorists of the Western world. Besides Dutch speakers, foreign speakers included Mason (RAF); Drew (SAASS); David Deptula (USAF); Robert Pape (US); Clodfelter (USAF); Robert Owen (USAF), the editor of the final report of the Air University Balkans Air Campaign Study; and Martin van Creveld (Israel).[161] These events provided the RNLAF with an opportunity to enrich its doctrinal thought with some of the most sophisticated and latest thinking on air power. They largely coincided with the publication of the RNLAF's first air power doctrine. Thereafter, it was predominantly major air campaigns or military operations that provided additional impetus for colloquia or symposia.

In the aftermath of Allied Force, the topic of coercion was especially emphasised to reflect NATO's air campaign. From 6 to 8 June 2000, for instance, the NDC organised a three-day colloquium. Besides Dutch speakers, scholars from the United States such as Benjamin Lambeth, Karl Mueller, and Hammond were invited.[162] In 2003 the 50th anniversary of the RNLAF and the role of air power in Operations Iraqi Freedom and Enduring Freedom provided the opportunity for a high-level air power symposium. Areas addressed included the RNLAF's history and significance, the military transformation process, and effects-based operations. Amongst the international guests was Gen Tommy Franks.[163] In many of its post–Cold War conferences, the RNLAF has managed to invite a wide range of well-established air power experts and practitioners from the United States, the UK, and elsewhere.

In terms of publications on air power doctrine and theory, the RNLAF has remained rather passive. Yet, it is interesting to point out that a former air power studies lecturer at the NDC wrote his thesis on John Boyd's theories and published it as a book.[164]

How Have Dutch Defence Planners Attempted to Maintain a Relevant Air Force in Light of Escalating Costs and Advanced Technologies?

The Netherlands has a modest defence industrial base. Apart from naval systems and some Army vehicles, the Netherlands does not have the capacity to produce complete main weapon systems such as major battle tanks or fighter aircraft.[165] As a rather small country, the Netherlands is in many ways dependent upon foreign defence manufacturers. To gain bargaining leverage with a particular supplier, the Netherlands has been interested in pooling its requirements with other nations.[166] This procurement policy has its roots in the early days of the Cold War when the Netherlands, together with Belgium, procured and produced under licence British Hawker Hunters in the mid-1950s and American F-104 Starfighters in the early 1960s.[167]

The Cold War era largely shaped the RNLAF's reliance upon American suppliers. Apart from the Hunters from Britain, the Netherlands selected American over French aircraft types on three occasions in the remainder of the Cold War era. According to a Swedish scholar, the outcomes were not coincidental since the Netherlands had always been more transatlantic than European in national security matters. However, in the second half of the 1960s, the Netherlands was actually on the brink of leaving its path of procuring and assembling American combat aircraft under licence. The Netherlands together with the other Western European F-104 Starfighter operators—the Federal Republic of Germany, Italy, and Belgium—held talks about designing an uncomplicated multipurpose aircraft as a successor for the F-104. As other partners joined the project, it became more and more difficult to determine common performance specifications. As a result, the Netherlands withdrew from the programme in May 1969. Subsequently, potential contenders for an F-104 Starfighter successor narrowed down to an American combat aircraft, the Cobra by Northrop; the Mirage F1 by Dassault; and the Swedish Viggen. In the mid-1970s, the Dutch decided to go with the F-16 by General Dynamics, which won a USAF competition against the Cobra. There were those, however, particularly in the Dutch aerospace industry, who would have preferred Dassault as a potential partner for politico-industrial reasons.[168]

In the post–Cold War era, the Netherlands has continued to buy primarily American as this has been perceived to deliver low unit costs combined with high quality. Yet, as the country has become almost exclusively dependent upon foreign suppliers, Dutch defence industrialists are concerned that purchasing off-the-shelf technology could reduce the Netherlands' research and development capabilities.[169]

Combat Aircraft

Along with Belgium, Denmark, and Norway, the Netherlands was one of the four initial European customers for the F-16 combat aircraft. Dutch F-16s were assembled at Fokker, and the first delivery to the RNLAF was in June 1979. Overall, the Netherlands procured 213 aircraft. As examined above, this number had to be considerably reduced in the post–Cold War era, and by 2003, all remaining operational F-16A/Bs had undergone an MLU programme.[170]

Initially, the aircraft's operational service life was estimated at 20 years. But developments in avionics technology made it possible to extend the operational life span of the F-16 aircraft considerably. Accordingly, the Netherlands, along with Belgium, Denmark, and Norway, opted for a thorough MLU rather than replacing the aircraft. For about 25 per cent of the original procurement costs, the operational life span of the Dutch F-16 aircraft was planned to be extended by 10 to 15 years in the late 1990s.[171] Planning for 2006 actually foresaw maintaining the capability of deploying at least one F-16 squadron up to 2020.[172] As has been previously noted, this MLU occurred under the EPAF framework. In 1993 the F-16 MLU contract was awarded to Lockheed Martin.[173] The RNLAF, remaining tightly involved in the programme, established a multinational operational test and evaluation centre at Leeuwarden Air Base in 1997.[174] Amongst the four European services involved in the multinational MLU programme, the RNLAF had the lion's share, with 138 of its 183 F-16 A/Bs being progressively refurbished and modernized.[175] For Allied Force, a number of upgraded F-16s were already available.[176] In the course of NATO's air campaign, F-16A/B MLU aircraft proved to be a valuable asset as modern multirole-capable combat aircraft were still in short supply in Europe.[177] In line with the Netherlands' push towards expeditionary warfare, the Ministry of Defence announced in 2003 that it would reduce the number of F-16A/B MLU aircraft by approximately 30 to a total of 108, with the aim of lowering operating costs and improving deployability of the remaining aircraft.[178] When further budget cuts were announced in mid-2007, the F-16 fleet was further reduced to 87 aircraft. Retirement of these remaining aircraft is currently foreseen for the period between 2015 and 2021, with an average airframe age of 31 years. The remaining frontline aircraft are planned to undergo further upgrade programmes.[179]

A relatively cost-effective way to improve a fighter aircraft's effectiveness is to attach avionics pods. The RNLAF made extensive use of this solution. In the wake of Operations Deny Flight and Deliberate Force, the RNLAF ordered 60 night-vision pods (forward-looking infrared or FLIR) and 10 targeting pods (low-altitude navigation and targeting infrared for night). Delivery of these systems was expected to start in 1999. Yet, pending delivery of the LANTIRN pods prior to Operation Allied Force, the commander in chief, General Droste, took the personal initiative to ask his American counterpart for pods from USAF arsenals. The USAF finally lent three LANTIRN pods to the RNLAF for the duration of Allied Force. To make optimum use of these pods, the Dutch normally flew in packages of two aircraft, with one plane fitted with a targeting pod providing buddy-lasing for the second F-16. For less dangerous missions, these packages grew bigger. In the aftermath of the air operations over Kosovo and Serbia, the Dutch increased their initial order of 10 LANTIRN pods to a total of 20.[180] In the meantime, the LANTIRN pods were supplemented with 22 Litening Block II advanced targeting pods.[181] The close relationship between the USAF and the RNLAF was especially highlighted by the unorthodoxy in making targeting pods available for Allied Force.

With the experience gained over the Balkans, the armament of the F-16 aircraft had to be constantly adapted to the latest demands. In the wake of Deliberate Force, it was decided to acquire a precision ground-attack capability by the procurement of AGM-65G Maverick air-to-ground missiles.[182] These would complement LGBs already available in RNLAF stocks. Originally, these LGBs were supposed to be released with RAF Buccaneer aircraft providing buddy-buddy lasing, as Dutch combat aircraft did not have laser-designator pods at their disposal in the early 1990s.[183] Continuing its path of acquiring PGMs, the Ministry of Defence announced in 2003 the introduction of the GPS-guided JDAM beginning in 2004 and the acquisition of long-range air-to-surface precision weapons beginning in 2007.[184] The acquisition of the latter was deferred and is currently foreseen to take place between 2012 and 2016.[185]

As regards air-to-air armament, the RNLAF made a major leap by the introduction of the "fire-and-forget" AIM-120 AMRAAM, its first medium-range air-to-air missile. In 1995 the contract was awarded to Hughes/Raytheon.[186] It was by means of an AMRAAM that a Dutch pilot downed a Serb MiG-29 during the first hours of Allied Force.[187]

While the F-16's lifespan was significantly extended, a potential successor was evaluated. Though no firm decision was taken in 2002, the Netherlands joined together with nine other nations to establish the partnership for the system development and demonstration phase of the F-35 Joint Strike Fighter. Besides the United States as lead nation, there are different levels of partnership. The US-led co-operative JSF programme is structured hierarchically according to different levels of participation (Levels I, II, and III). The Netherlands participates as a "Level II" partner. Level II partnership was restricted to no more than two countries, with Italy as the second Level II partner and the UK as the only "Level I" partner. In 2003 the total development cost of the JSF was estimated at about $26 billion. At the time, the Netherlands intended to contribute $800 million, of which the direct government contribution was expected to be limited to $200 million, with Dutch industry providing the remaining $600 million. In return, the Netherlands is supposed to receive a proportional share of levies on sales to third parties. Furthermore, the various international partners of the programme are entitled to receive their aircraft ahead of other potential customers.[188]

An F-35 Joint Strike Fighter manoeuvres during its first flight over Eglin AFB, FL, 23 April 2009. The jet is a stealth-capable multirole strike fighter.

A Dutch defence official confirmed in a 2004 interview with the author that contrary to European offers such as the Eurofighter or the French Rafale, only the JSF would offer full interoperability with the USAF. He added that American manufacturers were expected to provide the aircraft at low unit costs due to high production runs, as had already been the case with the F-16 combat aircraft. Thus, Dutch combat aircraft procurement decisions have been primarily driven by operational requirements and low unit costs rather than by the need to further a European defence industrial base.

It has to be reemphasised that the Dutch participation in the JSF project does not offer equal partnership as, for instance, a European cooperative venture might have done.[189] In mid-2009, RNLAF procurement planning foresaw the acquisition of 85 F-35 JSFs. At that point in time, the Netherlands government had committed to one aircraft for operational test and evaluation and was awaiting parliamentary approval of a second.[190] One year later, the decision on the second aircraft was still pending. Moreover, amidst the Dutch elections in mid-2010, some parliamentarians suggested a new aircraft evaluation, as the projected JSF's acquisition unit cost had soared significantly by over 50 per cent from its original baseline cost estimate.[191] As of June 2010, a decision on the Netherlands' continued participation in the system development and demonstration phase was expected to be part of a deal on a new coalition government.[192] Given Dutch industry's relatively strong involvement in this phase, the Netherlands is likely to continue its participation in the JSF programme.[193]

Courtesy Netherlands Institute for Military History; photo by Frank Visser

RNLAF F-16 after sunset at Kandahar Airfield in November 2008

Air Mobility

Up to the mid-1990s, the transport capacity of the RNLAF consisted of 13 light Fokker transport aircraft that had been built in the 1960s.[194] In accordance with the increasing Dutch involvement in peace support operations, the 1993 *Defence Priorities Review* identified a need for larger transport planes. All in all, two C-130 Hercules aircraft, four light Fokker utility transport aircraft, and two KDC-10 tanker/transport aircraft were to be procured.[195] In June 1992, two DC-10 aircraft were purchased from the Dutch airline Martinair. These DC-10s were modernised and converted to the KDC-10 standard. In April 1996, both aircraft were officially commissioned at Eindhoven Air Base.[196] In 2003 it was decided to acquire a third civilian DC-10. To make better use of the air-to-air refuelling capacity of the two available KDC-10 aircraft, this DC-10 is restricted to cargo and personnel.[197] In the same vein, the C-130 fleet was expanded to four aircraft by purchasing and modernising two former US Navy planes.[198]

According to a Dutch commentator, the transformation of the Dutch air mobility capacities is one of the most radical examples of how an air force adapted to the post–Cold War requirements in the 1990s. Yet the need to respond to a plethora of demands almost simultaneously—combined with a tight budget—led to a heterogeneous air mobility force, with the related logistical challenges. Particularly, the Dutch KDC-10 tankers are marred by interoperability problems. The modified commercial DC-10s have many non-standard features as compared to the USAF's KC-10A Extender. More importantly, RNLAF KDC-10 aircraft do not have hose-and-drogue refuelling facilities, making interoperability with air forces from the United Kingdom, France, or Germany impossible.[199]

Given the challenges of swiftly building up an airlift capacity with limited funding, Dutch defence planners have been open to unorthodox options compromising national sovereignty but offering cost-effective solutions. In 2000, for instance, the Dutch defence minister proposed contributing to the procurement costs of German A400M transport aircraft in exchange for freeing up German airlift capacities for Dutch requirements.[200] Moreover, the Netherlands is a member of NATO's Strategic Airlift Capability, centred on the shared ownership of three C-17 Globemasters. The Netherlands MOD plans to use 500 of the 3,500 C-17 flying hours projected annually.[201] The country is also a member of SALIS, which makes Ukrainian wide-body, long-range strategic transport aircraft available for both NATO and EU operations.[202] (Both cooperative ventures are highlighted in chap. 2.) In the medium term, the RNLAF's participation in the EATC is supposed to become a further force-multiplier in the field of air mobility.[203] As of mid-2009, with the Netherlands armed forces partially relying upon the chartering of IL-76 and An-124s for transport flights, these measures still needed to take full effect.[204]

In the field of rotary wing air mobility, the *Defence White Paper 1991* put a high premium upon battlefield mobility. Accordingly, it was decided to establish an air mobile brigade, which should be able to participate in crisis management operations outside the NATO treaty area.[205] In line with this, the 1993 *Defence Priorities Review* announced the procurement of 13 heavy CH-47 Chinook transport helicopters and 17 lighter transport helicopters, which later turned out to be French Cougar helicopters.[206] The order of 13 Chinook transport helicopters included seven secondhand Chinooks from Canada and six new machines from Boeing. The ex-Canadian helicopters underwent an MLU programme and have been operational since 1996. Delivery of the French Cougar transport helicopters started in 1996.[207]

After two Chinooks were lost in 2005, the MOD ordered six CH-47Fs to augment the RNLAF's Chinook fleet to 17 aircraft. As of 2009, it was planned to supplement the transport helicopter fleet with eight NH90s. As a consequence, the fleet will slightly fall short of the 2004 MOD capacity review, which foresaw a requirement of 20 Chinooks, 17 Cougars, and 10 NH90s.[208]

Combat Helicopters

These transport helicopters were to be augmented with combat helicopters. In November 1996, the RNLAF received 12 AH-64A Apaches on lease from the US Army. They would be operated until the delivery of 30 AH-64D Apaches had been completed. The leasing of US Army AH-64As allowed Dutch crews and mechanics to familiarise themselves with the system. Together with the Chinooks and Cougars, the Apaches form the RNLAF THG.[209] In July 1998, the first AH-64D arrived in the Netherlands, and by May 2002, the delivery of the new Apaches was completed.[210] During the combat helicopter evaluation process, the Netherlands received four offers, two from the United States and two from Europe. It finally boiled down to a decision between the AH-64D Apache and the Franco-German Tiger combat helicopter. Weighing heavily against the Tiger was that neither Germany nor France had yet signed production contracts. Finally, in early 1995, the cabinet decided in favour of the Apache. The earlier availability of the Apache was considered to be a decisive factor.[211] Moreover, the chances that in any future operation Dutch combat helicopters would be deployed alongside US forces were deemed to be highly likely.[212] To modernise the Dutch Apache fleet, the *Prinsjesdag Letter* proposed reducing the operational fleet of 30 aircraft to 24. Thanks to this measure, savings in operating costs were expected in order to free resources to modernise the remainder of the fleet with improved sensors.[213] As in the case of the F-16A/B MLU fleet, operational concerns outweighed quantitative aspects. Accordingly, the Netherlands defence minister argued in late 2003 that "the number of units on paper is not what counts, but rather the number of deployable units that can be fully committed at very short notice and with sufficient support."[214]

In 2006 efforts to sell the surplus aircraft had been of no avail. As a consequence, the government decided to keep them in storage as operational reserves. As of 2009, the RNLAF operated 16 AH-64Ds, with a further eight helicopters at the NATD at Hood Army Airfield for pilot training. In 2004 one combat helicopter was lost in Afghanistan.[215]

Courtesy Netherlands Institute for Military History; photo by Frank Visser

RNLAF AH-64D at Tarin Kowt in the province of Uruzgan in November 2008

The THG, combining transport and Apache combat helicopters, and the Royal Netherlands Army's 11th Air Mobile Brigade are integrated into the "blue-green" Air Manoeuvre Brigade. As of October 2003, the brigade reached full operational readiness in Poland.[216] It can be deployed as a whole or in smaller modules.[217] The well-balanced mixture of transport and combat helicopters gives the THG the ability to carry out diverse tasks across the entire spectrum of force, and it underlines the thrust towards integrated air-land operations.

C4ISTAR

Multinational cooperation in the domain of C4ISTAR is fully in line with the Netherlands' defence and alliance policy. Being closely integrated into NATO and the emerging ESDP, the Netherlands can benefit from international cooperation and thereby overcome the limits imposed by the size of its GDP. It is for exactly this reason that the Netherlands became a member of NATO's Airborne Early Warning and Control Force in the early 1980s.[218]

On a tactical level, reconnaissance pods have provided the RNLAF with an autonomous air-to-ground reconnaissance capability. The latest systems in the arsenal of the RNLAF are six Israeli-supplied Rafael RecceLite sensor pods equipped with electro-optical cameras and data links for providing real-time intelligence. The contract was signed on 17 November 2005.[219]

Due to financial restraints, however, the Netherlands forewent a number of bi- and multilateral projects in the field of reconnaissance and surveillance in recent years. As such, it bailed out of the French-led Helios 2 satellite programme as well as a Franco-Dutch MALE RPA project.[220] As of 2009, the Netherlands—against the logic of the Dutch cooperation rationale—is also no longer a member state in the multilateral NATO Alliance Ground Surveillance Core project.[221] Due to this lack of national intelligence assets, the Dutch Patriot units stationed in Turkey during Operation Iraqi Freedom relied upon American intelligence on potential missile launches, and the RNLAF fighter community as well had sufficient access to US intelligence in air campaigns, which was provided on a need-to-know basis.[222]

Ground Based Air Defence

For GBAD, including TBMD, the RNLAF has also focused upon quality and deployability. Systems no longer considered of immediate relevance in the current threat environment have been decommissioned to free resources for state-of-the art equipment. Accordingly, the RNLAF's radar-guided anti-aircraft artillery for point defence of air bases was phased out in 2000.[223] The decommissioning of the Hawk SAM batteries and the Stinger missiles followed suit in 2004 and 2007, respectively. The latter is still in service with the Royal Netherlands Army together with the Norwegian Advanced Surface-to-Air Missile System (NASAMS) II that entered Army service in 2009. NASAMS II is a medium-range air defence system based on the ubiquitous AMRAAM.[224]

In line with NATO's post–Cold War extended air defence concept embracing TBMD, the RNLAF has put a premium upon the modernisation of its Patriot batteries. For further improvement, PAC-3 radar modifications were ordered in 1996, with the intention of upgrading missiles and launchers to PAC-3 standard at a later stage.[225] Finally, in February 2004, the modernisation of the Patriot launchers and the procurement of an initial batch of 32 PAC-3 missiles were announced in Parliament.[226] Since deployability and modernisation of available systems for the TBMD role were deemed more important than increasing their number, plans for the acquisition of additional Patriot batteries from German surplus were finally shelved in 2003.[227] As examined in the section on employment ("Dutch Contributions to Western Air Campaigns"), the RNLAF has gained leverage through its TBMD capability in deployed operations. As of 2010, the RNLAF and Royal Netherlands Army GBAD units underwent a process of amalgamation.[228]

Conclusion

The Netherlands' changing security assessment, the redefinition of the tasks of the armed forces, and the shift away from territorial defence to expeditionary warfare have brought with them far-reaching impacts upon air power. Despite reducing its fighter aircraft fleet in response to its defence policy, the Netherlands has an enhanced ability to project power beyond NATO's boundaries.

In parallel, the RNLAF has embarked upon a number of cooperative ventures. In fact, they represent a response to the Air Force's limited size and shrinking resources. Dutch decision makers believe that air forces of a limited size without far-reaching cooperation would not remain relevant on the international stage. The establishment of the European Participating Air Forces' Expeditionary Air Wing is so far the most significant Dutch-led cooperative effort. Nowadays, the RNLAF has well-established links to the USAF. Simultaneously, it is well anchored in the emerging European defence architecture. This provides Dutch politicians with a broad array of political choices and deployment options. Therefore, it can be concluded that the RNLAF has grasped the uncertainties created by shifting defence and alliance policies as a chance to benefit from international synergies. Cooperation has not yet led to role specialisation, as this has been seen as unduly limiting deployment options. Nevertheless, certain capabilities have been kept at a minimum level, such as intertheatre airlift. Given its limited number of C-130 transport aircraft, the Netherlands hinges upon international cooperation, which could be seen as limited role specialisation.

Due to far-reaching transformation in the early post–Cold War era and its proactive stance in the field of multinational cooperation, the RNLAF has been able to punch above its weight. With its below average NATO defence budget, this is a remarkable achievement. However, the Netherlands' limited intertheatre airlift capacity might constrain its planned goal of deploying 36 F-16 combat aircraft in the context of NATO or EU operations. While the RNLAF proved capable of dispatching and sustaining up to 20 combat aircraft over a prolonged period for air operations over the Balkans, more distant theatres involving operations from bare bases might turn out to be too challenging. The RNLAF operated only contingents no larger than 10 combat aircraft in Central Asia. In this regard, an imbalance in the RNLAF's force structuring can be pointed out. While a number of advanced combat helicopters were purchased in the 1990s, the procurement of intertheatre air mobility was not given equal priority. The *Prinsjesdag Letter* addresses this imbalance by proposing a reduction of operational combat helicopters. In parallel, the RNLAF transport fleet gained two C-130 Hercules as well as one modified DC-10 aircraft by

2007. However, this imbalance could only be mitigated through these measures.

On an operational level, cooperation has fostered mutual trust. In the post–Cold War era, the RNLAF has gained a reputation of being a reliable partner, even in dangerous situations. The lessons of the "near deployment" to the Gulf area in 1990–91 were swiftly implemented, and the new emphasis on crisis intervention in the early 1990s led to a substantial Dutch contribution to the air campaigns over the former Yugoslavia. In the wake of the terrorist attacks in 2001, the RNLAF actively took on its share in the fight against terrorism. It provided the weapon of choice in operations at the upper end of the spectrum of military force and has made Dutch alliance solidarity visible. In the post–Cold War era, all of the RNLAF's major weapon systems have been deployed to out-of-area operations. The Dutch contributed to the air campaigns over the Balkans in terms of not only quantity but also quality. Dutch aircrews were prepared to execute dangerous frontline missions. RNLAF AH-64 combat helicopters and Patriot missile units provided sought after capabilities to deployed multinational operations. Due to its flexibility, the Dutch transport/tanker fleet has also shown an outstanding performance. Despite a below average budget, the RNLAF has proved to be a member of the A-team, partly explained by its can-do mentality. Even in cases where Dutch equipment did not meet the latest Western standards, the RNLAF achieved respectable results by making the best of the available means. For instance, devoid of an air-to-ground precision strike capability, Dutch F-16s released unguided air-to-ground weapons with great accuracy against ground targets during NATO's air operations over Bosnia.

In the final years of the Cold War, bottom-up attempts were already being undertaken to enhance education in air power theory and history. Some RNLAF officers understood that air power had to be mastered not only in technical terms but also in terms of theoretical and intellectual developments. This bottom-up approach was backed by the upper echelons of the RNLAF leadership and finally led to an institutionalised relationship with SAASS at Maxwell. It ensured that there has been a corps of educated Dutch officers throughout the post–Cold War era. This educational relationship has particularly been in line with RNLAF policy

Courtesy Netherlands Institute for Military History; photo by W. F. Helfferich

RNLAF CH-47 Chinook flying through a narrow valley in the former Yugoslavia

of putting a high premium upon interoperability with the USAF. Moreover, the RNLAF's participation in air operations over Bosnia led to the first formal strategic air power doctrine in the mid-1990s and to air power symposia conveying the close interrelationship between operations and air power precepts. These seminars can be considered an integral, though implicit, part of the RNLAF's lessons learned process. The RNLAF senior command has been receptive to sound education, perceiving it as one of the fundamentals for operational performance.

Since the Netherlands does not have the capacity to produce complete main weapon systems nationally, the RNLAF has procured its major weapon platforms from abroad, primarily from the United States. Buying American is in line with the Netherlands' transatlantic orientation, and it reflects the RNLAF's combat experience in air campaigns with the USAF being the lead air force. While this approach makes the Netherlands dependent upon foreign suppliers, the RNLAF's procurement policy can be directed more by combat experience than by politico-industrial aspects. Undoubtedly, this approach has benefited from short-term bargains—such as the procurement of secondhand Chinook transport helicopters from Canada—and has enhanced the RNLAF's combat effectiveness at relatively low costs. Yet, it does not contribute to the overarching goal of a capable European defence industrial base. Another factor contributing to the RNLAF's outstanding performance has been the reduction of certain key assets. The numbers of available F-16 aircraft, Patriot batteries, and Apache combat helicopters have been deliberately kept low or reduced to improve deployability of the available systems. Moreover, certain systems were completely phased out. Amongst the major GBAD systems from the Cold War era, only the Patriot has remained in service; this system also has a TBMD capability and is thus particularly suited for the current threat spectrum. Through this constant focus upon the requirements of deployed operations, the RNLAF has effectively played in the league of the major European air forces.

Notes

1. Lt Gen Dick L. Berlijn, commander in chief, RNLAF, "The Significance and Evolution of Air Power: The Royal Netherlands Air Force: One Team, One Mission," *Military Technology*, no. 6 (2003): 50.

2. IISS, *The Military Balance 2010* (Abingdon, Oxfordshire: published by Routledge for the IISS, February 2010), 149.

3. John Olsen, "Effects-Based Targeting through Pre-Attack Analysis," in *RNLAF Air Power Symposium*, 19 November 2003, 52.

4. MOD, Netherlands, *Defence White Paper 1991: The Netherlands Armed Forces in a Changing World*, abridged version (The Hague: MOD, 1991), 6, 18.

5. MOD, Netherlands, *Defence Priorities Review*, abridged version (The Hague: MOD, 1993), 4.

6. Ibid., 5.

7. MOD, Netherlands, *Defence White Paper 2000*, English summary (The Hague: MOD, November 1999), 11.

8. Ibid., 12.

9. MOD, Netherlands, *The Prinsjesdag Letter—Towards a New Equilibrium: The Armed Forces in the Coming Years*, English translation (The Hague: MOD, 2003), 13.

10. Ibid., 49.

11. MOD, Netherlands, *Defence White Paper 1991*, 6.

12. Ibid., 13.

13. MOD, *Defence Priorities Review*, 6.

14. MOD, Netherlands, *Defence White Paper 2000*, 9.

15. MOD, Netherlands, *Prinsjesdag Letter*, 9.

16. Ibid., 48.

17. MOD, Netherlands, *Defence White Paper 1991*, 15; and MOD, Netherlands, *Defence Priorities Review*, 2.

18. MOD, Netherlands, *Prinsjesdag Letter*, 23.

19. Ibid., 25–26.

20. Henk Kamp, minister of defence (address, Royal Netherlands Association of Military Science, Nieuwspoort Press Centre, The Hague, 1 March 2004).

21. Niels Lutzhöft, "Die Niederlande und die Wehrpflicht," interview, Lt Gen Hans Couzy, commander in chief, Royal Netherlands Army, *Europäische Sicherheit*, no. 6 (June 1996): 33; and MOD, Netherlands, *Defence Priorities Review*, 2.

22. Lt Gen Dirk Starink, commander in chief, RNLAF, briefing (presented at Shephard Air Power Conference, London, 25–27 January 2005).

23. Kees van der Mark, "Tall Ambitions for the Lowlands," *Air Forces Monthly* (July 2009): 40.

24. Jan Willem Honig, *Defense Policy in the North Atlantic Alliance: The Case of the Netherlands* (Westport, CT: Praeger, 1993), 229.

25. MOD, Netherlands, *Defence White Paper 1991*, 19.

26. Erwin van Loo, RNLAF History Unit, The Hague, to the author, e-mail, 9 September 2004.

27. MOD, Netherlands, *Defence White Paper 1991*, 30–33.

28. Ibid., 32–33.

29. Air Commodore P. J. M. Godderij, "The Evolution of Air Power Doctrine in the Netherlands," in *RNLAF Air Power Symposium* [15–19 April 1996] (The Hague: RNLAF, April 1996).

30. MOD, Netherlands, *Prinsjesdag Letter*, 20.

31. Dr. Sebastian Reyn (senior policy advisor, MOD, The Hague), interview by the author, 23 June 2004.

32. MOD, Netherlands, *Defence White Paper 1991*, 33–34.

33. Ibid., 19.

34. Godderij, "Evolution of Air Power Doctrine in the Netherlands."

35. IISS, *Military Balance 1990–91*, 75; and Ingemar Dörfer, *Arms Deal: The Selling of the F-16* (New York: Praeger Publishers, 1983), 109.

36. MOD, Netherlands, *Prinsjesdag Letter*, 41–42.
37. Van der Mark, "Tall Ambitions for the Lowlands," 47.
38. MOD, Netherlands, *Prinsjesdag Letter*, 26.
39. Van der Mark, "Tall Ambitions for the Lowlands," 44.
40. Ibid., 45.
41. MOD, Netherlands, *Prinsjesdag Letter*, 9–10.
42. Michael P. D'Abramo, *Military Trends in the Netherlands: Strengths and Weaknesses* (Washington, DC: Center for Strategic and International Studies, 31 August 2004), 5.
43. Alfred van Staden, "The Netherlands," in *The European Union and National Defence Policy*, ed. Jolyon Howorth and Anand Menon (London: Routledge, 1997), 92; and Honig, *Defense Policy in the North Atlantic Alliance*, 229.
44. IISS, *The Military Balance 1994–95* (London: Brassey's for the IISS, 1994), 59; and IISS, *Military Balance 2000–01*, 70.
45. Van der Mark, "Tall Ambitions for the Lowlands," 47; and Kees van der Mark, "Joint Helicopter Ops Dutch Style," *Air Forces Monthly*, August 2008, 66, 68.
46. MOD, Netherlands, *Prinsjesdag Letter*, 31; and IISS, *Military Balance 1994–95*, 59.
47. Question and answer session with Colonel Warden and Lt Col Osinga, RNLAF Air Power Symposium, 19 November 2003.
48. Janssen Lok, international editor, *Jane's International Defence Review*, and special correspondent, *Jane's Defence Weekly*, to the author, e-mail, 10 December 2004.
49. Ibid.
50. Jan G. Siccama, "The Netherlands Depillarized: Security Policy in a New Domestic Context," in *NATO's Northern Allies: The National Security Policies of Belgium, Denmark, the Netherlands, and Norway*, ed. Gregory Flynn (Totowa, NJ: Rowman & Allanheld, 1985), 117.
51. Ibid., 117.
52. Van Staden, "The Netherlands," in *European Union and National Defence Policy*, ed. Howorth and Menon, 96–98.
53. Ibid., 102–3.
54. MOD, Netherlands, *Prinsjesdag Letter*, 12–13.
55. MOD, Netherlands, *Defence White Paper 1991*, 15.
56. Joseph Soeters, professor for organisation studies and social sciences, Royal Netherlands Military Academy and War College, to the author, e-mail, 13 December 2004.
57. MOD, Netherlands, *Defence White Paper 2000*, 11.
58. MOD, Netherlands, *Prinsjesdag Letter*, 50.
59. Ibid., 25, 27.
60. Starink, briefing.
61. Wim H. Lutgert and Rolf de Winter, *Check the Horizon: De Koninklijke Luchtmacht en het conflict in voormalig Joegoslavië 1991–1995* (The Hague: Sectie Luchtmachthistorie van de Staf van de Bevelhebber der Luchtstrijdkrachten, 2001), 512–13.

62. Erwin van Loo, *Crossing the Border: De Koninklijke Luchtmacht na de val van de Berlijnse Muur* (The Hague: Sectie Luchtmachthistorie van de Staf van de Bevelhebber der Luchtstrijdkrachten, 2003), 749–50.

63. Droste, "Shaping Allied TBM Defence," 50.

64. Dave L. Orr, "The Benelux Deployable Air Task Force: A Model for EU/NATO Defense Force Integration," *Air and Space Power Journal* 17, no. 3 (Autumn 2003): 93.

65. GlobalSecurity.org, "F-16C/D Fighting Falcon," http://www.globalsecurity.org/military/systems/aircraft/f-16cd.htm, 19 March 2010.

66. Col Henk Bank, "Development of the EPAF [European Participating Air Forces] Expeditionary Air Wing" (presentation, European Air Chiefs Conference, Noordwijk Aanzee, Netherlands, 20 November 2003).

67. *Memorandum of Understanding between the Minister of Defence of the Kingdom of Belgium, the Minister of Defence of the Kingdom of Denmark, the Minister of Defence of the Kingdom of the Netherlands, the Minister of Defence of the Kingdom of Norway and the Minister of State and National Defence of the Portuguese Republic concerning the Establishment of the European Participating Air Forces' Expeditionary Air Wing* (Istanbul, Turkey, 28 June 2004), secs. 3.3, 4.1.

68. Bank, "Development of the EPAF Expeditionary Air Wing."

69. Maj Edwin Altena, RNLAF (Headquarters RNLAF, The Hague), interview by the author, 23 June 2004.

70. MOD, Netherlands, *Prinsjesdag Letter*, 50–53.

71. Droste, "Shaping Allied TBM Defence," 50–52.

72. Extended Air Defense Task Force, "History," http://www.eadtf.org; and Extended Air Defense Task Force, "Mission," http://www.eadtf.org.

73. European Air Group, "History of the EAG."

74. Nijean, "L'Armée de l'Air et l'Europe: gros plan—au coeur de l'Europe de la défense," 38.

75. Van der Mark, "Tall Ambitions for the Lowlands," 42, 48.

76. Ibid., 45, 48.

77. Horner quoted in Olsen, "Effects-Based Targeting," 52.

78. "For a New Dutch Defence Posture: Interview with Cornelis van der Knaap, State Secretary for Defence of the Netherlands," *Military Technology* 28, no. 2 (2004): 48; and Lok to the author, e-mail.

79. Col Henk Bank (former head, Policy Integration Branch, RNLAF), interview by the author, Royal College of Defence Studies, London, 4 May 2004.

80. Lt Col Peter Tankink, RNLAF, commander, 323 TACTESS, to the author, e-mail, 11 July 2004.

81. Kees van der Mark, "Firing Patriots in Greece," *Air Forces Monthly*, February 2010, 40.

82. Starink, briefing.

83. Lutgert and De Winter, *Check the Horizon*, 516.

84. Maj Gen Kees Homan, Royal Netherlands Marine Corps, retired (Netherlands Institute of International Relations, Clingendael), interview by the author, 22 June 2004.

85. Kamp, address.
86. Lutgert and De Winter, *Check the Horizon*, 500.
87. Ibid., 500–501.
88. Rinus Nederlof, *Blazing Skies: De Groepen Geleide Wapens van de Koninklijke Luchtmacht in Duitsland, 1960–1995* (The Hague: Sectie Luchtmachthistorie van de Staf van de Bevelhebber der Luchtstrijdkrachten, 2002), 534.
89. Ibid., 535.
90. Lutgert and De Winter, *Check the Horizon*, 503–4, 506, 508.
91. Ibid., 511.
92. Ibid., 514.
93. Ibid., 515.
94. Tankink to the author, e-mail, 21 January 2005; calculation of percentage is based upon figures from Sargent, "Weapons Used in Deliberate Force," 258, 265.
95. Owen, "Summary," 491; and John A. Tirpak, senior editor, "Deliberate Force," *Air Force Magazine* 80, no. 10 (October 1997), https://www.afa.org/_private/Magazine/Oct1997/1097deli.asp.
96. MOD, Netherlands, *Defence White Paper 1991*, 30.
97. Starink, briefing.
98. Lt Gen B. A. C. Droste, commander in chief, RNLAF, "Decisive Airpower Private: The Role of the Royal Netherlands Air Force in the Kosovo Conflict," *NATO's Nations and Partners for Peace*, no. 2 (1999): 129.
99. Lutgert and De Winter, *Check the Horizon*, 510.
100. Tankink to the author, e-mail, 11 July 2004; and Godderij, "Evolution of Air Power Doctrine in the Netherlands."
101. Lutgert and De Winter, *Check the Horizon*, 511.
102. Air Commodore J. L. H. Eikelboom, RNLAF (Headquarters RNLAF, The Hague), interview by the author, 22 June 2004.
103. Lt Gen H. J. W. Manderfeld, retired, commander in chief, RNLAF, 1992–95, to the author, e-mail, 27 December 2004.
104. Droste, "Decisive Airpower Private," 127–28; and Rolf de Winter and Erwin van Loo (RNLAF History Unit, The Hague), interview by the author, 24 June 2004.
105. MOD, UK, *Kosovo: Lessons from the Crisis*, Report to Parliament by the Secretary of State for Defence (London: The Stationery Office, 2000), annex f; calculation of percentage is based upon figures from Cordesman, *Lessons and Non-Lessons*, 4.
106. Clark, *Waging Modern War*, 197–98.
107. Adm Lukas Kroon, retired, chief of defence staff 1998–2004, Netherlands armed forces, to the author, e-mail, 5 January 2005 (forwarded by Maj Gen Kees Homan, Royal Netherlands Marine Corps, retired, Netherlands Institute of International Relations, Clingendael).
108. MOD, Netherlands, "Latest News of the Ministry of Defence," The Hague, 5 April 2002, accessed 14 June 2004, http://www.defensie.nl; and ibid., 27 June 2002, accessed 14 June 2004.

109. Ibid., 26 September 2002, accessed 14 June 2004; and Baldes, "Les Mirages français passent le relais," 34.

110. MOD, Netherlands, "Latest News of the Ministry of Defence," 6 October 2003, accessed 14 June 2004, http://www.defensie.nl.

111. Capt Inge van Megen, "Dutch and Norwegian F/16s Maintaining ISAF's Security Presence Round the Clock," *ISAF Mirror*, February 2006, accessed 27 February 2005, http://www.afnorth.nato.int.

112. Van der Mark, "Tall Ambitions for the Lowlands," 43.

113. MOD, Netherlands, "Latest News of the Ministry of Defence," 11 February 2003, accessed 14 June 2004, http://www.defensie.nl.

114. Ibid., 3 March 2003.

115. Soeters to the author, e-mail.

116. Maj Gen Kees Homan, Royal Netherlands Marine Corps, retired, Netherlands Institute of International Relations, Clingendael, to the author, e-mail, 5 January 2005; and Soeters to the author, e-mail.

117. David Charter and Tom Coghlan, "Dutch Confirm Afghan Troop Pullout Sparking Fears of Domino Effect," *The Times*, 22 February 2010, http://www.timesonline.co.uk/tol/news/world/ afghanistan/article7035719.ece.

118. Van Loo, *Crossing the Border*, 748.

119. De Winter and Van Loo, interview.

120. Government of the Netherlands, "Netherlands to Send Helicopters to Afghanistan," Government.nl, 30 January 2004, http://www.government.nl.

121. Anne-Claude Gouy, "Apache Helicopters Arrive in ISAF," NATO, 1 April 2004, http://www.nato.int/shape/news/2004/04/ i040401.htm.

122. MOD, Netherlands, "Latest News of the Ministry of Defence," 21 May 2004, accessed 14 June 2004, http://www.defensie.nl.

123. Van der Mark, "Tall Ambitions for the Lowlands," 43.

124. Lutgert and De Winter, *Check the Horizon*, 499.

125. De Winter and Van Loo, interview.

126. MOD, Netherlands, "Latest News of the Ministry of Defence," 29 December 2003, accessed 14 June 2004, http://www.defensie.nl.

127. Ibid., 22 January 2002, accessed 15 June 2004; and ibid., 20 February 2002.

128. Ibid., 5 February 2002.

129. Van der Mark, "Tall Ambitions for the Lowlands," 43.

130. Tankink to the author, e-mail, 8 January 2006.

131. Kuebart, "Air Policing Baltikum," 32.

132. "German Eurofighters Arrive for Baltic Air Policing Debut," 18.

133. Godderij, "Evolution of Air Power Doctrine in the Netherlands."

134. Lt Gen P. J. M. Godderij, RNLAF, military representative of the Netherlands to NATO, to the author, e-mail, 4 January 2006.

135. Godderij, "Evolution of Air Power Doctrine in the Netherlands."

136. *Royal Netherlands Air Force Air Power Doctrine* (The Hague: RNLAF, 1999). Originally published as *KLu Air Power doctrine voor het basis- en operationele niveau* [RNLAF basic and operational level doctrine] (The Hague: RNLAF, 1996).

137. Godderij to the author, e-mail.

138. *Royal Netherlands Air Force Air Power Doctrine*, table of contents.

139. Ibid., II-3–II-8.

140. Ibid., table of contents; AP 3000, *Royal Air Force Air Power Doctrine*, 1991, vii; and AP 3000, *Royal Air Force Air Power Doctrine*, 2nd ed., 1993, vii.

141. Col Peter Wijninga, RNLAF (head, Policy Integration Branch), telephone interview, 21 December 2004; and Col Peter Wijninga, RNLAF, head, Policy Integration Branch, to the author, e-mail, 20 July 2005 (forwarded by Lt Col Erik Wijers, RNLAF).

142. RNLAF, *Airpower Doctrine* (The Hague: RNLAF, April 2002), 1, 84, 86–87.

143. Ibid., 98–101.

144. MOD, Netherlands, *Netherlands Defence Doctrine*, trans. Netherlands Ministry of Defence Translation Service (Netherlands Defence Staff: The Hague, 2005), i.

145. Lt Col (Dr.) Marcel de Haas, Royal Netherlands Army (Netherlands Institute of International Relations, Clingendael), telephone interview by the author, 17 August 2007. During his most recent assignment at the Defence Staff, De Haas was responsible for the drafting of the first edition of the *Netherlands Defence Doctrine*.

146. Lt Col Eric A. de Landmeter, head, Joint Doctrine Branch, MOD, The Hague, to the author, e-mail, 30 June 2010.

147. Air Commodore J. T. Bakker, "Air Power in Peace and Humanitarian Operations: A Dutch Perspective" (paper presented at the RNLAF Air Power Colloquium, "Air Power: Theory and Application," The Hague, 24–28 November 1997).

148. Col Lex Kraft van Ermel, RNLAF, former director of studies, NDC, to the author, e-mails, 13 June and 22 June 2005.

149. Col Jan F. W. van Angeren, RNLAF, retired, to the author, e-mail, 9 January 2006.

150. Col Dennis M. Drew, USAF, retired, professor and associate dean, SAASS, to the author, e-mail, 9 July 2004.

151. Ibid., 21 December 2005; and Lt Col Vincent Scharrenberg, RNLAF, NDC, The Hague, to the author, e-mail, 9 January 2006.

152. Dr. Grant T. Hammond, director, Center for Strategy and Technology, USAF, Air University, Maxwell AFB, AL, to the author, e-mail, 1 July 2004.

153. Syllabus, Strategy, Doctrine, and Air Power Course, 2000–2001, NDC, The Hague, 2–4.

154. Ibid., 12.

155. Van Ermel to the author, e-mail, 13 June 2005.

156. Ibid.

157. Ibid.

158. Ibid.

159. For symposia during the Cold War, see, for instance, RNLAF Air Power Symposium, "The Future of Air Power," The Hague, 13–14 October 1988.

160. RNLAF Air Power Symposium, "Air power en doctrinevorming [Air power and doctrine development]," The Hague, 21 September 1995; and RNLAF

Air Power Symposium, "De rol van air power bij gezamenlijk optreden [The role of air power in joint operations]," The Hague, 27 September 1996.

161. *RNLAF Air Power Symposium*, 15–19 April 1996; and RNLAF Air Power Colloquium, 24–28 November 1997.

162. RNLAF Air Power Colloquium, "Coercive Air Strategies," The Hague, 6–8 June 2000.

163. *RNLAF Air Power Symposium*, 19 November 2003, table of contents.

164. Lt Col Frans Osinga, RNLAF (Netherlands Institute of International Relations, Clingendael), interview by the author, 24 June 2004; and Frans Osinga, *Science, Strategy and War* (London: Routledge, 2006).

165. "Sustaining Modern Armed Forces," interview with Dr. Jan Fledderus, Netherlands national armaments director, *Military Technology*, no. 8 (1997): 19.

166. "We want the Best Materiel for the Best Price," interview with Dr. Jan Fledderus, Netherlands national armaments director, *Military Technology* 24, no. 12 (December 2000): 26–27.

167. Dörfer, *Arms Deal*, 107.

168. Ibid., 106–8, 116, 122, 128–29.

169. D'Abramo, *Military Trends in the Netherlands*, 28–29.

170. De Winter and Van Loo, interview.

171. Maj Gen Marcel Wagevoort, "Materiel Projects in the RNLAF," Defence Procurement in the Netherlands, *Military Technology*, no. 8 (August 1997): 45.

172. Joris Janssen Lok, "RNLAF Concerns Increase over Targeting Pod Acquisition Being Slow to Hit the Mark," *Jane's International Defence Review*, April 2006, 6.

173. "Sustaining Modern Armed Forces," 13.

174. De Winter and Van Loo, interview.

175. *The Military Balance 2002–2003* (Oxford, England: Oxford University Press for the IISS, October 2002), 51.

176. Droste, "Decisive Airpower Private," 127.

177. MOD, Netherlands, *Prinsjesdag Letter*, 41.

178. Ibid., 41.

179. Van der Mark, "Tall Ambitions for the Lowlands," 44–45.

180. De Winter and Van Loo, interview.

181. Van der Mark, "Tall Ambitions for the Lowlands," 44.

182. Wagevoort, "Materiel Projects in the RNLAF," 45.

183. Tankink to the author, e-mail, 21 January 2005.

184. MOD, Netherlands, *Prinsjesdag Letter*, 40–41.

185. Van der Mark, "Tall Ambitions for the Lowlands," 44.

186. "Sustaining Modern Armed Forces," 13.

187. Van Loo, *Crossing the Border*, 679.

188. Dreger, "JSF Partnership takes Shape," 28–31.

189. See "Norway Commits Funding to Gripen Development," *Gripen News*, no. 1 (2007): 2, http://www.gripen.com.

190. John R. Kent, F-35 Lightning II Communications, Lockheed Martin Aeronautics Co., to the author, e-mail, 3 August 2009; and Hans de Vreij,

"Dutch Reveal Plan to Purchase Two JSF Test Aircraft," *Jane's Defence Weekly* 45, no. 11 (12 March 2008): 13.

191. Gerrard Cowan and Craig Caffrey, "Dutch Parliament Casts Doubt on F-35 Deal . . ," *Jane's Defence Weekly* 47, no. 21 (26 May 2010): 6.

192. Gerrard Cowan, "Future of Dutch Aerospace Sector 'Depends on JSF,'" *Jane's Defence Weekly* 47, no. 25 (23 June 2010): 15.

193. Col Robert J. Geerdes, RNLAF, F-16 Replacement Office, Defence Material Organisation, to the author, e-mail, 28 July 2010.

194. Wagevoort, "Materiel Projects in the RNLAF," 50.

195. MOD, Netherlands, *Defence Priorities Review*, 14–15.

196. Van Loo to the author, e-mail, 21 October 2004.

197. MOD, Netherlands, *Prinsjesdag Letter*, 42-43.

198. Van der Mark, "Tall Ambitions for the Lowlands," 47.

199. Joris Janssen Lok, "Revamping Dutch Airlift in a Post–Cold War Environment," *Jane's International Defence Review* 32 (February 1998): 27.

200. "We Want the Best Materiel for the Best Price," 24; and Lange, "Lufttransport," 40.

201. NATO, "Strategic Airlift Capability"; and Van der Mark, "Tall Ambitions for the Lowlands," 48.

202. BMVg, "Verlässlicher Zugriff."

203. See section on "Alliance Policy and Its Impact upon the Royal Netherlands Air Force."

204. Van der Mark, "Tall Ambitions for the Lowlands," 48.

205. MOD, Netherlands, *Defence White Paper 1991*, 27.

206. MOD, Netherlands, *Defence Priorities Review*, 15.

207. Wagevoort, "Materiel Projects in the RNLAF," 46.

208. Van der Mark, "Joint Helicopter Ops Dutch Style," 65; and Van der Mark, "Tall Ambitions for the Lowlands," 47.

209. Van Loo, *Crossing the Border*, 742.

210. Van Loo to the author, e-mail, 21 October 2004.

211. Air Commodore Dirk Starink, "The Apache Programme," *Military Technology: Defence Procurement in the Netherlands*, no. 8 (August 1997): 53–55.

212. Ian Bustin, "Apache Country: The Selection of the AH-64 Apache by the Royal Netherlands Air Force," *Military Technology*, no. 7 (July 1995): 30.

213. MOD, Netherlands, *Prinsjesdag Letter*, 43.

214. Henk Kamp, minister of defence, opening address, in *RNLAF Air Power Symposium*, 19 November 2003, 10.

215. Van der Mark, "Joint Helicopter Ops Dutch Style," 65, 67.

216. MOD, Netherlands, "Latest News of the Ministry of Defence: 11 Air Manoeuvre Brigade makes the Grade," 29 October 2003, accessed 16 October 2004, http://www.defensie.nl.

217. Kamp, opening address, 10.

218. NATO, "NATO Airborne Early Warning and Control Force," 1.

219. Lok to the author, e-mail; and Lok, "RNLAF Concerns Increase," 6.

220. MOD, Netherlands, *Prinsjesdag Letter*, 31; and Mercier to the author, e-mail.

221. MOD, Netherlands, *Prinsjesdag Letter*, 41; and NATO, "NATO's Allied Ground Surveillance Programme."

222. Lok to the author, e-mail.

223. Van Loo to the author, e-mail, 21 October 2004.

224. Van der Mark, "Firing Patriots in Greece," 42.

225. Wagevoort, "Materiel Projects in the RNLAF," 50.

226. MOD, Netherlands, "Latest News of the Ministry of Defence," 16 February 2004, accessed 16 October 2004, http://www.defensie.nl.

227. MOD, Netherlands, *Prinsjesdag Letter*, 44.

228. Col Peter Wijninga, RNLAF, commander, De Peel AFB, to the author, e-mail, 30 July 2010 (forwarded by 1Lt Ton Steers, RNLAF).

Chapter 6

Swedish Air Force (Flygvapnet)

In June 1925, the separate Army and Navy air corps were uni-fied in an independent air force, the then Royal Swedish Air Force.[1] Despite its early birth, the Swedish Air Force could only muster slightly more than 100 operational aircraft at the dawn of World War II. Due to lack of spare parts from the United King-dom, a large number of fighters were grounded.[2] This deplorable situation was a wake-up call and resulted in an intensive rearma-ment process. At the end of the war, the Air Force had more than 800 aircraft at its disposal. The domestic aircraft industry continued to deliver new aircraft after the war, resulting in the SwAF becom-ing the fourth largest air force in the Western world, operating some 1,000 aircraft.[3]

The 1958 long-term Defence Act put a clear emphasis upon air power in the Swedish overall defence posture. As a consequence, the Navy had to absorb significant budget cuts at the hands of the SwAF.[4] The naval budget shrank from 18 to 13 per cent of the total defence budget, as combat aircraft were considered to be more versatile and less vulnerable than surface ships.[5]

The Swedish defence rationale of the late 1950s considered the national territory of secondary priority for potential enemies.[6] It was assumed that no major power would and could allocate major defence resources for an attack against Sweden and, therefore, that a potent Swedish defence would have a sufficient deterrent effect. The Swedish armed forces were geared up towards a multilayered anti-invasion defence, which aimed at striking a potential aggres-sor outside Swedish territory.[7] This was primarily the task of the SwAF and the submarine fleet. The main task of the Army was to defend the land border with Finland as well as southern Sweden.[8] During the 1950s, the acquisition of nuclear weapons for the Swedish armed forces was also debated but never realised. With the signature of the Non-Proliferation Treaty in 1968, this debate came to an end.[9]

Whereas Swedish defence doctrine concentrated exclusively upon defence against full-scale invasion in the early periods of the

Cold War, defence planners during the 1970s started to envisage the possibility of limited surprise attacks directed against Sweden's key infrastructure. Soviet submarine incursions into Swedish coastal waters during the early 1980s corroborated this concern. Henceforth, preparation for defence against limited attacks was given particular emphasis.[10] Simultaneously, adherence to a defence-in-depth doctrine was reiterated. Sweden's ability to engage invasion forces beyond the national border and coast by combat aircraft and submarines was considered important.[11]

Courtesy Gripen International; photo by Stefan Kalm

JAS 39D Gripen dual-seat combat aircraft

Notwithstanding the country's neutral status, Sweden played a key role in Western defence plans of Scandinavia.[12] A large and powerful SwAF was regarded as pivotal by NATO defence planners, given the relatively weak Norwegian Air Force. In the later stages of the Cold War, Sweden possessed indeed one of the largest and best equipped air forces in Europe.[13] Other impressive features were its modern early warning and command systems, an extensive dispersed basing system, and its indigenous combat aircraft designs, such as the advanced JA 37 Viggen, tailored to Sweden's defence requirements.[14]

Given Sweden's proximity to Russia and its considerable autonomous defence effort throughout the Cold War, the Swedish defence establishment only slowly adapted to the altered security paradigms of the post–Cold War era. Sweden's Cold War legacy exerted significant inertia upon defence reforms throughout the 1990s.[15] It also had consequences on a technical and tactical level. Interoperability was hampered by equipment, particularly in the field of command and control. For these reasons, the SwAF could not be deployed to out-of-area operations during the 1990s, with the exception of its transport unit.[16]

Only in the late 1990s did a defence reform process set in, with the vast Cold War defence complex beginning to be dismantled.[17] Nowadays, the SwAF finds itself in a transformation process, which is leading to enhanced interoperability and deployability. Sweden's EU membership since 1995 has been a major catalyst in this process.

How Has the Swedish Air Force Adapted to the Uncertainties Created by Shifting Defence and Alliance Policies?

This section first analyses Swedish post–Cold War defence policy and the SwAF's response to it. It then analyses Sweden's alliance context and its influence on the SwAF.

Defence Policy

Sweden's changing defence posture in the post–Cold War era is closely related to its threat and risk perception. Over the centuries, Russia was the dominant factor in the Swedish strategic calculus. It took Sweden almost a decade to adapt to the post–Cold War realities, and the Nordic state finally shifted away from a predominantly Russia-centric view to an international one, which resulted in a significant impact upon its Air Force.[18] Sweden's relatively slow defence policy transformation can to a large degree be ascribed to its geostrategic location. Accordingly, integration of the Baltic states into NATO and the European Union was of major significance for Swedish decision makers. Yet, the Russian intervention in Georgia as of August 2008 led to an increased emphasis on national defence again, while not reversing previous developments in Sweden's defence-political opening process.

Threat and Risk Perception. During the Cold War, Soviet-Swedish relations reached periods of considerable tension. As described later, Soviet fighters shot down at least one Swedish aircraft over the Baltic Sea in 1952, and in the early 1980s, Soviet submarines intruded into Swedish waters. In 1990 Swedish authorities urged caution in the Swedish response to the events unfolding in Eastern Europe. They were indeed very suspicious of the permanence of the developments in the Soviet Union. A setback in Soviet reforms by communist hardliners was still regarded as a possibility.[19] Only after the dissolution of the Soviet Union in late 1991 did the process of gradually transforming the Swedish defence posture start. While the immediate threat of a major armed invasion across coastal or land borders was ruled out from then onwards, such a scenario was required to be met after one year of preparation. Yet, it was deemed necessary that the armed forces be capable of dealing with limited strategic surprise attacks against key targets in Sweden at any time.[20]

In 1996–97 the Swedish government concluded that the security situation had further evolved in a positive way. This development was mainly attributed to the stabilising function of an enlarging European Union.[21] However, it was only in 1999 that the threat of an invasion ceased to be the dominant factor in shaping Sweden's defence architecture. Henceforth, a full-scale invasion of Swedish territory did not seem feasible within the time span of 10 years, provided that Sweden had a basic defence capability at its disposal. These fundamental improvements in the security assessment initiated a development towards smaller but qualitatively advanced armed forces, with the aim of effectively using them both nationally and internationally. However, limited surprise attacks, primarily from the air, were still deemed a potential threat that could occur any time with the aim of coercing Swedish decision makers.[22]

Government Bill 1999/2000:30, The New Defence, thus represented a milestone step in Swedish defence policy.[23] According to the former director of the Defence Commission, Michael Mohr, reduced financial resources played a significant role in this reorientation in 1999. A balance had to be struck between tasks and resources. As a result, Swedish decision makers were forced to accept the realities of a changed security environment.[24]

In the wake of the terrorist attacks on 11 September 2001, non-state actors shifted into the focus of Swedish security policy. It was emphasised that conflicts with local origins could have repercussions far outside the actual area of conflict, endangering global stability as a whole. Regardless, Sweden's regional security was deemed positive.[25] Finally in 2003, any form of armed attack on Sweden by another state was not expected to be likely within the space of at least 10 years.[26] This positive assessment can be related to the Baltic states being integrated into NATO and the EU, which was perceived to strengthen decisively Sweden's own security.[27] With the integration of these countries into established European structures, the Baltic region—the buffer zone between Sweden and Russia—was no longer seen as a grey zone in Swedish security policy, which fundamentally changed the strategic situation in the immediate vicinity. These developments reinforced the shift in Sweden's security policy from one primarily emphasising territorial defence to one aiming at robust contributions to international security.

The positive security assessment within Sweden's neighbourhood stood in stark contrast to other regions of the world. Swedish Government Bill 2004/05:5, Our Future Defence, underlines the concerns about stability in the wider world. Regional conflicts were not believed to be restricted to a single country or region; instead, they could produce ripple effects. Leni Björklund, minister of defence at the time, concluded that by participating in international operations for peace and security, Sweden would enhance its own security.[28] This is a significant conclusion, given Sweden's legacy of neutrality. The fact that Sweden wanted to be perceived as a reliable partner amongst Western states certainly contributed to a stronger Swedish involvement in international security affairs, as is examined in the section on alliance policy.

However, the 2008 Russian military intervention in Georgia was received by Swedish decision makers with a certain degree of concern and gave new impetus to the importance of a capability to counter limited but rapidly occurring military crises. In 2009 a new government bill, A Functional Defence, was issued to meet these concerns. While it argued that both NATO and the EU exerted a stabilising influence upon the Baltic region and that the threat of invasion from a previous superpower had disappeared, the risk of military incidents either on or around the Baltic Sea could not be discounted.[29]

Tasks of the Armed Forces. Throughout the post–Cold War era, the changing threat perception has had a major influence upon the formulation of the tasks for the armed forces. Accordingly in 1991, the tasks of the armed forces were primarily relegated to defensive tasks against various kinds of military threats to Sweden, ranging from full-scale invasion to limited strategic attacks. Safeguarding Sweden's neutrality was considered pivotal. In particular, the integrity of Swedish airspace was emphasised.[30] Persistent emphasis upon territorial defence reflects the doubts of Swedish decision makers who—despite the considerable relaxation in the East-West relationship—reckoned with a possibility that the situation might deteriorate again.

Only the 1999–2000 government defence bill brought about a significant shift towards out-of-area operations. Broad participation in European defence cooperation and contributions to crisis management were henceforth regarded as essential means to safeguard Sweden's own security, reflecting the 1999 turning point in Swedish defence policy.[31] Finally in 2004–5, involvement in international operations was given first priority in relation to other tasks. The Ministry of Defence stated that the "objective for Sweden's total defence is to preserve the country's peace and independence by helping to manage and prevent crises in the world around us, asserting our territorial integrity, defending Sweden against armed attack, [and] protecting the civilian population and safeguarding the most important societal functions."[32]

Accordingly, Swedish decision makers, both political and military, were very clear in the aftermath of the 1999–2000 reform on the need for restructuring the armed forces and making international operations the number one priority.[33] The 2009 government bill, however, partially reversed this one-sided emphasis on the international dimension. In particular, it was decided to significantly improve the readiness of Sweden's defence resources to counter a limited but rapidly unfolding military crisis lasting only a couple of days—a reference to Russia's 2008 military intervention in Georgia that lasted five days and was decided in two. Yet at the same time, the government also announced a significant increase in deployed personnel on peace support operations.[34]

Courtesy Swedish Air Force

Swedish C-130 Hercules in Afghanistan

Towards Expeditionary Warfare and Deployable Armed Forces. Sweden had already developed a tradition of contributing to UN peacekeeping operations in the Cold War. For instance, it participated in the first UN mission—the ceasefire supervision mission in the Middle East—commencing in 1948. It sent infantry units to Suez in 1956, and in 1960–64, Sweden engaged at the upper end of the spectrum of military force in the former Belgian Congo, representing the first Swedish combat operation since 1814.[35] Yet, this militarily robust deployment remained an exception throughout the Cold War. Despite a relatively strong commitment to UN missions, anti-invasion defence was by far the most important concern for the Swedish armed forces. The political leadership proved to be restrictive in using military power as a security policy instrument abroad.[36]

Throughout the 1990s, international operations gradually gained in importance. As a first step towards a more robust international commitment, the 1996–97 defence bill put forward the creation of an international command, the purpose of which was to organise, train, and support international missions.[37] But as the establishment of a separate command reveals, the international dimension remained divorced from the main bulk of the Swedish armed forces.

Only since 1999 have Swedish international ambitions intensified significantly. Regarding European security, two major events took place during this year, serving as catalysts for Sweden's defence transformation. First, the Kosovo campaign showed the inadequacies of European defence postures. Second, at the European Council in Cologne in June 1999, EU member states asserted the need for the EU to become a strategic actor. In line with these developments, the 1999–2000 government defence bill stated that the defence organisation was to be restructured for both national and international missions. Particular reference was made to the conflict in Kosovo by concluding that Sweden's ability to participate in an international operation rapidly and with interoperable equipment was to be improved.[38] Sweden's contribution to the development of a European rapid reaction capacity was seen as the foremost factor shaping Swedish defence transformation in the ensuing years.[39] At the Council of Defence Ministers in Brussels on 22 November 2004, it was declared that Sweden, Finland, Norway, and Estonia intended to establish a multinational battle group based on the EU battle group concept, with Sweden being the lead nation. Out of 1,500 troops, Sweden contributes approximately 1,100. The Nordic Battle Group's (NBG) first standby period lasted from 1 January to 30 June 2008.[40] The EU battle group concept is likely to have a profound impact upon the entire Swedish Army. The 2009 government bill announced that the main bulk of the ground units were to be organised in battle groups, which can be employed nationally or multinationally.[41]

Since the late 1990s, preparedness of Swedish decision makers to engage in actual combat has increased significantly. In 2003 the Defence Commission stated that the ability to engage in armed conflict, both nationally and internationally, was the single most key factor governing the development of Swedish military capabilities, implying a shift towards more robust peace support operations, including peace enforcement.[42]

Despite these ambitions towards increased power projection, the Swedish defence establishment continued to underline its adherence to conscription. In the 1999–2000 defence bill, it was explicitly stated that the units used for both territorial defence and international operations would be based on conscription. The bill also mandated that the number of conscripts should be kept to the lowest possible level.[43] In the same vein, the supreme commander of the Swedish armed forces argued in 2004 that the willingness to serve in international missions ought to be taken into consideration in the recruitment of new conscripts.[44] Particularly the establishment of the Swedish-led EU battle group called for modifications in the recruiting system, as it required Sweden to have military personnel available at very short notice. Accordingly, the 2004–5 defence bill foresaw national servicemen being employed for a number of years.[45] As described later, the SwAF set up rapid reaction units depending upon contracted personnel.

Only in 2009 was the abolishment or temporary annulment of compulsory conscription seriously considered. As an alternative, the Ministry of Defence proposed voluntary military service.[46] This shift towards a voluntary force can be explained by the 2009 government bill's goal to strengthen Sweden's defence capability by creating immediately employable armed forces. The bill stated that "the entire operational organisation of some 50,000 people will be able to be used within a week after a decision on heightened alert."[47] Moreover, readily available armed forces also allow Swedish decision makers to increase deployed personnel on peace support operations.

Finally, on 19 May 2010, the Swedish Parliament formally suspended compulsory military service. On 1 July 2010, the Ministry of Defence initiated the transition to a 100 per cent professional and contract force.[48] The Swedish armed forces are projected to total 55,200 personnel by 2014, including civilians and 22,000 in the home guard units, largely comprised of part-time soldiers. The planning figure for the SwAF—including the Armed Forces Helicopter Wing—is 3,350 personnel, approximately 650 of whom are part-time professionals.[49]

Defence Policy and Its Impact upon the Swedish Air Force. At the outset of the post–Cold War era, the SwAF operated one of the largest combat aircraft fleets in Europe, numbering more than 400 aircraft all of domestic manufacture.[50] Sweden's threat and risk perception has determined the size as well as structure of the SwAF in the post–Cold War era.

During the Cold War, the Swedish armed forces were structured in a way that would have allowed them to conduct a defence in depth. The SwAF attack aircraft fleet, like the Swedish submarine force, was intended to attack invading forces far out in the Baltic Sea. Moreover, the coastal defences would have been supported by maritime strikes, and the limited ground attack capacity—primarily in the air interdiction role—would have been concentrated in northern Sweden.[51]

In the later stages of the Cold War, the attack aircraft fleet consisted of AJ 37 Viggen and SK 60 aircraft. The SK 60 was primarily used as a trainer, but it also had a limited attack and photoreconnaissance capability. Aerial reconnaissance was largely the function of the SF 37 and SH 37 Viggens. The former was a dedicated photoreconnaissance aircraft, whereas the latter was a maritime reconnaissance aircraft. Air defence was primarily carried out by the JA 37 Viggen, an all-weather interceptor that first entered service in 1979. Furthermore, older but modified J 35J Draken aircraft supported the JA 37 Viggens in the air defence task into the 1990s. To enhance survivability in a high-intensity conflict scenario, dispersed basing of its squadrons was a crucial pillar in the SwAF's defence doctrine. Besides the main operating bases with their large underground storage facilities, the SwAF also maintained wartime reserve air bases. At the end of the Cold War, the SwAF had an extensive network of as many as 30 main and reserve air bases and approximately 50 additional operating sites, including civilian airports, at its disposal.[52]

According to Swedish post–Cold War security assessments made in the early 1990s, the size and structure of the SwAF continued to be determined by a potential invasion and also by the threat of limited aerial surprise attacks. Subsequently, priority was placed upon air defence, and it was planned that the SwAF should be capable of better countering a limited surprise attack against Swedish territory by 1997. In line with this goal, the defence planning programme for 1993–98 emphasised enhancing the recognised air picture. Particularly, the introduction of an airborne radar surveillance system before the turn of the century was expected to enhance considerably Swedish air defence capabilities. It was furthermore decided to retain all eight JA 37 Viggen air defence squadrons. Additionally, the acquisition of a new medium-range air-to-air missile was to improve the air defence capability. In contrast, the acquisition of new reconnaissance pods was to be delayed until well after the turn of the century. Furthermore, the SwAF pursued a modification programme for its AJ 37 attack and reconnaissance Viggens to AJS 37 standard. These aircraft would

Courtesy Gripen International; photo by Frans Dely

ASC 890 Erieye airborne early warning and control aircraft escorted by a Gripen C and D. The Erieye significantly enhances the SwAF's air defence capabilities.

bridge the gap till the introduction of 16 planned JAS 39 Gripen squadrons was completed.[53] In the mid-1990s, the number of planned Gripen squadrons was further reduced to 12.[54]

In accordance with the strategic security assessment, the Government Defence Bill 1996/97 reiterated the emphasis upon air defence. Therefore, the bulk of the JAS 39 Gripen aircraft was planned to be configured in the air defence role. Furthermore, the JA 37 air defence Viggen squadrons were to be retained. In contrast, the AJS 37 attack and reconnaissance Viggen squadrons were planned to be reduced significantly in the ensuing years.[55]

Force modernisation programmes throughout the 1990s were an integral part of the Flygvapnet 2000 (Swedish Air Force 2000) concept. It was conceived in the late 1980s and early 1990s, with the purpose of turning the SwAF into a network-based air force, implying free flow of information across multiple levels.[56] From a material point of view, this meant concentration upon a new generation of systems.[57] Besides the combat aircraft fleet, this programme particularly hinged upon a modernisation of the SwAF's C2 structure (see section on maintaining a relevant air force). In parallel to these developments, the command structure was streamlined. In the early 1990s, the four air defence sectors were superseded by three air commands, and in July 2000, these air commands were replaced by a single air force command.[58] The SwAF thus shifted from a decentralised command structure to a centralised one. Moreover, the command and control battalions were reduced from 11 in 1990 to three in 2005, with only one planned for 2006.[59] The streamlining process within the SwAF was part of a larger streamlining process of the Swedish armed forces as a whole. In 1993–94 a major organisational reform was initiated which elevated the position and authority of the supreme commander and relegated the formerly independent peacetime commanders of the single services to inspectors. For this purpose, the joint armed forces headquarters was established in September 1993, replacing the separate staff functions of the three services.[60]

Initiating Sweden's most fundamental defence reform in the post–Cold War era in 1999 had a significant impact upon the organisation of the SwAF. Yet already prior to 1999, a certain shift away from the primacy of air defence can be noticed. As such, the inspector general of the SwAF announced in 1998 that a planned midlife update of approximately 60 JA 37 air defence Viggen fighters was likely to be reduced or even cancelled. Instead, requirements for international operations would be investigated.[61] In 1999 plans for the reorganisation and reduction of the wartime and peacetime air bases were decided upon.[62] Moreover, the ground support units were to experience considerable cuts.[63] A particularly significant step was the disbandment of all JA 37 air defence Viggen squadrons by 2005, which led to a substantial reduction of the air defence capability in quantitative terms.[64] These cuts can be directly related to the positive security assessment in Sweden's immediate neighbourhood.

With Sweden's increasing thrust towards international operations across the spectrum of military force, interoperability became a specifically important issue and had a far-reaching impact upon the SwAF's force inventory. Only Gripen squadrons equipped with the NATO interoperable JAS 39C/D version have been considered operational from 2005 on. The remaining squadrons, purely equipped with the JAS 39A/B version, lost operational squadron status and were either disbanded or have been used for conversion training and tactical evaluation. Accordingly, the number of operational Gripen squadrons was reduced from seven in 2004 to three in 2005 and increased to four again in 2006, with more JAS 39C/Ds delivered.[65] These squadrons are to be fully equipped with the JAS 39C/D version.[66] Moreover, while the air defence capability had lost in relative importance, the CAS capability gained unprecedented emphasis, as it was regarded as a critical capability for robust peace support operations.[67] This shift in priorities went hand in hand with Sweden's efforts in setting up a Nordic EU battle group.

Yet against the backdrop of a reasserting Russia, the 2009 government bill, A Functional Defence, declared that "the air force must primarily develop the capability to operate in our region. It should also be able to participate in air operations together with other countries, in Sweden and within and outside our region."[68] With the SwAF and the Swedish Navy readopting a more national focus, the Swedish Army was expected to carry the main bulk of international missions.[69] As regards the SwAF, these developments represented a shift away from the rather one-sided thrust towards deployed operations. Yet, interoperability remained the cornerstone of the SwAF's transformation. Moreover, this did not preclude potential deployed operations. The Navy, with its similar regional focus, was increasingly involved in the EU's anti-piracy mission off Somalia, Operation Atalanta, and in April 2010, a Swedish rear admiral became the force commander of the European Union Naval Force. This maritime effort was supported by helicopters from Sweden's Armed Forces Helicopter Wing.[70]

While table 5 (next page) clearly reveals the prioritisation of air defence in the 1990s, it also shows the massive force reductions and thrust towards a homogenous combat aircraft fleet after 1999. These reductions can be attributed to a perceived improvement in security in Sweden's close neighbourhood and the demands of potential international operations and interoperability. Deployable combat aircraft units require larger investments than units tasked solely for territorial defence. For instance, the JAS 39C/D rapid reaction units demand intense training, particularly in the areas of planning and executing multinational operations and close air support.[71] Prior to the SwAF's significant force reductions, such intense focus upon training was not conceivable. In the later stages of the Cold War, the average flying hours of a fully trained pilot were significantly below NATO's mini-mum requirement. This situation was further aggravated by a very low pilot-to-aircraft ratio.[72] A parliamentary report pub-lished in 1992 highlights serious shortcomings in training, mod-ern air-to-air and air-to-ground weapons, and spare parts in general.[73] Difficulties in generating enough training hours per combat pilot persisted up to 2001.[74] Hence, it can be argued that the SwAF's force structure had actually been overstretched and only became more stable against the backdrop of Sweden's in-creasing commitment to out-of-area operations.

Table 5. SwAF combat aircraft inventory

Aircraft Type	1990	1997	2001	2005	2012 (planned)
J 35 Draken (air defence)	72	60	-	-	-
JA 37 Viggen (air defence)	139	133	91	-	-
AJ/S/AJS/SK 37 Viggen (attack, recce, standoff jamming)	147	104	32	19	-
JAS 39A/B Gripen (multirole)	-	37[a]	95	124	-
JAS 39C/D Gripen (multirole)	-	-	-	27	100
SK 60 (training, recce, CAS)	105	105[b]	105	103	?

Adapted from IISS, *The Military Balance 1990–91* (London: Brassey's for the IISS, 1990), 93; IISS, *The Military Balance 1997–98* (London: Oxford University Press, 1997), 96; IISS, *The Military Bal-ance 2001–2002* (Oxford, England: Oxford University Press, 2001), 100; IISS, *The Military Balance 2005–2006* (London: Oxford University Press, 2005), 128; and MOD, Sweden, "A Functional De-fence," fact sheet, March 2009.

[a] Primarily employed in the air defence role. MOD, *Regeringens proposition 1996/97:4: Totalförsvar i förnyelse—etapp 2* [Government defence bill 1996/97:4: renewal of total defence—stage 2] (Stock-holm: MOD, 12 September 1996).

[b] In 1996 the SK 60 ceased to be used in the reconnaissance and close air support role. Col Bertil Wennerholm, SwAF, retired, Stockholm, to the author, e-mail, 11 February 2008.

Emphasis upon Air Power. A 1991 RAND analysis considered the SwAF to be the most important branch of the Swedish armed forces, as it would have provided the first line of defence in a high-intensity conflict scenario with the Soviet Union.[75] Viewed from a

Nordic perspective, Swedish emphasis upon air power made sense. While Finland was expected to become bogged down in close quarter battles due to its large land border with the Soviet Union, the SwAF together with the Swedish submarine fleet constituted a manoeuvre element across the Baltic Sea. Not only military considerations but also politico-industrial considerations explain Sweden's emphasis upon air power. With the end of the Cold War, this emphasis upon air power remained, as the single service budget allocations from 1989 to 2002 reveal (see fig. 5).

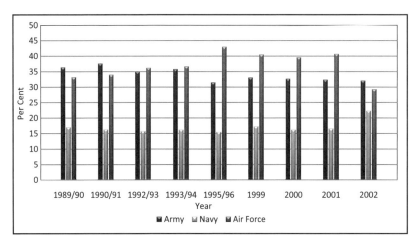

Figure 5. Budget allocation to the single services (Sweden). The overall defence budget further includes expenditure on the central administration (support and command), military assistance including UN peacekeeping, and other. Until 1995–96, Sweden had a fiscal year from 1 July to 30 June. (Author developed figure based on percentages from Bengt-Göran Bergstrand [senior analyst, Swedish Defence Research Agency (FOI), Defence Expenditures Project, Stockholm], interview by the author, 11 February 2005.)

According to the broad approach to air power which this study takes, the air power assets of the Swedish Army and Navy have to be included as well. From 1994 to 2002, the Army operated up to 600 air-defence guns plus a considerable number of SAM systems, mostly MANPADS but also a small number of upgraded Hawk systems and other advanced systems. Furthermore, the Army and Navy air arms operated up to 100 training, liaison, rescue, transport, anti-tank, and anti-submarine warfare helicopters. If these systems are taken into account, air power has featured even more prominently in the Swedish defence architecture and accounted for significantly more than a third of the Swedish defence expenditure.[76]

This book argues that, against the backdrop of an increasing shift towards out-of-area operations, it became increasingly difficult for the SwAF to retain its prominent position within the Swedish defence architecture. Apart from the SwAF's transport unit, Swedish air power was basically relegated to control of the Swedish airspace and to territorial defence up to the year 2000.[77] Air power could not be integrated into multinational air campaigns due to the lack of interoperability, nor was the Navy in a position to participate in the embargo operations against the former Yugoslavia in the absence of appropriate blue-water assets.[78] In the 1990s, this did not really matter, as Swedish authorities still reckoned with the possibility of strategic surprise attacks, the countering of which demanded a sophisticated air defence capability. Yet with Sweden's increasing involvement in European security from 1999 onwards, readiness to participate in multinational operations had become a key factor. So far, the main burden of international missions has been carried by the Army. Army contingents up to battalion strength served in Lebanon, the former Yugoslavia, and Liberia.[79] With the setting up of the Nordic Battle Group, this trend continued. Hence, it was in the SwAF's own interest to focus upon international missions and to redress this imbalance. However, recent events, particularly the Russian intervention in Georgia, partially reversed this logic. The SwAF, together with the Navy, might again be seen as the first line of defence in a potential military incidence in the Baltic area.

Alliance Context

Prior to the dissolution of the Warsaw Pact and the Soviet Union, Swedish neutrality policy was defined as military non-alignment in peace, implying neutrality in war. In early 1992, Swedish neutrality in war was no longer deemed an imperative. Instead, a flexible approach was adopted, whereby neutrality was regarded as one option amongst others. Henceforth, the purpose of being non-aligned has been to enable Sweden to remain neutral in the event of a conflict in its vicinity, should Sweden wish to do so.[80] Despite the policy of non-alignment, the UN, the EU, and NATO have played a significant role in Sweden's defence policy. The UN is regarded as the ultimate responsible body for maintaining international peace and security.[81] As such, a UN mandate is regarded as imperative for Swedish participation in peace support operations.[82] Since the late 1990s, the EU has exerted an unprecedented influence upon the Swedish defence architecture. In contrast to the UN, the ESDP has had an immediate impact upon Swedish military transformation.

Sweden joined the EU in 1995 and embraced the 1999 Cologne Declaration stating that the EU should acquire the necessary civilian and military capacities to engage in crisis management operations. Due to its policy of non-alignment, Sweden did not endorse an EU solidarity clause, calling for collective defence.[83] To reconcile ESDP membership and the policy of military non-alignment, Swedish officials identified elements incompatible with this policy early in the process.[84] Especially, Sweden together with Finland pushed for the ESDP to be focused upon the Petersberg Tasks, which include crisis management and exclude collective defence arrangements.[85]

The embedding into the ESDP changed the basis of Sweden's defence policy.[86] Accordingly, the 2004–5 defence bill argues that the new Swedish focus should be on making a relevant contribution to the EU's power projection capabilities, particularly in the domain of rapid reaction capabilities.[87] The setting up of a multinational Nordic Battle Group has been one of Sweden's greatest military endeavours. Though there were several reasons for a defence reform, the ESDP process has no doubt been important in speeding up the process of military transformation.[88] In early 2004, the supreme commander of the Swedish armed forces expressed his intention to give ever more weight to European cooperation: "I see the steadily growing European cooperation as a basic starting point for our continuing defence reform."[89]

Throughout the 1990s, special attention was also given to NATO and to the US presence in Europe. Both were seen as fundamental pillars for Swedish security. The social democrats openly acknowledged that Swedish non-alignment policy depended upon a clear and unambiguous American commitment to European security.[90]

This development stood in clear contrast to the Cold War era, when a linkage between Sweden's security and the American-dominated NATO had been carefully avoided in public statements.[91] Swedish NATO membership was, however, not aspired to during the 1990s. A reason was that Swedish or Finnish NATO membership would with certainty have been viewed negatively by Russia and therefore would not have contributed to stability in the Baltic Sea region.[92]

Nevertheless, Swedish rapprochement towards NATO took place, particularly when the country joined the Partnership for Peace in May 1994 and when it established a formal national delegation, in fact a NATO embassy, within the NATO headquarters in Brussels in the late 1990s.[93] The tight bonds between Sweden and NATO were again underlined in the 2004–5 defence bill that called for a continuation and reinforcement of Sweden's cooperation with the alliance through PfP.[94] In relation to NATO, Sweden's ambition is to be seen as a non-ally that is ready to "give and not only to take."[95]

When the Baltic states were invited as NATO membership candidates, this development was clearly welcomed by Swedish decision makers. The enlargement of NATO in 2004 was considered to be conducive to sustaining a long-term American commitment to northern Europe, which had been a key issue for Sweden throughout the 1990s.[96]

Against the backdrop of a reasserting Russia, the Swedish Ministry of Defence declared in early 2009 that the country would show solidarity with other EU member states or Nordic countries in case of a military attack, and it implicitly stated that Sweden would expect reciprocal assistance.[97] This represents a cautious attempt at formulating an informal Nordic collective defence framework.

Alliance Policy and Its Impact upon the Swedish Air Force

Sweden's alliance policy has indeed influenced the SwAF. Nowadays, it can be regarded as one of the most important factors dominating force transformation. With Sweden's commitment to PfP and especially the ESDP, Swedish combat aircraft have been made available for international operations since the turn of the century. Furthermore, the SwAF started to participate in and to conduct international exercises. This has enhanced interoperability with other air forces on a tactical level. Sweden's alliance policy and

Courtesy Gripen International; photo by FMV

FAF C-135FR tanker refuelling Gripen C and D in midair

increased commitment to out-of-area operations have also had direct repercussions upon the SwAF at the technical level.

ESDP and PfP Context. In 2001 Minister of Defence Björn von Sydow stated that the internationalisation process permeated the Air Force organisation as a whole, from doctrine, training, exercises, staff procedures, procurement, and language training to the air base and C2 structure.[98] Immediately following the 1999 defence reform, 2000 and 2001 were indeed key years for the SwAF. In June 2000, the Swedish government decided to set up a dedicated Air Force rapid reaction unit for international peace support operations within the framework of the UN, the EU, or NATO's PfP. The SwAF's rapid reaction force became operational on 1 January 2001 and consisted of two units—the Swedish Air Force AJS 37 Rapid Reaction Unit (SWAFRAP AJS 37) and SWAFRAP C-130.[99] Both units were on a 30-day standby for international contingencies.[100] SWAFRAP AJS 37 was a tactical reconnaissance unit consisting of six AJS 37 photoreconnaissance Viggens and approximately 220 professional and contracted personnel. Potential deployment could last up to a six-month period.[101] At the same time, the SwAF transport unit readied four C-130 Hercules aircraft and the major portion of its personnel for rapid deployment in the context of SWAFRAP C-130.[102]

Though SWAFRAP AJS 37 was never deployed for real operations, it allowed the SwAF to gain international experience through participation in various exercises.[103] SWAFRAP AJS 37 was replaced by a new rapid reaction unit, SWAFRAP JAS 39, which became operational on 1 January 2004. The unit was originally equipped with eight JAS 39A Gripen aircraft, and the contracted and professional personnel numbered approximately 260.[104]

Against the backdrop of the NBG going on standby during the first half of 2008, NATO interoperable JAS 39C Gripen combat aircraft provided part of an air component. This NBG Gripen detachment replaced SWAFRAP JAS 39 and was designated Stridsflygenhet 01 (SE01—Combat Aircraft Unit 01). Its mission spectrum encompassed air defence, reconnaissance, and precision strike. CAS was especially regarded as an important capability. The SwAF contribution to the NBG was eight JAS 39C Gripen multirole fighters, four Hercules transport aircraft, and an air base unit. Moreover, the Armed Forces Helicopter Wing was ready to dispatch seven transport helicopters. SE01 could also be deployed outside the framework of the NBG, effectively taking over the role of SWAFRAP JAS 39.[105] In the near future, all operational JAS 39C/D Gripen squadrons are planned to be deployable.[106]

The SwAF's NBG contribution was also listed in the *EU Force Catalogue 2006*, which earmarked deployable military forces for major EU operations.[107] Moreover, the SwAF's S 102B Korpen signals intelligence aircraft can also be made available for international operations, and a possible future deployment of Swedish airborne early warning assets was discussed in 2005.[108] With a retrofit (see "C4ISTAR" section), two S 100 Argus airborne early warning aircraft were readied for international deployments. Hence in the future, all the SwAF's major weapon systems are likely to be available for international operations.

Cooperation on an Operational Level. During the Cold War, it was very unusual for the SwAF to participate in international exercises, and the mindset of SwAF officers was almost exclusively geared up for national defence scenarios. In this regard, the post–Cold War era brought about a significant shift. To facilitate the SwAF's thrust towards increased interoperability, English became the language of tactical aircraft-to-aircraft communication. Furthermore, PfP provided a platform for participation in multinational exercises.[109] The SwAF's rapid reaction units have particularly been subject to rigorous training abroad. From its very be-

Courtesy Gripen International; photo by Gunnar Åkerberg

JAS 39B Gripen at the multinational Exercise Cold Response in Norway

ginning, SWAFRAP AJS 37 participated in NATO exercises, and SWAFRAP C-130 crews were able to improve their skills in integrated exercises such as Maple Flag in Canada.[110] Nowadays, JAS 39 Gripen units participate in various multinational exercises in Europe, primarily northern Europe.[111]

In accordance with the SwAF's evolving international commitment, the SwAF's rapid reaction units first concentrated on low-intensity PfP exercises.[112] With SWAFRAP JAS 39 becoming operational in early 2004, Swedish combat aircraft have participated in several multinational integrated air exercises, such as the Nordic Air Meet in Sweden, Frisian Flag in Holland, and Cold Response in Norway. Moreover, participation in Exercise Red Flag in the United States or the NATO Air Meet in Europe was considered in 2007. Against the backdrop of the NBG, Swedish JAS 39C/D Gripen combat aircraft were also deployed to Alaska to practise CAS under realistic conditions.[113]

Besides Sweden, Hungary and the Czech Republic are European JAS 39 Gripen operators. The SwAF has been providing training assistance for Czech and Hungarian Gripen pilots and technicians. This kind of cooperation has fostered the SwAF's international expertise. For instance, training with the Czechs took place in English, and all training material was translated into English for this purpose.[114] Furthermore, Swedish Gripen aircrews had the opportunity to participate in exercises in Hungary.[115] This close training cooperation led to considerations of a more formalised partnership amongst the European Gripen operators, which could cover aspects such as training and logistics.[116]

How Has the Swedish Air Force Responded to the Challenges of Real Operations?

Sharing the Baltic Sea with the Soviet Union led to a number of military incidents during the Cold War. For the SwAF, probably the most significant incidents were the June 1952 alleged shoot-down of a DC-3 signals intelligence aircraft over the Baltic Sea and the ensuing Soviet fighter downing of a Swedish Catalina flying boat that was searching for the DC-3 crew. These events marked the beginning of a comprehensive alert service involving round-the-clock radar surveillance of the Swedish airspace, with fighter as well as reconnaissance aircraft in a high state of readiness.[117]

During the Cold War, Swedish aircraft on numerous occasions had to scramble to intercept foreign aircraft. Aerial readiness for interception missions varied over the years. During the 1950s, 1960s, and 1970s, it was particularly high, and readiness was maintained around the clock. Contacts with foreign military aircraft over the Baltic Sea were common for Swedish pilots. During certain years, there were as many as 250 aerial contacts. These occasions were particularly valuable for information gathering, and it was a way to show Swedish determination. Furthermore, reconnaissance aircraft were on standby to identify Soviet and Western military vessels and to monitor foreign maritime exercises. Aerial readiness placed a considerable burden upon the SwAF, which was shared among the various wings. In the 1980s, the SwAF often used mixed pairs consisting of one JA 37 air defence Viggen and one AJ 37 attack Viggen for interception missions. The reason for this was to diminish the burden on the JA 37 interceptor wings. At the same time, the attack Viggens provided a very good maritime picture.[118]

Probably the tensest situation with the Soviet Union in the latter half of the Cold War occurred during the submarine incident in 1981. When a Soviet submarine grounded in the Swedish archipelago, Soviet surface vessels closed in on the Swedish coast to cover a rescue attempt. The SwAF answered by having JA 37 attack Viggens constantly in the air.[119]

Yet, Swedish combat and reconnaissance aircraft were not purely relegated to border watch missions and territorial defence tasks. They also actively participated in UN missions, as in 1958 when five Swedish aircraft were sent to Lebanon for surveillance missions.[120] Soon afterwards, Swedish combat aircraft were readied for a UN mission in the former Belgian Congo. After the declaration of the country's independence, the Katanga province also declared independence and separated itself from the rest of the country, which led to a civil war. To stabilise the situation, the Swedish government deployed combat aircraft in late 1961. On two occasions, Swedish pilots conducted successful offensive counterair missions against the Katanga Air Force, wiping out substantial parts of it on the ground. Moreover, Swedish aircraft protected UN convoys and gave fire cover to UN ground forces. First supported by Indian and Ethiopian combat aircraft, the Swedish unit operated on its own in the autumn of 1962, when the situation became tense again. The mission lasted until 1963.[121] Not only Swedish combat aircraft but also transport aircraft contributed their share during the Cold War. In 1952 the SwAF's transport unit conducted its first international relief flights to Holland, and very soon, it expanded its missions to numerous disaster areas around the world.[122]

In the post–Cold War era, the SwAF has continued to be ready for real operations. With regard to international operations, the Air Force's contributions have so far been confined to the lower spectrum of military force, given the country's legacy of neutrality.

Swedish Aerial Readiness and Potential Future Scenarios

The SwAF has continued to patrol the Swedish airspace and the Baltic Sea in the post–Cold war era. Swedish combat aircraft have remained ready to scramble in case of an emergency. While the number of military flights over the Baltic Sea has considerably decreased, 11 September underlined the importance of aerial readiness.[123] In the immediate aftermath of the attacks against New York and Washington, Swedish fighters patrolled the sky over Stockholm.[124] The SwAF also maintained combat air patrols during important events such as Pres. George W. Bush's visit to the EU summit in Göteborg, Sweden, from 13 to 15 June 2001, or during the memorial service for the minister of foreign affairs,

Annah Lindh, in late 2003.[125] Yet as of 2004, rules of engagement did not allow for rogue planes to be downed.[126] The SwAF also maintained aerial readiness together with the Norwegian Air Force, as was the case during the Euro-Atlantic Partnership Council Security Forum in Åre, Sweden, from 24 to 25 May 2005. Besides fighter aircraft, Swedish S 100 Argus airborne early warning aircraft were on station.[127]

Courtesy Swedish Air Force; photo by Peter Liander

In the early 1990s, 13 SF 37 reconnaissance Viggens were upgraded to the AJSF 37 standard.

Despite the rather relaxed situation, missions over the Baltic Sea were not without risk in the 1990s. At the end of the Cold War, the Soviets lost a fighter aircraft following a Swedish reconnaissance Viggen, and on 16 October 1996, the SwAF lost a fighter aircraft between the island of Gotland and Latvia. Executing a low-level pass for identification of a Russian warship, the Viggen collided with the water surface.[128]

In the future, Swedish combat aircraft might not only be employed in the aerial readiness role. With JAS 39 Gripens getting readied for out-of-area interventions, Swedish aircraft are likely to conduct operations across the spectrum of military force within the framework of robust peace support operations. In the post–Cold War era, Swedish decision makers have already proven that they are willing to share risks. In mid-2003, during the EU operation Artemis in the Democratic Republic of Congo, some 70–80 Swedish special forces conducted missions at the upper level of the spectrum of military force together with their French counterparts. This Franco-Swedish force provided the EU task force with a highly effective capability to engage and neutralize armed threats.[129] Again in 2004, the Swedish government deployed heavily armed troops to Liberia. The unit consisted of 231 soldiers with armoured personnel carriers, acting as a standby force for quick intervention in particularly difficult situations.[130] Hence, it can be concluded that it might only be a question of time until JAS 39 fighter-bombers will engage in deployed operations, despite the new emphasis upon regional scenarios in the wake of the crisis in Georgia. This assumption is corroborated by a 2004 opinion poll. According to this survey, half of the Swedish population supported robust military operations at the request of the UN or the EU, even if they entailed fatalities.[131]

The option of potentially employing offensive forces in deployed operations was pursued incrementally. In 2001 the minister of defence, Björn von Sydow, declared that SWAFRAP AJS 37 would be relegated to reconnaissance and air-defence missions. Ground-attack missions were regarded as too risky, involving a high probability of collateral damage. The minister, however, added that this position might be reconsidered in the future.[132] Already in 2003, the Swedish Defence Commission argued that the capacity to engage in deployed armed conflict is considered to be a key competence of the Swedish armed forces.[133] In an interview with the author in February 2006, the state secretary for defence, Jonas Hjelm, declared that once JAS 39 fighter-bombers are able to deliver LGBs, this capability could in principle be employed within a robust peace support operation, provided there is a UN mandate and tight political control of potential air strikes.[134] Gaining political leverage might certainly have played a major role in Sweden's decision to consider joining in the early entry phase of an operation. Moreover, it has become an international norm for troops involved in peace support operations to have a robust escalation capacity at their disposal.

Humanitarian Operations, Troop Deployments, and Subsidiary Tasks

Since the delivery of the first C-130 in 1965, Swedish C-130 Hercules transport planes have participated in relief operations around the world, mostly for the International Committee of the Red Cross and the UN. The first major relief operations were conducted by the SwAF transport unit at the formal request of the Red Cross in Ethiopia, when the country was struck by successive famine disasters in 1985 and 1988.[135] The task of the SwAF transport unit was to provide intratheatre airlift for delivering food and aid packages from the ports to the more remote and mountainous regions of the country.[136] The SwAF transport unit also supported the UN Transition Assistance Group mission in Namibia in 1989. Moreover, throughout the 1980s, Swedish Hercules aircraft resupplied Swedish UN contingents in Cyprus and Lebanon.[137]

The first significant troop and materiel deployment supported by the air transport unit in the post–Cold War era occurred during Desert Storm, when Swedish C-130s airlifted a Swedish field hospital with 500 staff from Sweden over Cyprus to Riyadh, Saudi Arabia. The operation lasted three weeks and pushed the organisation, a total of eight aircraft and approximately 150 personnel, to its limits. A major lesson learned was to keep enough spare parts in storage. In 1993 the transport unit again transported a Swedish field hospital to Mogadishu.[138]

Soon after Desert Storm, the Swedish transport unit again became involved in a major operation. In mid-1992, one Swedish C-130 transport aircraft, based at Zagreb International Airport in Croatia, was contributing to the UN air bridge to Sarajevo from the very beginning of the operation. Yet when it became known that the warring parties tracked transport planes with air defence radars, and after the Swedish C-130 Hercules came under mortar fire at the airport of Sarajevo, it was decided to withdraw the Swedish contingent after less than two months of operations and prior to the shootdown of an Italian aircraft on 3 September 1992. After its C-130 Hercules fleet had been retrofitted with self-defence systems, the SwAF rejoined the airlift effort in 1994, operating from Ancona, Italy. Whereas in 1992 the SwAF detachment had to rely upon Canadian intelligence, a Swedish intelligence officer actively contributed to the intelligence gathering process in 1994. Besides supply sorties, the Swedish unit also conducted medical evacuation missions.[139]

In the post–Cold War era, the Swedish air transport unit has also facilitated Swedish power projection. In this task, the C-130

Hercules aircraft have been complemented by civilian airliners.[140] The Swedish transport unit also flew personnel and supplies for the Austrian, Danish, and Finnish armed forces to the Balkans or Afghanistan.[141] The Swedish C-130s were, however, not considered adequate for intertheatre airlift in the context of the NBG (see section on maintaining a relevant air force).

Africa has remained a hot spot for the Swedish transport unit. In 1996, for instance, one Swedish C-130 Hercules was based in Nairobi, Kenya, for missions on behalf of the International Committee of the Red Cross in Rwanda, Burundi, and the Democratic Republic of Congo.[142] But also countries such as Honduras, El Salvador, and Thailand received logistical support from Swedish C-130s after having been struck by natural disasters.[143]

The Swedish air transport unit has a long tradition in assisting civilian authorities. In November 2004, for instance, it conducted evacuation flights out of the civil-war-torn Ivory Coast.[144] Within Sweden itself, the SwAF transport unit has executed relief missions in case of natural disasters.[145]

The SwAF has played a crucial role in providing logistical support for UN operations as well. Within the framework of UN operation Mission de l'Organisation des Nations Unies en République

Courtesy Swedish Air Force

Swedish C-130 Hercules in Afghanistan

Démocratique du Congo (UN mission in the Democratic Republic of Congo), the SwAF took on responsibility for the UN's airport service in the town of Kindu from June 2003 to June 2004. The Swedish unit consisted of professional and reserve officers, as well as civilians, who provided the main bulk of personnel and had previously worked at different civilian airports in Sweden. The tasks of the Swedish unit were comprehensive, ranging from air traffic control to air terminal security.[146]

The deployment of a Swedish airport support unit to the Democratic Republic of Congo was in certain aspects a forerunner to the future Swedish base battalion concept. While base battalions were originally designed to service Swedish combat and transport aircraft in Sweden only, from 2008 onwards, Swedish base battalions are to operate autonomously in an international environment. With the NBG on standby in the first half of 2008, deployable air base units were a critical requirement.[147]

Signals Intelligence

A rather veiled Swedish contribution to international operations is SIGINT. Within the scope of Sweden's participation in the UN missions in Bosnia and Kosovo, S 102B Korpen SIGINT aircraft conducted electronic intelligence missions.[148] As of 2005, these aircraft were made available only for PfP and ESDP operations within the closer vicinity of Europe, yet this was expected to change.[149] To what degree signals intelligence is shared with partner nations remains confidential. Hence, the international leverage gained by these operations cannot be assessed.

How Has the Swedish Air Force Responded to the New Intellectualism in Air Power Thinking and Doctrine?

Milestone steps in the evolution of air power doctrine development and military education have been closely related to the steps Sweden made towards opening up its defence policy in the late 1990s. As such, Sweden has increasingly cooperated with other countries, and its armed forces have become more interoperable (e.g., SwAF's rapid reaction units). According to a Swedish scholar, the demands of international operations required the military to rethink the Swedish approach to doctrine and education. To foster interoperability, it was considered necessary to embed Swedish military thinking into a common Western—or rather Anglo-Saxon—discourse.[150]

Air Power Doctrine

In 1996–97, the doctrinal development process within the SwAF started as a bottom-up approach, when a mid-career officer attending the Swedish National Defence College (SNDC) wrote a draft of an air power doctrine. This draft was well received by the higher echelons, and a small study group was set up within the SwAF to elaborate on it. Yet officers in charge could only work part-time on the project, and it soon faded out.[151] According to the officer who started this bottom-up approach, one of the main obstacles was the very limited number of officers educated in air power theory and doctrine.[152] Another unsuccessful attempt was reportedly undertaken in 2001. Only in 2002 did the doctrinal development process within the SwAF receive a solid institutional foundation, when the Air Tactical Command decided to implement a project team that would work full-time on doctrinal issues for a year in order to produce a draft version of a national air power doctrine.[153] It can be argued that this reorientation was closely linked to the publication of an overarching military-strategic doctrine. Published in May 2002, it already anticipated joint and single service doctrines.[154] Hence, an exogenous factor provided the necessary impetus for properly institutionalising the doctrine development process within the SwAF.

The doctrine team consisted of three officers working full-time and a number of other officers, researchers, and staff personnel helping out on a part-time basis. Officers returning from foreign staff colleges gave their input on an ad hoc basis. In contrast to the earlier attempts at developing a national air power doctrine, a steering committee headed by a general as well as an expert panel had taken on responsibility for the work in progress.[155] As a result, the doctrine development process was well anchored and monitored within the institution and led to a published air power doctrine in early 2004.

The chapters of the 2004 air power doctrine were structured according to a genuine logic, which started out with the air environment and air power's abilities and principles. As a further step, the single mission types were analysed. To conclude, the international dimension of air operations and an outlook for the future were presented.[156] Moreover, the 2004 edition introduced a broad panoply of topics, from interwar air power thought in Europe and the United States, through Western air power thought in the aftermath of World War II, to modern air power theory. The section on modern air power was dominated by American thinkers and concepts. As such, it referred to concepts by John Warden and John Boyd and also dealt with the concept of parallel warfare.[157] With regard to Swedish specifics, a particular reference was made to the provision of indigenous equipment.[158]

Given Sweden's defence political opening process, it was an explicit aim to harmonise the national air power doctrine with NATO doctrine. This thrust towards doctrinal interoperability had already been underlined by Sweden's Military-Strategic Doctrine of 2002, which stated that "our participation in multinational operations within the framework of the UN, NATO and European crisis management entails certain harmonisation with multinational doctrine, primarily NATO's *Allied Joint Doctrine* (AJP-01)."[159] In certain areas, such as the categorisation of air operation or the CJTF concept, the alliance's air power doctrine, *Joint Air and Space Operations Doctrine*, AJP-3.3, served as an example, though Swedish specifics were taken into account. For instance, so-called strategic air operations were not introduced as a separate topic.[160]

Air power seminars held annually at the SNDC served as a platform for input from a wider audience during the doctrine development phase. During the first seminar in 2002, for instance, the essence and value of doctrine were discussed in general terms. Representatives from the military strategic, joint, and single service doctrine working groups discussed the progress of their work. During the 2003 seminar, an entire day was exclusively devoted to the outline and progress of the Swedish air power doctrine. In a number of workshops, its contents were discussed and suggestions to improve the draft were given.[161] Undoubtedly, these phases of reflection prevented the SwAF from merely copying the structure and outline of already existing Anglo-Saxon air power doctrines.

The 2004 Air Power Doctrine was bestowed only draft status, as the decision was taken that the single service doctrine had to be harmonised with the joint doctrine that was still being developed, primarily by Army officers. To comply with the demands for harmonisation, the 2004 air power doctrine was amended without further delay, and an updated edition was published in early 2005 alongside the joint doctrine. The preliminary version of 2004 served as a basis for the revisions.[162] Despite its unofficial status, it nevertheless represented a milestone step for the SwAF.

The updating process focused primarily upon rearranging existing material. As such, the structure of the revised air power doctrine was brought completely in line with the joint doctrine.[163] Yet, topics such as interwar air power thought or American concepts, explicitly referred to in the 2004 edition, were heavily reduced or completely omitted.[164] The harmonisation process occurred against the backdrop of the general Swedish view that there was actually only one doctrine with several volumes rather than different doctrines.[165] This thrust can be considered an integral part of the ambition to create a joint mentality.

The doctrine development process within the Swedish armed forces in general and within the SwAF in particular followed the major defence reform in 1999. It can therefore be argued that Sweden's increasing engagement in international operations as well as the shift from a threat-driven defence towards a capability-based

Courtesy Gripen International; photo by Katsuhiko Tokunaga

Split from Gripen dual-seat cockpit

defence represented the key motivations for grasping doctrinal issues. Yet, the revision of the first edition represents an institutional evolution and not an intellectual one.

Teaching of Air Power Thought

A SwAF career officer undergoes three courses in which air power history and theory are taught in greater depth. The first course is the seven-month junior staff officer (JSO) programme. Next, majors attend the general staff officer programme I and II (GSO I & II) at the SNDC, each of them taking one year. The JSO and GSO programmes were established in the 1960s, though with different names. Until the late 1990s, these courses put a premium upon the study of tactics and staff work. While the courses were almost exclusively intraservice, GSO II allocated brief interludes for joint training and education.[166]

Since the late 1990s, which marked the starting point of an educational reform, the SNDC has been evolving from a purely military institution to an institution embracing academic standards, education, and research. In particular, in 2002 completely new curricula putting a premium upon military theory and history were introduced.[167] This crucial step was facilitated by a Norwegian professor, Nils Marius Rekkedal, who was employed at the Department of War Studies at the SNDC in 2001.[168] The air faculty within that department had rediscovered the classical and modern air power theorists in 1999 and included them into the curricula and reading lists for the JSO and GSO courses. The 2002 reform thus underlined and reinforced the approach the air faculty had already taken three years before.[169]

The JSO curriculum and reading list serve to introduce SwAF officers to military theory in general and air power theory in particular. In terms of air power, the reading list focuses upon American authors such as Warden and Phillip S. Meilinger. Moreover, the SNDC's own air power publications are part of the required reading list. Besides these theoretical writings, doctrinal writings form an important part of the JSO reading list. Air Force officers study not only Swedish air power doctrine but also military strategic and joint doctrine. Furthermore, NATO doctrine publications such as AJP-3.3 or the NATO *Guidelines for Operational Planning* are part of the readings, which clearly reveals the SwAF's thrust towards interoperability on a doctrinal level.[170]

The GSO reading list elaborates on the JSO reading programme, reiterating the general topics as well as broadening and deepening the contents. Particular attention is devoted to military theory, the classics of military thought, grand strategy, and security policy. Furthermore, the programme exposes the students to the basic theories of their sister services. With regard to air power theory, the programme again puts a clear emphasis upon Anglo-Saxon literature. Books, publications, and articles by authors such as Tony Mason, Robert A. Pape, David A. Deptula, and James Corum are studied. The students are also introduced to topics such as the evolution of air power theory, John Boyd's decision-making concept, effects-based operations, air power and coercion, and air power in low-intensity conflict, as well as Russian military aviation.[171] Regarding the dissemination of doctrinal thought, Swedish air power doctrine is also taught at the GSO level. While it is introduced in rather rudimentary form at the JSO level, students at the GSO level are encouraged to reflect upon air power doctrine, to analyse it in its entire depth, and even to present alternative suggestions to the "absolutes" stated in the SwAF's formal doctrine.[172]

Since the beginning of the reform in the late 1990s, the development of a joint culture and mindset has been a major ambition at the SNDC.[173] The curricula for the JSO and GSO programmes are divided into interservice and intraservice sequences, facilitating a mutual understanding and at the same time furthering professionalism in one's own branch.[174] In the same vein, the study of Swedish joint doctrine can be regarded as a fundamental element on the way towards an enhanced interservice culture and joint mindset.

Sweden's security political reform had an impact not only upon joint education but also upon international cooperation. In the late 1990s, the SwAF started to send officers to the Air War College or the Air Command and Staff College at the Air University at Maxwell AFB.[175] Today, about 10 selected students annually complete their GSO II programmes at non-Swedish military institutions such as the Finnish Military Academy, the Staff College in Oslo, the Führungsakademie in Hamburg, the JSCSC in the UK, or—as just mentioned—AU.[176] Upon their return from foreign staff colleges, some SwAF officers serve in the staff of the air faculty at the SNDC. Through their first-hand experience of foreign views and nuances in air power thinking and theory, they are supposed to improve the quality of the SNDC curriculum. Moreover, approximately 10 foreign students study at the SNDC each year. The basic requirement is a fairly good grasp of a Nordic language, which obviously limits the number of potential candidates. Over the past few years, some British, American, and German, but primarily Norwegian, Danish, and Finnish, students took advanced courses at the SNDC. Particularly, the Norwegian students offer an insight into NATO procedures.[177] In general, links with the Norwegian defence establishment are strong. As such, Col John Olsen, the dean of the Norwegian Defence University College and a prolific air power scholar, is also a visiting professor of operational art and tactics at the SNDC.

An educational goal of the SNDC air faculty is to form officers capable of informed judgement in the field of air power and beyond. It is assumed that academic warriors have a better chance of accomplishing their objectives due to a broader range of insights. After having attended the GSO programmes, airmen are expected to be capable of accomplishing their tasks in combined and joint air operations. This goal is supported by combining the teaching of the operational art with historical case studies from the Second World War up to the latest air campaigns and conflicts. The concept of defensive counterair operations, for instance, is exemplified by studying the Battle of Britain, while offensive counterair operations are introduced by scrutinising the air campaign of Desert Storm. Operation Allied Force for its part is deemed an adequate example for the study of multinational operations. Furthermore, the teaching of air power theory heavily draws upon the historical evolution. A particular emphasis is put upon air power thinking in the interwar years, with particular reference to the situation in Sweden.[178]

Dissemination of Air Power Thought

The annual air power seminars held at the SNDC became an important vehicle to further the intellectual air power debate within the SwAF. Furthermore, publications on air power theory produced by the SNDC underline the SwAF's thrust towards achieving a professional grasp of the new intellectualism in air power thinking.

From 2002 to 2005 air power seminars were organised annually. The 2002 seminar dealt with the nature and essence of military doctrine. It was basically run by Swedes.[179] In the following years, the air power seminars received an international or rather Anglo-Saxon character. Speakers such as Mason, Meilinger, Warden, or Philip Sabin were invited to share their latest views upon the application of Western air power in a dynamically changing environment.[180] The seminars lasted two days, with the first day conducted in English and the second in Swedish. Whereas the first day was dominated by Anglo-Saxon speakers, the second day was reserved for internal debates. To debate and to transfer the Anglo-Saxon dimension into a Swedish context, workshops and panel debates are held. In 2004, for instance, the future development of Swedish air power capabilities was pondered. The year after, the status and the mission types of the SWAFRAP JAS 39 squadron as well as the threat of "air" terrorism were debated.[181] Hence, foreign—particularly Anglo-Saxon—ideas are put into a Swedish context and not just copied without any further reflection.

In addition to the annual air power seminars, the SNDC started to produce its own publications on air power theory, with a particular focus on the application of air power in the post–Cold War era.[182] Though not with an explicit air power focus, the SNDC has also published volumes that have a strong relevance for air power theory, such as a publication on effects-based operations.[183]

A particularly interesting publication is *Daidalossyndromet: Om luftmakt 2005* (*The Daedalos Syndrome: Thoughts on Air Power 2005*).[184] It is composed of essays written by previous SNDC students and presentations given by international speakers at the annual air power seminars. According to the head of the SNDC air faculty, this publication was supposed to be food for thought for a doctrinal air power debate. In particular, it relates to the sort of missions the Swedes expect to be dispatched to as part of the EU Battle Group concept, revealing the intertwined relationship between air power education and Sweden's changing defence political orientation.[185] Hence it can be concluded that, though the air power debate in Sweden is dominated by Anglo-Saxon ideas, continuous efforts are made to generate genuine Swedish thinking on air power.

How Have Swedish Defence Planners Attempted to Maintain a Relevant Air Force in Light of Escalating Costs and Advanced Technologies?

Throughout the Cold War, Sweden maintained a vast defence industrial complex capable of producing complete main weapon systems. A particular concern was the development and production of indigenous fighter aircraft. The roots of the Swedish aircraft and aerospace industry reach back before World War II. In 1937 Svenska Aeroplan was founded and later became known as Saab. Originally, activities were relegated to building foreign aircraft under license, but being cut off from foreign supplies during World War II forced the Swedes to design and manufacture their own aircraft. As a result, since the 1940s, the procurement of domestically designed, developed, and manufactured combat aircraft has been considered an important pillar of the Swedish defence posture. Throughout the Cold War era, the Swedish aerospace industry proved its capacity to turn out very sophisticated aircraft such as the Draken or the Viggen. Yet, components threatening to overstretch Sweden's industrial base, such as jet engines or air-to-air missiles, were built under licence. The American C-130 Hercules is a rare example where an off-the-shelf procurement policy was adopted.[186] In line with Sweden's policy of neutrality, the goal was to be as self-reliant as possible in the supply of defence materiel. The Swedish defence industry was able to cover 85 to 90 per cent of the Swedish armed forces' acquisition programmes in the later stages of the Cold War.[187]

With the end of the Cold War and with an increasing need for Western power projection in peace support operations, the need for international interoperability increased dramatically. Consequently, the legacy of the autonomous Swedish defence architecture of the Cold War era led to inertia. Since equipment was not compatible with NATO standards, air power could not be deployed to out-of-area operations.[188] To remedy this situation, solutions have also been sought on an industrial level.

Throughout the 1990s, a major reform within the Swedish defence industrial complex took place. It was acknowledged that autarchy in armaments production was no longer economically and technically reasonable. Sweden therefore began to pursue a process of consolidating and rationalising its national defence industrial base, thereby focusing upon core competencies and promoting arms exports. This process was accompanied by a vigorous thrust towards international cooperation.[189] It was even argued by Swedish decision makers that international cooperation would help to ensure reliable access to defence material in the event of a deteriorating security situation—an argument standing in stark contrast to Sweden's Cold War approach to defence procurement.[190] In the same vein, government subsidies ceased in the 1990s, and large parts of the defence industrial base were sold to foreign investors.[191] In 1995 Saab and British Aerospace, now BAE Systems, joined together in order to export the JAS 39 Gripen.[192] In 2003 British BAE Systems owned 35 per cent of Saab. Moreover, there is no state ownership in the Swedish defence industry anymore.[193] This thrust towards international cooperation was also embraced by the Swedish armed forces. In early 2004, the supreme commander of the Swedish armed forces, Gen Hakan Syren, stated that the future modernisation of equipment would be based on procuring systems together with other countries with similar requirements and needs. In this context, the European Defence Agency was expected to play an important role.[194]

Combat Aircraft

Undoubtedly, the development, manufacture, and introduction of the JAS 39 Gripen into the SwAF were the most important achievements of the Swedish defence industry in the 1990s. To bridge the gap between earlier aircraft designs and the Gripen, the SwAF pursued a path of upgrading operational aircraft designs. Since the delivery of the first JA 37 interceptor Viggen to the SwAF in 1978, the aircraft had been constantly upgraded, primarily in the area of software, and a data link allowing aircraft-to-aircraft communication was integrated in 1986.[195] Moreover, between 1993 and 1995, 86 attack and reconnaissance Viggens were modified to AJS 37 standard. The AJS 37 had a limited multirole capability and was designed to carry all JAS 39 air-to-ground weapons

Courtesy Gripen International; photo by Katsuhiko Tokunaga

The Gripen is a versatile weapons platform. For ground and maritime strike missions, the Gripen can carry the indigenously designed Saab Bofors Dynamics RBS 15F anti-ship missile and the ubiquitous AGM-65 Maverick air-to-ground missile.

planned at the time. Hence, familiarisation with the AJS 37 platform reduced conversion time on the Gripen.[196]

In the second half of the 1990s, the JAS 39 Gripen was introduced into operational service. The Gripen is the result of project studies that began in the early 1970s. In the late 1970s, alternatives to a future Swedish aircraft design were considered. These included the purchase of foreign aircraft such as the F-16 or F-18. Yet in June 1982, the development of a Swedish multirole aircraft was approved by the Swedish parliament. Despite being an indigenous programme, the JAS 39 Gripen is an expression of Sweden's defence industrial opening process. A substantial part of the aircraft's components have been procured from abroad, which also helped to break the cost spiral.[197]

The aircraft has a swing-role capability, allowing it to switch between interception, ground attack, and reconnaissance tasks within one mission. The more advanced C and D versions offer a crucial improvement in that they meet not only domestic but also international requirements. They are fully NATO interoperable, have an in-flight refuelling capability, an integrated GPS navigation suite, and the latest NATO standard IFF equipment. In response to international requirements, all system information is presented in English. In the tradition of Swedish combat aircraft designs, the Gripen is able to operate from very short strips and thus fits into the Swedish doctrine of dispersed basing. Furthermore, the aircraft is easy to maintain and delivers a high readiness and sortie rate.[198] A total of 204 Gripens were ordered for the SwAF in the mid-1990s—140 JAS 39A/B models plus 64 C/D models.[199] As it finally turned out, more JAS 39C/D and fewer JAS 39A/B aircraft than originally planned were procured. Against the backdrop of the fundamental reorganisation of the SwAF in the wake of the 1999 defence reform, the number of Gripens being procured proved too great. As a consequence, the SwAF has been attempting to sell surplus aircraft. Twenty-eight JAS 39C/D aircraft were, for instance, delivered to the Czech Republic and to Hungary, while the operational SwAF squadrons have been equipped with the more advanced C/D version. Currently, approximately 30 JAS 39A/B Gripens are being upgraded to achieve the force goal of 100 JAS 39C/D aircraft by 2012.[200]

A combat aircraft's versatility depends to a large degree upon the array of weapons it can deliver. In this regard, the Gripen is indeed a very versatile weapons platform. For ground and maritime strike missions, the Gripen can carry a variety of standoff weapons. Among them are the indigenously designed Saab Bofors Dynamics RBS 15F anti-ship missile and the ubiquitous AGM-65 Maverick air-to-ground missile.[201] With Sweden's increasing commitment to out-of-area operations, the SwAF also saw the need to acquire LGBs for close air support missions. In 2005 Mark 82

Paveway II LGBs as well as Litening III laser-designator pods were procured for the Gripen fleet.[202] In the air-to-air role, the JAS 39 Gripen can be fitted with a variety of advanced weapons such as the IR-guided AIM-9L Sidewinder and the AIM-120 AMRAAM active radar-guided missile.[203] Currently, the AIM-9L Sidewinder is being supplanted by the IRIS-T.[204] In the coming years, the Gripen will be equipped with the Meteor for medium and long ranges. Sweden has been involved in the development of both air-to-air weapons, the IRIS-T, and the Meteor.[205]

In the future, the JAS 39 Gripen system is likely to be complemented by remotely piloted combat aircraft. Together with five partner nations, Sweden is currently participating in the French-led Neuron remotely piloted combat aircraft programme, providing a significant budget share.[206]

Courtesy Gripen International

SwAF C-130 refuelling Gripen C and D in midair

Air Mobility

The SwAF was the first European air force to buy the C-130 Hercules and took delivery of its first C-130 in February 1965. It bought a total of eight aircraft, the bulk of them in the early 1980s.[207] During their lifespan, the Swedish C-130 aircraft have been continuously upgraded. The transport unit's engagement in the air bridge to Sarajevo was a major catalyst for an upgrade with self-protection suites. After the Swedish C-130 had been pulled out of Croatia in 1992, the upgrades were implemented without any further delay. Politicians recognised that force protection in real operations was of the highest importance, and the required budget was approved immediately.[208] A major midlife update was scheduled for 2007, with the SwAF participating in a major USAF C-130 modernisation programme. The upgrade programme was supposed to focus primarily upon aircraft avionics.[209] Moreover, one C-130 Hercules was converted into an air-to-air refuelling aircraft in 2004 so that Gripen pilots can train in air-to-air refuelling procedures.[210]

As the lead nation of the NBG, Sweden is responsible for intertheatre deployment. Despite continuous upgrades of the Swedish C-130 fleet, these aircraft are not to be used for intertheatre airlift but for providing intratheatre airlift.[211] Since the buildup of an autonomous strategic airlift capability would go beyond Sweden's national resources, Sweden has been engaging in two multinational NATO and EU projects. On 23 March 2006, Sweden became a member of the Strategic Airlift Interim Solution, and it is also participating in NATO's Strategic Airlift Capability, centred on the shared ownership of C-17 Globemasters. Sweden's role in the latter is particularly prominent, as the SwAF provides the deputy commander for the multinational military structure operating the C-17s.[212] (Both initiatives were examined in chap. 2.)

Insufficient power projection capabilities also became apparent against the backdrop of the establishment of the SwAF's rapid reaction units. Though all necessary support functions were contained within the unit, SWAFRAP JAS 39 was dependent upon lead nation support in deployed operations, especially when operating from so-called bare bases.[213]

The most important Swedish rotary-wing programme of the post–Cold War era is the acquisition of 18 Eurocopter NH90s.[214] In the late 1990s, the helicopters from the three services were united in the Armed Forces Helicopter Wing. The wing is composed of a broad array of helicopters, from transport, search and rescue, anti-submarine warfare, and liaison to training helicopters.[215] The streamlining of the helicopter force avoids unnecessary redundancies between the services and hence is a cost-saving factor.

C4ISTAR

Due to Sweden's strategic position and proximity to the Soviet Union, early warning and rapid response were crucial for Swedish defence preparations throughout the Cold War era. Hence, a sophisticated intelligence gathering, processing, and response system was indispensable. For this purpose, a highly integrated C2 system, the Stril 60, was procured in the 1960s and early 1970s.[216] Stril 60 contained up to 40 surveillance radar stations, both stationary and mobile, and other surveillance devices. Furthermore, it could be fed with data from reconnaissance and patrol Viggens. By means of data links, Swedish combat aircraft could be directed under conditions of radio silence. Stril 60 was also connected with Army GBAD batteries, the coastal surveillance and Navy C2 systems, and the national civil defence organisation.[217] As such, Stril 60 was an early attempt at modern network-centric warfare concepts.

Given Sweden's emphasis upon the ability to counter limited strategic attacks throughout the 1990s, the need for situational awareness was given particular priority in the aftermath of the Cold War. The Flygvapnet 2000 concept was specially designed for this purpose, allowing for a free data flow between the different systems, with the combat aircraft itself serving as an intelligence collection platform.[218] As a consequence of these developments, the integrated C2 system Stril 60 was supplanted by a new system, the Stric, which became operational in October 1999.[219] Simultaneously, a new data link system, the Taras, was introduced, particularly designed to comply with the demands for an extension of data transfer and the demands of electronic warfare–resistant communications.[220] Moreover, Taras was supposed to have a joint dimension by linking naval and other platforms with Air Force assets.[221]

The Flygvapnet 2000 concept also foresaw the introduction of an autonomous Swedish airborne early warning capability, particularly suited for detecting targets at low altitudes and at long distances. The Swedish airborne radar surveillance system was called FSR 890, while the complete system including the aircraft was known as S 100 Argus.[222] According to the SwAF's network-based defence approach, the FSR 890 had been designed to be fully integrated into the Stric C2 system. By means of the Taras data link, the airborne FSR 890 radar could be remotely controlled from ground-based Stric C2 posts, with the recognised air picture being generated on the ground.²²³ The SwAF operated a total of six S 100 Argus airborne early warning aircraft.²²⁴

Yet, given Sweden's fundamental strategic reorientation in the late 1990s and the changing security environment in the Baltic in the ensuing years, the Flygvapnet 2000 concept had to be adapted. With the thrust towards out-of-area operations, it was discussed in 2006 whether to put four of the SwAF's Saab S 100 Arguses up for sale and to upgrade the two remaining aircraft to airborne early warning and control aircraft with permanently installed operator workstations and a Link-16 data link. This allows the upgraded Argus to operate outside the country's national air defence ground environment and to be deployed on multinational operations.²²⁵ By mid-2009, two aircraft, designated S 100B Argus or ASC 890 Erieye, had been upgraded and handed over to the SwAF, with the remainder of the original S 100 Arguses progressively sold abroad or decommissioned.²²⁶ This development again shows Sweden's reorientation. While two advanced and NATO interoperable airborne early warning aircraft do not allow the country to maintain 24/7 surveillance of its airspace over an extended period of time, they provide a key capability in multinational operations. In this regard, the 2009 government bill also put a premium upon interoperability in regional defence scenarios.²²⁷ Since these scenarios conceived limited and rapidly unfolding military crises of short duration, a fleet of two airborne surveillance and control aircraft might have been deemed sufficient.²²⁸

To increase the situational awareness of its combat pilots, the SwAF was amongst the pioneers of data links. Already during the 1960s, the first Swedish combat aircraft were equipped with a simple ground-to-air data exchange system. In the 1980s, more advanced links were introduced in the JA 37 air defence Viggen fleet.²²⁹ Hence, the SwAF could draw upon a pool of experience, when network-centric or network-based approaches came to the fore in the post–Cold War era. Yet given Sweden's legacy of neutrality, a major problem that emerged in the wake of Sweden's strategic reorientation in the late 1990s was how to make the SwAF's advanced C2 structure interoperable with partner nations.²³⁰ As such, Sweden's increased engagement in the ESDP and PfP had a direct impact upon the data-link architecture of the SwAF. The

Courtesy Gripen International; photo by Frans Dely

ASC 890 Erieye airborne early warning and control aircraft escorted by Gripen fighter aircraft

JAS 39C/D Gripens are not being equipped with the Taras, the latest Swedish fighter link, but with the NATO interoperable Link-16 terminal MIDS.[231] Furthermore, the aircraft are receiving a NATO-compatible IFF.[232] This necessity for interoperability on a technological level was backed by the highest Swedish decision makers. As such the 2004–5 defence bill argues that Swedish network-based defence developments were to be coordinated with international developments.[233]

Since a network-based defence depends upon control of the electromagnetic spectrum, the SwAF also made efforts to retain its expertise in airborne standoff jamming. For this purpose, 10 two-seat Viggens were modified into dedicated electronic warfare platforms in the late 1990s. These so-called SK 37E Viggens were phased out in mid-2007. As a next step, the JAS 39D Gripen might be further developed into an electronic warfare platform. Yet the project does not have the highest priority for the SwAF.[234] Against the backdrop of European shortfalls in airborne standoff jamming, however, the SwAF potentially could gain significant leverage through its long-standing experience in this area. The SwAF also uses the electromagnetic spectrum as a means for intelligence gathering. Since 1974 two Sud Aviation Caravelles had been in SwAF service as SIGINT aircraft. These aircraft were replaced by two modified Gulfstream G-IVSPs, with the SwAF designation S 102B Korpen.[235] The S 102B systems were commissioned in 1997 and 1998.[236]

In the area of tactical air-to-ground reconnaissance, the SwAF had two dedicated reconnaissance Viggen versions at its disposal during the 1980s, the SF 37 Viggen (reconnaissance-land) and the SH 37 Viggen (reconnaissance-sea). In the early 1990s, 13 SF 37 Viggens and 25 SH 37 Viggens were upgraded to AJSF 37 and AJSH 37 standard, respectively. To carry out their reconnaissance tasks, the AJSF 37 Viggens used optical cameras and the AJSH 37 Viggens air-to-ground radar, particularly suited for maritime reconnaissance.[237] With the introduction of the JAS 39 Gripen combat system, the SwAF entered into an era where specifically dedicated reconnaissance aircraft are no longer required. The JAS 39 Gripen's radar is particularly suited for maritime reconnaissance, and for ground reconnaissance missions, the Gripen received in 2007 an indigenously developed modular reconnaissance pod system, fitted with a secure data link to provide real-time information to ground stations.[238] Up to the introduction of this new reconnaissance pod, 10 AJSF 37 reconnaissance Viggens were kept in the SwAF's inventory.[239] In the future, the SwAF's reconnaissance capability might be augmented by RPAs. In this domain, the Swedish aerospace industry is pursuing a cooperative approach.[240]

Conclusion

During the post–Cold War era, the SwAF has come a long way from an air force almost exclusively geared up to autonomous territorial defence to an air force that can operate in an international environment. Yet, in an era where power projection and combined operations have become increasingly important, Sweden's Cold War legacy exerted significant inertia upon reforms. Only in the late 1990s was the Swedish defence establishment ready to forego fundamental principles such as the focus upon autonomous territorial defence. The defence reform, commencing in 1999, had far-reaching consequences for the SwAF. As such, it underwent massive force cuts. In fact, the SwAF seems to have reached a critical force level for autonomous territorial defence scenarios. Such drastic reductions were particularly justified in light of NATO's expansion to the Baltic, which was perceived to have a stabilising effect upon the region. Moreover, the SwAF accepted a setback in its very advanced course towards an autonomous network-based defence approach when it forewent the indigenous Taras data link and instead opted for a Link 16–compatible terminal for its combat aircraft. Particularly, the decision to reduce the operational S 100 Argus fleet from six to two upgraded and NATO-interoperable airborne early warning and control aircraft underlines the importance that Swedish defence policy attaches to the multinational dimension. Defence is no longer understood in purely national terms. The 2009 government bill made it clear that the SwAF should be able to conduct combined air operations in the Nordic region. Immediately available and interoperable resources to counter rapidly unfolding military crises of short duration are considered more important than a large autarchic defence that requires a significant time span to reach operational readiness.

Despite drastic force cuts and the abandonment of an advanced Swedish data link, the SwAF has—in the context of territorial defence—remained a full-spectrum air force, allowing it to reconstitute in a time of heightened tension. For instance, unlike many other European air forces, it has acquired airborne SIGINT and retained an expertise in airborne standoff jamming. Moreover, the planned number of 100 JAS 39C/D Gripens is still significant, given the size of the Swedish state and compared, for instance, to the GAF, which originally planned to acquire 180 Eurofighters.

Besides the 1999 defence reform and a reduced defence budget, the SwAF's considerable force cuts can be related to the requirements of deployed operations. Whereas the Cold War was characterised by deterrence postures and hence dominated by how many squadrons an opponent could potentially muster, the ability to generate air power has become much more dependent upon other factors than primarily airframes. Already suffering from overstretch during the Cold War and throughout the 1990s, the SwAF could no longer sustain an oversized combat aircraft fleet. To operate across the spectrum of military force in deployed scenarios, sophisticated training and professionalism could no longer be compromised in order to retain an excessive aircraft fleet. SwAF

rapid reaction units regularly participated in multinational integrated exercises and exclusively drew upon professional and contracted personnel for potential deployed operations. Moreover,
the demands of interoperability dictated the SwAF's force structure. Only NATO interoperable combat aircraft are retained in the
operational squadrons. Force reductions have been partly made
up through force multiplying factors such as increased situational
awareness as well as swing-role-capable combat aircraft and advanced armament. It can be concluded that Sweden's alliance policy
and emphasis upon power projection were the driving factors behind the SwAF's transformation since 1999. Yet, against the backdrop of a reasserting Russia, Nordic defence scenarios in cooperation with other neighbouring states will most likely become the
cornerstone of the SwAF's continuing adaptation.

During the Cold War and throughout the 1990s, international
operations were completely divorced from Sweden's main defence
effort. The SwAF's implicit response to Desert Storm was not to
emulate its Western counterparts but to improve its air defence for
countering the new capabilities of precision air power. Today, it is
the declared aim to generate defence resources which are poly-
valent and can be used in both a national and international context. This dual-use approach combined with the indigenous combat aircraft industry has restricted the buildup of adequate force
enablers for deployed operations. A relatively large strategic airlift
fleet only makes sense in the international context, and Sweden's
aircraft industry required economy of scale to avoid excessive
costs. Therefore, given Sweden's limited defence budget, the imbalance between strike assets and force enablers could not substantially be redressed. Through cooperative ventures such as
SALIS or NATO's Strategic Airlift Capability initiative, these
shortcomings can most likely only be mitigated. As in the case of
SWAFRAP JAS 39, it can be assumed that the NBG Gripen detachment is still dependent upon foreign assistance in deployed
scenarios, particularly when operating from bare bases.

Sweden's shift towards enhanced deployability also had an impact upon the prioritisation of the SwAF's air power roles. Whereas
air defence was the predominant mission in the 1990s, the ability
to influence events on the ground has gained in importance. This
shift in importance is closely related to the changing threat and

risk perception. Limited strategic attacks against Swedish key infrastructure were considered a major threat scenario during the 1990s. To counter them, a good air defence and early warning capability were considered crucial. Yet, with the thrust towards international operations at the end of the century, air-to-ground capabilities have become more important. Particularly against the backdrop of the buildup of rapidly deployable Army units, the SwAF had to develop from scratch an advanced CAS capability.

Due to its tradition of neutrality and the fact that most of its air power assets were not interoperable throughout the 1990s, the SwAF contributed only in a limited way to deployed operations. It did this mainly through the provision of airlift at the lower level of the spectrum of military force, which nevertheless entailed significant risks for the crews involved. Since the turn of the century, the SwAF has readied rapid reaction units for international operations, also involving fighter-bombers. While SWAFRAP AJS 37 was primarily relegated to reconnaissance, the NBG Gripen detachment is able to deliver CAS for ground troops. The capability to act across the spectrum of military force in deployed operations became a key criterion for the Swedish armed forces in general and for the SwAF in particular. These requirements demanded a high degree of professionalism, which was met by the SwAF's rapid reaction units.

Not only with regard to hardware and software but also in terms of the intellectual mastery of air power, the SwAF has become increasingly interoperable. Both the revision of the air power curriculum at the SNDC and the air power doctrine development process coincided with Sweden's defence reform and hence reflected the shift away from an emphasis upon autonomous territorial defence towards a combined and expeditionary mindset. The defence reform itself provided the necessary top-down backing for these intellectual responses to the changing security environment. Though the teaching of air power theory has heavily drawn upon Anglo-Saxon concepts, the SNDC has been slowly emancipating itself and has produced its own publications. It can therefore be concluded that the SwAF has proactively responded to the new intellectual challenges since the late 1990s. Though this process has required top-down backing, individuals and personalities at lower levels have played a key role.

Notes

1. Kurt Karlsson, "The History of Flygvapnet," in *Flygvapnet: The Swedish Air Force*, ed. Raymond Andersson, Kurt Karlsson, and Anders Linnér (Stockholm: Swedish Air Force, 2001), 85.

2. Ibid., 106.

3. Ibid., 124–25; and Gunnar Åselius, "Swedish Strategic Culture after 1945," *Cooperation and Conflict: Journal of the Nordic International Studies Association* 40, no. 1 (2005): 27.

4. Bertil Wennerholm, *Fjärde flygvapnet i världen? Doktrinutveckling i det svenska flygvapnet i försvarsbesluten 1942–1958. Underlag, beslut och genomförande i nationellt och internationellt perspektiv* [The fourth Air Force in the world? Doctrine development in the Swedish Air Force in the long-term parliamentary defence acts of 1942–1958. Foundations, resolutions and implementation in a national and international perspective] (Stockholm: SNDC, 2006), 263; and Wolfgang A. Schmidt, *Die schwedische Sicherheitspolitik: Konzeption:—Praxis—Entwicklung, Beiträge und Berichte*, monograph no. 88 (St. Gallen, Switzerland: Hochschule St. Gallen, December 1983), 146–47.

5. Lars Wedin, "La pensée navale suédoise après 1945," Institut de Stratégie Comparée, http://www.stratisc.org/PN7_Wedin.html.

6. Schmidt, *Die schwedische Sicherheitspolitik*, 147.

7. Wedin, "La pensée navale suédoise après 1945."

8. Schmidt, *Die schwedische Sicherheitspolitik*, 147.

9. Ibid., 147–48.

10. Wedin, "La pensée navale suédoise après 1945."

11. Schmidt, *Die schwedische Sicherheitspolitik*, 168–69; and Bitzinger, *Facing the Future*, 2.

12. Robert W. Komer, "Sweden's Role in the Security System of the High North," in *Proceedings of a Symposium on Changing Strategic Conditions in the High North*, ed. Ingemar Dörfer and Lars B. Walin (Stockholm: The Swedish National Defence Research Establishment, 1990), 77.

13. Bitzinger, *Facing the Future*, 7, 13.

14. Ibid., 2.

15. See National Defence Academy, *Report: Austrian and Swedish Security Policy*, series 6/98 (Vienna: Schriftenreihe der Landesverteidigungsakademie, 1998), 65, 70.

16. Ingemar Dörfer, *The Nordic Nations in the New Western Security Regime* (Washington, D.C.: Woodrow Wilson Center Press, 1997), 42.

17. Åselius, "Swedish Strategic Culture after 1945," 25.

18. Ibid., 39.

19. Bitzinger, *Facing the Future*, 32–33.

20. MOD, Sweden, "The New Defence," summary of Government Bill 1999/2000:30, presented to Parliament on 25 November 1999, http://www.sweden.gov.se/sb/d/574/a/25639, 4; and Dörfer, *Nordic Nations*, 40.

21. MOD, Sweden, *Regeringens proposition 1996/97:4: Totalförsvar i förny-else—etapp 2* [Government defence bill 1996/97:4: renewal of total defence—stage 2] (Stockholm: MOD, 12 September 1996), 33.

22. MOD, Sweden, "New Defence," 3–5.

23. See also Eriksson, "Sweden and the Europeanisation of Security," 128.

24. Michael Mohr (director, Defence Commission, MOD, Stockholm), interview by the author, 10 February 2005.

25. MOD, Sweden, "Continued Renewal of the Total Defence," fact sheet, October 2001.

26. Defence Commission, *Summary: Secure Neighbourhood—Insecure World* (Stockholm: MOD, 27 February 2003), 12.

27. Defence Commission, *Defence for a New Time: Introduction and Summary* (Stockholm: MOD, 1 June 2004), 3.

28. MOD, Sweden, *Our Future Defence: The Focus of Swedish Defence Policy 2005–2007*, Summary of the Swedish Government Bill 2004/05:5 (Stockholm: MOD, October 2004), 5.

29. MOD, Sweden, "A Functional Defence—with a Substantially Strengthened Defence Capability," press release, 19 March 2009, http://www.regeringen.se/sb/d/10448/a/123010.

30. Parliamentary Defence Committee, *Försvarsutskottets betänkande 1991/92 FöU12: Totalförsvarets fortsatta utveckling 1992/93–1996/97* (Stockholm: Parliamentary Defence Committee, 1992), 2–3.

31. MOD, Sweden, "New Defence," 3–6.

32. MOD, Sweden, *Our Future Defence*, 10.

33. Lars Wedin, "Sweden in European Security," in *Strategic Yearbook 2004: The New Northern Security Agenda—Perspectives from Finland and Sweden*, ed. Huldt et al. (Stockholm: SNDC, 2003), 330.

34. MOD, Sweden, "A Functional Defence: Government Bill on the Future Focus of Defence," fact sheet no. 2009.06, March 2009, http://www.sweden.gov.se/content/1/c6/12/30/22/3ed2684c.pdf; and MOD, Sweden, "Functional Defence," press release.

35. Pertti Salminen, "'Nordic Power Projection' and International Operations: Aspects from the Past and New Opportunities," in *Strategic Yearbook 2004*, ed. Huldt et al., 181, 202.

36. Eriksson, "Sweden and the Europeanisation of Security," 127.

37. MOD, Sweden, *Regeringens proposition 1996/97:4*, 87.

38. MOD, Sweden, "New Defence," 3, 5–6.

39. Defence Commission, *Defence for a New Time*, 5.

40. Organisation for Security and Cooperation in Europe, *Annual Exchange of Information on Defence Planning: The Kingdom of Sweden* [*According to the*] *Vienna Document 1999* (Stockholm: The Kingdom of Sweden, March 2005), 9; and Andersson, *Armed and Ready?*, 39.

41. MOD, Sweden, "Functional Defence," fact sheet.

42. Defence Commission, *Summary: Secure Neighbourhood*, 13.

43. MOD, Sweden, "New Defence," 1, 9.

44. Gen Hakan Syren, supreme commander of the Swedish armed forces, "The Swedish Armed Forces Today and towards the Next Defence Decision" (presentation, National Conference on Folk och Försvar [people and defence], Sälen, Sweden, 21 January 2004).

45. MOD, Sweden, *Our Future Defence*, 16.

46. MOD, Sweden, "Functional Defence," fact sheet.

47. MOD, Sweden, "Functional Defence," press release.

48. Bruce Jones, "Sweden Abolishes Conscription," *Jane's Defence Weekly* 47, no. 21 (26 May 2010): 10.

49. Mattias Robertson, Communications and Public Affairs Directorate, Swedish armed forces, to the author, e-mail, 14 June 2010.

50. Clifford Beal, "The JAS 39 Gripen: Last Swedish Fighter?," *International Defence Review* 24, no. 10 (1991): 1,067.

51. Bitzinger, *Facing the Future*, 9–11.

52. Ibid., 24.

53. *Programplan för det militära försvarets utveckling 1993*–98 (ÖB 93), huvuddokument, ÖB skr 1992-09-15 PLANL 483:62700 (Stockholm: Armed Forces Headquarters, 1992), 33–35.

54. "European Military Aircraft Programmes Revisited," 16.

55. MOD, Sweden, *Regeringens proposition 1996/97:4*, 75–77.

56. Jan Andersson, "Flygvapnet 2000—a Vision Fulfilled," in Andersson, Karlsson, and Linnér, *Flygvapnet*, 12.

57. Maj Gen Staffan Näsström, chief, Air Force Materiel Command, "The Swedish Air Force 2000—Materiel Aspects," *Military Technology: Defence Procurement in Sweden*, special issue, 1996, 35.

58. Karlsson, "History of Flygvapnet," 144; and Christer Salsing, "New Tasks and Requirements for a New Command System," in Andersson, Karlsson, and Linnér, *Flygvapnet*, 16.

59. Bertil Wennerholm and Stig Schyldt, *1990-talets omvälvningar för luftstridskrafterna: Erfarenheter inför framtiden* [The air power revolution of the 1990s: experiences for the future]: *Krigsvetenskapliga Forskningsrapporter* [War scientific research report] no. 3 (Stockholm: SNDC, 2004), 61; and Organisation for Security and Cooperation in Europe, *Annual Exchange of Information*, 2005, annexe 3, 1.

60. Åselius, "Swedish Strategic Culture after 1945," 40; and Karlsson, "History of Flygvapnet," 144.

61. Joris Janssen Lok, "Defense Review to Refocus Sweden's Military Forces," *Jane's International Defence Review* 31 (October 1998): 6.

62. Maj Gen Jan Jonsson, chief of staff, SwAF, "The Future of Air Power (II): Sweden," *Military Technology*, no. 7 (1999): 15.

63. Organisation for Security and Cooperation in Europe, *Annual Exchange of Information on Defence Planning: The Kingdom of Sweden* [*According to the*] *Vienna Document 1999* (Stockholm: The Kingdom of Sweden, March 2001), annexe 3, 1; and Ibid., 2005, annexe 3, 1.

64. Organisation for Security and Cooperation in Europe, *Annual Exchange of Information on Defence Planning*, 2005, annexe 3, 1.

65. Lt Col Anders P. Persson, SwAF, Armed Forces Headquarters, Stockholm, to the author, e-mail, 28 October 2005; Organisation for Security and Cooperation in Europe, *Annual Exchange of Information on Defence Planning*, 2005, annexe 3, 1; and ibid., 2006, annexe 3, 1.

66. MOD, Sweden, "Functional Defence," press release.

67. Lt Col Christer Björs, SwAF (chief, Air Operations [A3], Swedish Air Component Command), telephone interview by the author, 3 August 2005.

68. MOD, Sweden, "Functional Defence," fact sheet.

69. Gareth Jennings, "Sweden Rethinks Strategy in Wake of Georgia Crisis," *Jane's Defence Weekly* 46, no. 25 (24 June 2009): 16.

70. European Union, "EUNAVFOR [European Union Naval Force] Somalia," http://www.consilium.europa.eu/showPage.aspx?id=1518&lang=en; and "Swedish A109s Deployed to Horn of Africa for Anti-Piracy Mission," *Air Forces Monthly*, July 2010, 30.

71. Jan Jorgensen, "Battle Group Gripens," *Air Forces Monthly* (August 2007): 45–46.

72. Bitzinger, *Facing the Future*, 26.

73. Parliamentary Defence Committee, *Försvarsutskottets betänkande 1991/92*, 48.

74. Joris Janssen Lok, "Budget Crunch May Hit Swedish Air Operations," *Jane's Defence Weekly*, 5 September 2001, accessed 25 May 2007, http://search.janes.com.

75. Bitzinger, *Facing the Future*, 2.

76. International Institute for Strategic Studies, *Military Balance 1994–95*, 101–2; International Institute for Strategic Studies, *Military Balance 2000–01*, 103–4; and International Institute for Strategic Studies, *Military Balance 2002–03*, 80.

77. Stefan Helsing, Swedish Air Force Rapid Reaction Unit (SWAFRAP) AJS 37 information officer, "Pilots for Peace," in Andersson, Karlsson, and Linnér, *Flygvapnet*, 62.

78. Dörfer, *Nordic Nations*, 42.

79. Per Klingvall, chief press and information officer, Joint Forces Command, Sweden, to the author, e-mail, 7 February 2005.

80. Mike Winnerstig, "Sweden and NATO," in *Finnish and Swedish Security: Comparing National Policies*, ed. Bo Huldt et al. (Stockholm: SNDC and the Programme on the Northern Dimension of the CFSP conducted by the Finnish Institute of International Affairs and the Institut für Europäische Politik, 2001), 77–79.

81. MOD, Sweden, *Our Future Defence*, 7.

82. Leni Björklund, minister for defence, "Swedish Defence Policy in Times of Change" (speech, Bern, Switzerland, 22 October 2004).

83. Mikael af Malmborg, "Sweden in the EU," in Huldt et al., *Finnish and Swedish Security*, 48.

84. Eriksson, "Sweden and the Europeanisation of Security," 130.

85. Alison J. K. Bailes, "European Security from a Nordic Perspective: The Roles for Finland and Sweden," in Huldt et al., *Strategic Yearbook 2004*, 61.

86. Björklund, speech.

87. MOD, Sweden, *Our Future Defence*, 8.

88. Eriksson, "Sweden and the Europeanisation of Security," 129.

89. Syren, "Swedish Armed Forces Today."

90. Af Malmborg, "Sweden in the EU," 44.

91. Wedin, "Sweden in European Security," 328.

92. Örjan Berner, "Sweden and Russia," in Huldt et al., *Finnish and Swedish Security*, 136–37.

93. Dörfer, *Nordic Nations*, 50; and Winnerstig, "Sweden and NATO," 86.

94. MOD, Sweden, *Our Future Defence*, 8–9.

95. Bailes, "European Security from a Nordic Perspective," 60.

96. Defence Commission, *Summary: Secure Neighbourhood*, 3; and Bailes, "European Security from a Nordic Perspective," 73.

97. MOD, Sweden, "Functional Defence," press release.

98. Byörn von Sydow, minister of defence, "Flygvapnet and Network Defence," in Andersson, Karlsson, and Linnér, *Flygvapnet*, 14.

99. Jorgensen, "Battle Group Gripens," 42.

100. Maj Gen Mats Nilsson, commander, SwAF, "Air Force Revival," in Andersson, Karlsson, and Anders, *Flygvapnet*, 9.

101. Helsing, "Pilots for Peace," 65; Jorgensen, "Battle Group Gripens," 42; and Persson to the author, e-mail, 10 May 2005.

102. *Air Transport Unit: SWAFRAP C-130*, leaflet (Satenäs, Sweden: Skaraborg Wing F 7, 2004); and Lars Eric Blad, commanding officer, SWaF Rapid Reaction Unit C-130, "A Load-Carrying Member in International Service," in Andersson, Karlsson, and Linnér, *Flygvapnet*, 67.

103. Jorgensen, "Battle Group Gripens," 43.

104. Ibid., 43–44; and Peter Liander, "On Alert: Continued Focus on the New Gripen and New Helicopters," in *The Swedish Armed Forces Forum: Insats & Försvar*, English ed., no. 1 (2004): 23–24.

105. Jorgensen, "Battle Group Gripens," 42, 44–46.

106. Lt Col Johan Swetoft, SwAF, F-17 Wing, Ronneby, to the author, e-mail, 13 June 2006; and Lt Col Anders P. Persson, SwAF, chief of staff, Air Combat Training School, Uppsala, Sweden, to the author, e-mail, 25 August 2009.

107. Col Staffan F. Sjöberg (Swedish deputy military representative to the EU, Swedish Embassy, Brussels), interview by the author, 1 February 2007.

108. Persson to the author, e-mail, 10 May 2005.

109. Peter Liander (editor, *Joint Armed Forces Magazine*, SNDC, Stockholm), interview by the author, 11 February 2005; and Jonsson, "Future of Air Power (II): Sweden," 17.

110. Helsing, "Pilots for Peace," 64; and Blad, "Load-Carrying Member in International Service," 68.

111. Liander, "On Alert," 25.

112. Janssen Lok, "Budget Crunch May Hit Swedish Air Operations."

113. Jorgensen, "Battle Group Gripens," 44, 46.

114. Peter Liander, editor, *Joint Armed Forces Magazine*, to the author, e-mail, 9 May 2005.

115. Liander, "On Alert," 25.

116. Persson to the author, e-mail, 10 May 2005.

117. Karlsson, "History of Flygvapnet," 132.

118. Brig Gen Göran Tode, SwAF, retired, to the author, e-mail, 7 June 2005.

119. Ibid.

120. Karlsson, "History of Flygvapnet," 134.

121. Ibid., 134–35.

122. Ibid., 132.

123. Liander, "On Alert," 22–23.

124. Ministry of Foreign Affairs, Stockholm, "EAPC Security Forum 2005—Media Advisory," press release, 24 March 2005, accessed 30 September 2005, http://www.sweden.gov.se/sb/d/5314/a/41398 (site discontinued).

125. Tode to the author, e-mail, 7 June 2005.

126. Liander, "On Alert," 23.

127. Björs, interview.

128. Ibid.; and Tode to the author, e-mail, 14 June 2005.

129. United Nations, *Operation Artemis*, 13.

130. Adam Folcker, "Commanding Officer in Liberia," *Insats & Försvar*, English ed., no. 1 (2004): 52–53.

131. Göran Stütz, *Opinion 2004* (Stockholm: Styrelsen för Psykologiskt Försvar, 2004), 125.

132. Brig Gen C-G Fant, "Luftstridskrafterna i ett internationell perspektiv: Referat från International Air Power Symposium," in *Kungl Krigsvetenskapsakademiens Handlingar och Tidskrift*, no. 1 (Stockholm: The Royal Swedish Academy of War Sciences, 2002), 51.

133. Defence Commission, *Summary: Secure Neighbourhood*, 13.

134. Jonas Hjelm (state secretary for defence, Sweden), telephone interview by the author, 15 February 2006.

135. Brig Gen Åke Svedén, SwAF, commander, Air Transport Unit, 1983–90, to the author, e-mail, 22 January 2006.

136. Capt Åsa Schön, SwAF, deputy chief of public affairs, Skaraborg Wing, to the author, e-mail, 5 July 2005.

137. Svedén to the author, e-mail.

138. Lt Col Pepe Brolén, SwAF, retired (commander, Air Transport Unit, 1990–1998), telephone interview by the author, 2 November 2005.

139. Ibid.

140. Anders Hedgren, head, Public Affairs Department, Swedish armed forces, to the author, e-mail, 8 April 2005.

141. Lt Col Lars-Eric Blad, SwAF, commander, Air Transport Unit, 1998–2003, to the author, e-mail, 6 September 2005.

142. Brolén, interview.

143. Schön to the author, e-mail.

144. Ibid.
145. Lt Col Bertil Höglund, SwAF, commander, Air Transport Unit, to the author, letter, 2 August 2005.
146. Lt Col Torbjörn A. Olsson, SwAF, commanding officer, Airport Support Unit, to the author, e-mail, 18 June 2005.
147. Ibid.
148. Claes Winquist, "We See with Our Ears," in Andersson, Karlsson, and Linnér, *Flygvapnet*, 33–34.
149. Persson to the author, e-mail, 10 May 2005.
150. Åselius, "Swedish Strategic Culture after 1945," 40.
151. Maj Tobias Harryson, SwAF, to the author, e-mail, 12 October 2005.
152. Brig Gen Anders Silver, SwAF, commander, Air Component Command, to the author, e-mail, 5 February 2006 (forwarded by Maj Tobias Harryson, SwAF, 8 March 2006).
153. Harryson to the author, e-mail, 12 October 2005.
154. *Military-Strategic Doctrine*, English translation (Stockholm: Swedish Armed Forces, 2002), 11.
155. Harryson to the author, e-mail, 12 October 2005.
156. *Doktrin för luftoperationer 2004* (Stockholm: Swedish Armed Forces, 2004), table of contents; and Maj Tobias Harryson, SwAF (SNDC, Stockholm), interview by the author, 11 February 2005. Major Harryson, together with another major and lieutenant colonel, formed the air power doctrine project team, set up in August 2002.
157. *Doktrin för luftoperationer 2004*, 51–68.
158. Ibid., 33.
159. *Military-Strategic Doctrine*, 11.
160. Harryson to the author, e-mail, 24 April 2006; *Doktrin för luftoperationer 2004*, 164, 197, 207; and AJP-3.3, *Joint Air and Space Operations Doctrine*, 3-2–3-6, 4-1–4-25.
161. Lt Col Jan Reuterdahl, SwAF, head, air faculty, Department of War Studies, SNDC, Stockholm, to the author, e-mail, 17 February 2005; Luftstridsseminarium [Air power seminar] 2002, SNDC, Stockholm, 29–30 January 2002, accessed 7 November 2005, http://www.fhs.se; and Air Power Seminar 2003, SNDC, Stockholm, 27–28 January 2003, accessed 7 November 2005, http://www.fhs.se.
162. Harryson to the author, e-mails, 12 October 2005 and 29 March 2006.
163. *Försvarsmaktens doktrin för luftoperationer* (Stockholm: Swedish Armed Forces, 2005), 4–6; and *Försvarsmaktens doktrin för gemensamma operationer* (Stockholm: Swedish Armed Forces, 2005), 4–5.
164. *Försvarsmaktens doktrin för luftoperationer*, 4–6.
165. Harryson to the author, e-mail, 12 October 2005.
166. Reuterdahl to the author, e-mail, 9 June 2005.
167. Prof. Nils Marius Rekkedal, Department of War Studies, SNDC, Stockholm, to the author, e-mail, 10 January 2006.
168. Reuterdahl to the author, e-mails, 9 and 10 June 2005.

169. Ibid., 9 June 2005.
170. Prof. Nils Marius Rekkedal, Department of War Studies, SNDC, Stockholm, to the author, e-mail, "JSO curriculum (2005)," 2 November 2005.
171. Prof. Nils Marius Rekkedal, Department of War Studies, SNDC, Stockholm, to the author, e-mail, "GSO curriculum (2005)," 2 November 2005.
172. Reuterdahl to the author, e-mail, 17 February 2005.
173. Ibid., 9 June 2005.
174. Rekkedal to the author, e-mails, "JSO curriculum" and "GSO curriculum."
175. Silver to the author, e-mail.
176. Reuterdahl to the author, e-mail, 9 June 2005.
177. Ibid.
178. Lt Col Jan Reuterdahl, SwAF (head, air faculty, Department of War Studies, SNDC, Stockholm) interview by the author, 8 February 2005.
179. Air Power Seminar 2002.
180. Air Power Seminar 2003; Air Power Seminar 2004, SNDC, Stockholm, 26–27 January 2004, accessed 7 November 2005, http://www.fhs.se; and Air Power Seminar 2005, SNDC, Stockholm, 9–10 February 2005, accessed 7 November 2005, http://www.fhs.se.
181. Reuterdahl to the author, e-mail, 17 February 2005.
182. Bo Hofvander and Nils Marius Rekkedal, eds., *Luftmakt: En antologi* [Air power: an anthology] (Stockholm: SNDC, 2003); Bo Hofvander and Nils Marius Rekkedal, eds., *Luftmakt: Teorier och tillämpningar* [Air power: theory and application] (Stockholm: SNDC, 2004); Jan Reuterdahl, ed., *Tillbaka till framtiden? Tankar om luftoperationerna över Irak 2003* [Back to the future? Thoughts on air operations over Iraq 2003] (Stockholm: SNDC, 2004); and Wennerholm and Schyldt, *1990-talets omvälvningar för luftstridskrafterna.*
183. Johan Elg, ed., *Effektbaserade operationer: Teori, planering och applicering* [Effects-based operations: theory, planning and application] (Stockholm: SNDC, 2005).
184. Jan Reuterdahl, ed., *Daidalossyndromet: Om luftmakt 2005* [The Daedalus syndrome: thoughts on air power 2005] (Stockholm: SNDC, 2006).
185. Reuterdahl to the author, e-mail, 11 November 2005.
186. Bitzinger, *Facing the Future*, 17.
187. Schmidt, *Die schwedische Sicherheitspolitik*, 160.
188. Dörfer, *Nordic Nations*, 42.
189. Richard A. Bitzinger, *Towards a Brave New Arms Industry?* Adelphi Paper no. 356 (Oxford, England: Oxford University Press for IISS, 2003), 53.
190. MOD, Sweden, "Continued Renewal of the Total Defence."
191. Åselius, "Swedish Strategic Culture after 1945," 39.
192. Bitzinger, *Towards a Brave New Arms Industry?*, 57.
193. Martin Lundmark, "Nordic Defence Materiel Cooperation," in Huldt et al., *Strategic Yearbook 2004*, 220.
194. Syren, "Swedish Armed Forces Today"; and MOD, Sweden, *Our Future Defence*, 20.

195. Sven-Olof Hökborg, "The Swedish Air Force's Procurement Programmes," *Military Technology: Defence Procurement in Sweden*, special issue, 1993, 59; and Bengt Andersson (senior analyst, Swedish Defence Research Agency [FOI], Stockholm), interview by the author, 11 February 2005.

196. Karlsson, "History of Flygvapnet," 140–41; and Anders Annerfalk, communications manager, Gripen International, to the author, e-mail, 7 July 2005.

197. Ahlgren, Linnér, and Wigert, *Gripen*, 12–14, 19, 52–53.

198. Ibid., 24–26, 35–36.

199. Annerfalk to the author, e-mail; and Joris Janssen Lok, "JAS 39 Gripen Set for Major Upgrades," *Jane's International Defence Review* 30 (December 1997): 50.

200. Persson to the author, e-mails, 28 October 2005 and 27 November 2005.

201. Ahlgren, Linnér, and Wigert, *Gripen*, 41–42.

202. Persson to the author, e-mail, 10 May 2005.

203. Ahlgren, Linnér, and Wigert, *Gripen*, 39.

204. FMV, "Short-Range Missile Ready for Gripen," 12 November 2009, http://www.fmv.se/WmTemplates/page.aspx?id=4931.

205. Ahlgren, Linnér, and Wigert, *Gripen*, 43.

206. Defense Industry Daily, "nEUROn UCAV Project Rolling Down the Runway," 14 June 2007, accessed 27 December 2007, http://www.defenseindustrydaily.com.

207. Svedén to the author, e-mail.

208. Brolén, interview.

209. Höglund, letter.

210. "Airborne Fuelling Station," *The Swedish Armed Forces Forum: Insats & Försvar*, English ed., no. 1 (2004): 29.

211. Andersson, *Armed and Ready?*, 42.

212. BMVg, "Verlässlicher Zugriff"; and NATO, "Strategic Airlift Capability."

213. Jorgensen, "Battle Group Gripens," 44.

214. Lundmark, "Nordic Defence Materiel Cooperation," 218.

215. International Institute for Strategic Studies, *The Military Balance 1999–2000* (Oxford, England: Oxford University Press for the IISS, 1999), 99.

216. Näsström, "Swedish Air Force 2000," 41.

217. Bitzinger, *Facing the Future*, 22–23.

218. Näsström, "Swedish Air Force 2000," 35.

219. Jonsson, "Future of Air Power (II): Sweden," 16.

220. Näsström, "Swedish Air Force 2000," 41.

221. Hewish and Lok, "Connecting Flights," 41.

222. Näsström, "Swedish Air Force 2000," 41.

223. Martin Bergstrand, "The Earth Is Round . . ," in Andersson, Karlsson, and Linnér, *Flygvapnet*, 28; and Jonsson, "Future of Air Power (II): Sweden," 16.

224. Organisation for Security and Cooperation in Europe, *Annual Exchange of Information on Defence Planning*, 2005, annexe 8, 2.

225. Joris Janssen Lok, "Sweden Focuses on Erieye Fleet Changes," *Jane's International Defence Review* 39 (August 2006): 4.

226. Persson to the author, e-mail, 25 August 2009.

227. MOD, Sweden, "Functional Defence," fact sheet.

228. MOD, Sweden, "Functional Defence," press release.

229. Näsström, "Swedish Air Force 2000," 41; and Andersson, interview.

230. Wedin, "Sweden in European Security," 330.

231. Persson to the author, e-mail, 10 May 2005; and "Link 16 Added to Gripen's Datalink Systems," *Gripen News* no. 1 (2007): 7.

232. Persson to the author, e-mail, 28 October 2005.

233. MOD, Sweden, *Our Future Defence*, 15.

234. Lars-Åke Siggelin, "Electric Saabs," *Air Forces Monthly*, October 2007, 40–41.

235. Winquist, "We See with Our Ears," 34.

236. Näsström, "Swedish Air Force 2000," 41.

237. Annerfalk to the author, e-mail.

238. Ahlgren, Linnér, and Wigert, *Gripen*, 42; and Jorgensen, "Battle Group Gripens," 43.

239. Organisation for Security and Cooperation in Europe, *Annual Exchange of Information on Defence Planning*, 2005, annexe 8, 1.

240. Annerfalk to the author, e-mail.

Chapter 7

Conclusion

As the preliminary findings demonstrate, each air force has responded to the air power challenges of the post–Cold War era according to its context. The following discussion synthesises the responses of the air forces examined across the four guiding challenges—shifting defence and alliance policies, real operations, new intellectualism in air power thinking and doctrine, and procurement.

Shifting Defence and Alliance Policies

Since the threat and risk spectrum of the post–Cold War era was no longer determined by two opposing blocs, the military-strategic focus shifted from territorial defence to deployed operations. Power projection in support of the UN, the ESDP, or NATO has become the determining factor in shaping Continental European air power. Amongst the four countries, the Netherlands was the first to accept these new realities; consequently, its military transformation was the most advanced in the period studied. Its rapid shift towards expeditionary warfare allowed the RNLAF to contribute to deployed operations disproportionately to the size of the country. France, under the presidency of Jacques Chirac, came to terms with the new realities in the mid-1990s. Whereas the country's military strategic doctrine had long been dominated by the nuclear force, conventional power assumed centre stage, offering the possibility to prevent, contain, or stabilise regional crises.

Germany and Sweden, on the other hand, only began to accept the new strategic realities at the end of the 1990s, significantly delaying the transformation of their air forces. Germany's historical legacy and the vestiges of Sweden's neutral tradition exacerbated these delays. Moreover, both nations had been frontline states, and their decision makers found it difficult to relax after the Cold War. The GAF had a unique experience in the early 1990s, when it was tasked to disband the East German Air Force and to integrate some of its personnel and equipment. This process did not lead to a hybrid between Western and Eastern concepts—instead, West German concepts were extended across Germany. Carefully screening former East German officers who aspired to be reemployed was in part responsible for this outcome. The costs of German reunification and the so-called peace dividend led to severely reduced defence resources, causing delays and reappraisals of multinational acquisition programmes such as the Eurofighter. Reunification efforts probably also caused the German armed forces to be more preoccupied with themselves than with deployed operations. It can be concluded that the Netherlands entered the post–Cold War era with the least baggage, enabling it to adapt almost immediately to the altered post–Cold War strategic environment.

The sharp increase in deployed operations caused the four countries examined to initiate a shift towards professional armed forces. France and the Netherlands abolished conscription in the 1990s—primarily due to the requirements of deployed operations across the spectrum of military force. Sweden suspended conscription in mid-2010, while, as of 2010, Germany still held onto the tradition of conscription. The suspension of conscription is scheduled to be implemented in July 2011.

Unlike armies, air forces are to a larger degree dependent on equipment. Given the exponential rise in the cost of air power equipment, national economic performance can significantly impact a nation's air force. To put national defence resources and the size of the four examined air forces into context, the following tables compare gross national products (GNP)/GDP and defence expenditures (table 6), force reductions measured by manpower and squadrons (table 7), and aircraft numbers (table 8).

As this study shows, the size of the defence budget is only one determinant in shaping effective defence. Equally important is an air force's approach to real operations, multinational cooperation, and procurement, as well as the political will to make contributions across the spectrum of military force.

Resource allocations to the air forces vary across the four countries. In 2002 France allocated the air force 17.7 per cent of its overall defence budget—including expenditures for the national paramilitary police force (*Gendarmerie Nationale*) and research and development; according to a 1999 assessment, Germany appropriated 29 per cent.[1] In the Netherlands, this figure varied

Table 6. Tabular comparisons of GNP/GDP and defence expenditures

	1991[a]			2003[b]		
	GNP[c]	Defence expenditure	%[d]	GDP[e]	Defence expenditure	%[f]
France	1,199.1	42.433	3.5	1,745.5	45.384	2.6
Germany	1,690.1	39.517	2.3	2,483.0	34.762	1.4
Netherlands	287.0	7.246	2.5	512.4	8.199	1.6
Sweden	230.3	5.543	2.4	297.3	5.352	1.8

[a]Stockholm International Peace Research Institute (SIPRI), *SIPRI Yearbook 1993: World Armaments and Disarmament* (Oxford, England: Oxford University Press, 1993), 384. Calculation of defence expenditure as a percentage of GNP is based upon GNP and defence expenditure data. Hence, the figures enable an approximate comparison across the states in 1991.

[b]SIPRI, *SIPRI Yearbook 2005: Armaments, Disarmament and International Security* (Oxford, England: Oxford University Press, 2005), 360–61, 366–67. Calculation of GDP is based upon defence expenditure and defence expenditure as a percentage of GDP data. Hence, the figures enable an approximate comparison across the states in 2003.

[c]GNP and defence expenditures for 1991 are in US $b. at constant 1991 prices.

[d]Defence expenditure as a percentage of GNP.

[e]GDP and defence expenditures for 2003 are in US $b. at constant 2003 prices.

[f]Defence expenditure as a percentage of GDP.

between 23 and 28 per cent from 1989 to 2002, and in Sweden, the figure ranged from 29.3 to 43 per cent over the same time span (see fig. 4, "Budget allocation to the single services [Netherlands]," chap. 5; and fig. 5, "Budget allocation to the single services [Sweden]," chap. 6). Shrinking defence budgets, combined with escalating air power equipment costs, led to a reduction of operational squadrons and manpower, as examined in table 7.

In terms of flying frontline squadrons and air force personnel, the four air forces examined differ significantly. Yet, as is argued in this book, the operational potential does not directly correlate with the number of available squadrons but with the number of deployable and employable squadrons.

Single European states have been unable to acquire the capacities of a full-spectrum air force capable of autonomously staging an operation of the size of Allied Force. As table 8 shows, even five major European air forces combined did not get close to American capabilities in 1990 or 2000. This discrepancy in manned air assets is also mirrored in the domain of MALE and HALE RPAs. According to a JAPCC report, European NATO members (Turkey excluded) operated only 64 MALE/HALE RPAs in 2008 as opposed to the United States, which had more than 400.[2]

Table 7. Force reductions measured by flying frontline squadrons and air force personnel

| | 1991[a] | | 2003[b] | |
	Flying frontline squadrons[c]	Manpower[d]	Flying frontline squadrons	Manpower
FAF	28	92,900 (incl. 34,000 conscripts)	16	64,000
GAF	~29	103,700 (incl. 30,800 conscripts)	17	67,500 (incl. 16,100 conscripts)
RNLAF	9	16,000 (incl. 3,500 conscripts)	6	11,050
SwAF	19	7,500 (incl. 5,500 conscripts)	8	5,900 (incl. 1,500 conscripts and 1,600 active reservists)

[a]IISS, *The Military Balance* 1991–92 (London: Brassey's for the IISS, 1991), 55, 57, 60, 67, 94.
[b]IISS, *The Military Balance* 2003–4 (Oxford, England: Oxford University Press for the IISS, 2003), 39, 41, 43, 51, 80.
[c]Data on flying frontline squadrons is relegated to fixed-wing fighter/fighter-bomber aircraft in the air-to-air, air-to-ground, and reconnaissance roles. For the FAF, strategic and prestrategic nuclear squadrons, which can also be used in a conventional role, are included.
[d]Excluding reservists.

Table 8. Differential air power

| | 1990[a] | | 2000[b] | | Beyond 2010 (new-generation combat aircraft only) | |
	Combat aircraft	Transport/ tanker	Combat aircraft	Transport/ tanker	Combat aircraft	Transport/ tanker
UK	790	100	520	80	245[c]	43[d]
France	630	100	530	100	286[e]	67[f]
(West) Germany	710	90	490	90	143[g]	60[h]
Netherlands	230	-	157	4	85[i]	7[j]
Sweden	460	8	260	8	100[k]	8[l]
Total	2,820	298	1,957	282	859	185
					1,320[m] (Europe total)	301[n] (Europe total)
United States	6,800	1,600	4,900	1,600	3,259[o]	1,208[p]

Note: The above data is based upon approximate figures and also includes naval and marine aviation. While aircraft in store are included, jet trainers are excluded, apart from cases where they had a dual training/combat role as in the case of Germany and Sweden in 1990. Transport aircraft below the size of a C-160 Transall are excluded.

[a]IISS, *The Military Balance 1990–91* (London: Brassey's for the IISS, 1990), 17, 21–24, 65–66, 68, 75, 84–85, 93.

[b]IISS, *The Military Balance 2000–01* (Oxford, England: Oxford University Press for the IISS, 2000), 28, 30–31, 59–60, 62–63, 70, 81–82, 104.

[c]Figure takes into account a reduction in the UK's projected number of Eurofighters in service from the original order of 232 to 107 and includes 138 F-35 JSFs. Yet, the UK's *Strategic Defence and Security Review* 2010 foresees a reduction of the planned number of JSFs to be acquired. The exact figure has not been publicly released to date. Tim Ripley, "UK Plans to Axe a Third of Typhoon Force by 2015," *Jane's Defence Weekly* 47, no. 28 (14 July 2010): 5; John R. Kent, F-35 Lightning II Communications, Lockheed Martin Aeronautics Co., to the author, e-mail, 3 August 2009; and *Securing Britain in an Age of Uncertainty: The Strategic Defence and Security Review*, presented to Parliament by the Prime Minister by Command of Her Majesty (London: The Stationery Office, 2010), 27.

[d]14 Airbus A330 future strategic transport and tanker aircraft, 7 C-17s, and 22 A400Ms. The UK's C-130J Hercules transport aircraft are planned to be withdrawn from service by 2022. *Securing Britain in an Age of Uncertainty*, 25, 27.

[e]286 Rafales. See Xavier Pintat and Daniel Reiner, *Avis présenté au nom de la commission des Affaires étrangères, de la défense et des forces armées (1) sur le projet de loi de finances pour 2009, adopté par l'Assemblée Nationale*, Tome 5, *défense—équipement des forces*, no. 102, (Paris: Sénat, 20 November 2008), chap. 2, sec. 4, "Engagement et combat, B. Frapper à distance ," http://www.senat.fr/rap/a08 -102-5/a08-102-5.html.

[f]Includes 14 MRTTs, 50 A400Ms, and 3 A310-300s. Pintat and Reiner, *Avis présenté au nom de la commission des Affaires étrangères*, Tome 5, chap. 2, sec. 3, "Projection, mobilité, soutien, A. La projection vers un théâtre d'opération."

[g]Figure takes into account a potential reduction in Germany's Eurofighter order from 180 to 143. See Thomas Newdick, "A Change in the Air," *Jane's Defence Weekly* 47, no. 22 (2 June 2010): 30.

[h]7 A310s (4 MRTTs, 3 MRTs) and 53 A400Ms. Yet the GAF might only commission 40 A400M aircraft. See IISS, *The Military Balance 2010* (Abingdon, Oxfordshire, England: published by Routledge for the IISS, February 2010), 136; and "Frankreich attackiert Deutschland wegen A400M," *Handelsblatt*, 31 January 2011, http://www.handelsblatt.com/unternehmen/industrie/airbus-frankreich -attackiert-deutschland-wegen-a400m;2743458.

[i]85F-35 JSFs. Kent to the author, e-mail.

[j]2 KDC-10s, 1 DC-10, and 4 C-130Hs. Kees van der Mark, "Tall Ambitions for the Lowlands," *Air Forces Monthly*, July 2009, 47.

[k]100 Gripen C/Ds. IISS, *Military Balance 2010*, 193.

[l]8 Hercules. Ibid.

[m]Figures from above, plus planned or current fleets of F-35 JSFs, JAS-39C/D Gripens, and Eurofighters in the rest of Europe: JSFs—Italy (131), Norway (56), and Denmark (48); Gripens—Hungary (14) and Czech Republic (14); Eurofighters—Spain (87), Austria (15), and Italy (96). Figure takes into account a reduction in Italy's Eurofighter order from 121 to 96. Kent to the author, e-mail; Air Force Technology, "Gripen Multirole Fighter Aircraft, Sweden," http://www.airforce-technology .com/projects/gripen; Air Force Technology, "Eurofighter Typhoon Multirole Combat Fighter, Europe," http://www.airforce-technology.com/projects/ef2000; and Tom Kington, "Italy, France Protect Vital Future Capabilities," *Defense News*, 6 September 2010, 11.

[n]Figures from above, plus 116 transport aircraft of the size of a C-130 and above in Austria, Belgium, Denmark, Greece, Italy, Luxembourg, Norway, Poland, Portugal, Romania, and Spain. See IISS, *Military Balance 2010*, 121, 127, 139, 144, 152, 154, 156, 158, 163, 175; and Airbus Military, "A400M," accessed 4 February 2008, http://www.airbusmilitary.com.

[o]Includes 2,443 F-35 JSFs, 187 F-22 Raptors, 515 F/A-18E/F Super Hornets, and 114 EA-18G Growlers. Kent to the author, e-mail; James Ludes, "The F-22 Has a Future," *Defence News* (20 April 2009), http://www.defensenews.com/story.php?i=4045998; and John Reed, "U.S. Navy Awards Super Hornet Contract," *Defense News*, 28 September 2010, http://www.defensenews.com/story.php?i= 4799993&c=AIR&s=AME.

[p]Includes 174 C-17s, 47 C-5Bs, 441 C-130E/H/Js, 34 KC-130Js, 59 KC-10As, and 453 KC-135s. Exact number of AAR aircraft to replace KC-135 fleet has not been decided upon. IISS, *Military Balance 2010*, 46–47.

While including other select European air forces would increase the total number of combat aircraft, they would not add substantially to air mobility. The UK and France, for instance, had the only significant European tanker fleets and were the only air forces to have a national and deployable airborne early warning component in the 1990s.[3] Moreover, the USAF long-range bomber and strategic airlift fleets had no equivalent in Europe, and American combat aircraft were on average more advanced than their European counterparts. Partly through reductions in numbers, the Continental Europeans attempted to generate more interoperable and deployable air power. In terms of mere numbers, the UK, France, and Germany were comparable. Yet unlike France, for instance, the UK had already focused upon complex conventional air power during the Cold War, and export potentials influenced aircraft specifications to a lesser degree.[4] Moreover, as Desert Storm proved, the RAF was the only European air force capable of playing a leading role in a major combined air campaign at the outset of the post–Cold War era.

For the foreseeable future, the United States will retain its air power dominance. Though Europe's currently planned 1,320 new-generation combat aircraft seem impressive, they are by no means all usable in deployed operations, particularly due to a lack of force enablers. European defence resources turned out to be too scarce for the buildup of a tanker fleet that is even remotely comparable to that of the USAF. Air bases and equipment for deployed operations are also lacking, as has been demonstrated by NATO Response Force packages. Up to the time of writing, deployable air bases have remained a bottleneck, and only the RAF and FAF have been able to provide deployable air bases on a national basis. On the positive side, programmes such as the A400M are under way to remedy the situation. This airbus is expected to become the backbone of the multinational European Air Transport Command and is likely to considerably enhance European intertheatre airlift capacities.

Shortcomings in materiel can partly be compensated for through bi- and multinational cooperation, as the RNLAF has proven. Through embedding its Air Force into multilateral approaches, combined with its willingness to share risks across the spectrum of military operations, the Netherlands has counterbalanced a limited defence budget. By virtue of its Air Force's military prowess, the Netherlands has established itself as a trustworthy alliance partner—particularly in the view of the United States—that could make meaningful contributions to combined operations. It can be argued that the RNLAF's limited size—combined with its natural inclination to play a leading role in deployed operations and its conducive political environment—bolstered its adaptation to the organisational and doctrinal challenges of the post–Cold War era.

In France and Germany, the challenges of deployed operations have fostered a readiness to transfer national authority into cooperative ventures that have enhanced air power's effectiveness, particularly in air mobility. Promoted by these two countries, the EATC represents the culmination of European air mobility integration and is likely to appreciably cut administrative overheads. For rapid deployments, European nations need to be able to summon substantial airlift capabilities. The EATC provides member countries with an unprecedented surge capability. Another example of a successful cooperative venture was the establishment of a multinational European F-16 expeditionary air wing in 2004. So far, the trend has gravitated towards cooperation and pooling, while role specialisation has been viewed as limiting potential deployment options. Yet, role specialisation has occurred by default rather than by design. For instance, when its Jaguar fleet was phased out, the FAF also lost its dedicated SEAD capability. It was subsequently announced that the FAF would rely upon partners for this particular role in multinational air operations.

As regards alliance policy, American influence in European security matters has been welcomed by some, while others have attempted to minimise it. The Netherlands has chosen a strong transatlantic link, allowing it to generate effective segments of air power at a relatively low price. While the RNLAF has versatile and potent multirole combat aircraft and helicopter fleets at its disposal, it lacks essential functions such as deployable air operations centres, deployable air bases, or sufficient strategic airlift. The RNLAF is geared up for effectively plugging into American-led operations. In contrast, France has sought to generate independent European air power, which is nevertheless interoperable with US air power.

Unlike France, Germany has never had an ambition to supplant NATO by an independent European approach. Nonetheless, it strives to make the transatlantic partnership more equal, requiring the buildup of genuine European capabilities. Given Germany's close embedment into alliance structures, efforts such as NATO's Defence Capabilities Initiative or the Helsinki Headline Goal—launched in the same year—have considerably influenced force transformation. Consequently, the GAF has developed into a more balanced force. Moreover, Franco-German cooperation might be gaining in importance. Their combined armed forces are expected to cover a broad range of aerospace power capabilities in the medium term—including airborne early warning, SEAD, theatre ballistic missile defence, MALE/HALE RPAs, deployable air operations centres, combat search and rescue, air-to-air refuelling, and strategic airlift—without recourse to commonly owned NATO assets. This leads to a further conclusion highlighting the importance of European cooperation for generating relevant air power. A Franco-German core of European aerospace power capabilities will be coherent and will allow smaller European nations to plug and play.

Courtesy German Air Force; photo by Ulrich Metternich

Airbus A310 MRTT refuelling GAF Tornado and Swiss Air Force F/A-18D Hornet in midair

European countries do not have to be capable of autonomously dealing with major contingencies in deployed operations, such as a future conflict involving a major regional military power. In these circumstances, it is—according to the British strategic expert Lawrence Freedman—"inconceivable that European governments would act independently of the US."[5] However, the EU must be equipped to autonomously undertake peace support operations across the spectrum of military force, especially battle-group-sized early-entry operations in failed states. After severe disagreements over Iraq in early 2003, it is difficult to imagine that the EU could have drawn easily upon NATO resources for Operation Artemis in mid-2003.

Rapid intervention operations in failed states require solid strategic airlift, ISTAR, and CAS capabilities, the latter providing escalation dominance. While a NATO C-17 airlift fleet might prove useful in these scenarios, ultimately, US consent will be required—a right Americans have put a premium upon to secure their influence upon European security.[6] In contrast, cooperative arrangements that can be made equally available to NATO and ESDP operations, such as the EATC, offer assured availability for European operations.

NATO and the ESDP complement each other. The former guarantees a strong transatlantic link and provides for collective defence. The latter is particularly suited to respond to active requests by the UN secretary general against the backdrop of crises in the developing world. In many cases, the ESDP is the natural choice for the conduct of non–Article 5 crisis management scenarios. ESDP-led operations are seen in certain theatres as more benign than potential NATO operations. This specific reality was underlined by a French Air Force officer with ample experience in the African theatre at a 2008 NATO conference.[7] So far, the EU has indeed become a major actor in African security. Yet, while the ESDP provides a suitable framework for peace support operations throughout the spectrum of military force, it is less likely to provide an adequate framework for conventional high-intensity warfare in deployed operations.

With regard to European cooperation, the SwAF, for instance, still retains a coherent force structure for national defence scenarios and might be in a position to contribute key niche capabilities such as airborne early warning and control. Yet, the Netherlands and Sweden have been unable to build up capacities in resource-intensive areas, such as intertheatre airlift, similar to those of Germany and France. The SwAF has been forced to redress the imbalance between an overstretched force structure and operational requirements. As in the case of Germany, European headline goals have had a significant impact upon Sweden's military transformation—with interoperability and rapid deployability becoming the determining factor for the SwAF's force structure. Whereas during the Cold War the SwAF was able to maintain a large aircraft fleet through cost-effective national approaches to air power and the use of conscripts, deployed operations impose a number of minimum requirements such as participation in integrated exercises.

Challenge of Real Operations

Simultaneously with the end of the Cold War, Desert Storm proved to be a major catalyst for the evolution of Western air power. According to their national circumstances, the four air forces examined reacted differently to the air war in the Gulf. For the FAF, Desert Storm changed paradigms about interoperability and international integration and generated a new emphasis upon advanced conventional weapon systems. German decision makers felt uneasy with offensive deployed operations and continued to emphasise territorial alliance defence throughout the 1990s. This outlook led to a gap in doctrinal innovation in the GAF, which was only overcome with the 2002–3 defence reform. The Netherlands entered the post–Cold War era with the least baggage, and the country—closely aligned with the United States—swiftly implemented lessons on a political level from its "near-fighter-deployment" to the Gulf area during Operations Desert Shield and Desert Storm. But on a military level, particularly for developing doctrine and acquiring an air-to-ground precision capability, Operations Deny Flight and Deliberate Force were more instrumental. In Sweden, lessons were not explicitly formulated. Yet throughout the 1990s, the capacity for countering limited strategic surprise attacks was particularly emphasised—basically an attempt to offset conventional precision air power. It can be concluded that amongst the four air forces examined, Desert Storm had the biggest impact upon the FAF, both politically and militarily. This is not surprising since the FAF was directly involved in the operation. There is a difference between "being there" and "watching an operation."

While Desert Storm was a watershed for the FAF, operations at the upper end of the spectrum of military force were nothing new for the French—the FAF had been deployed to African theatres in the days of the Cold War. In the case of the RNLAF, the political leadership's willingness to employ it across the spectrum of force has become a significant leverage for Dutch air power. The RNLAF delivers proof that smaller nations can make a significant contribution to multinational operations disproportionate to their size when political resolve is combined with adequate military capabilities to carry out missions across the entire spectrum of military force. This is partly because the RNLAF entered the post–Cold War era with the least baggage amongst the four air forces examined. Yet, even with the Netherlands, an overtly aggressive stance in the use of offensive air power had to be avoided. To circumvent arousing public concerns, the Dutch supreme commander deliberately imposed a low silhouette upon the RNLAF's performance in Operation Allied Force. This was particularly the case after a Dutch F-16 downed a Serb MiG-29 in the first night of operations, the only European air-to-air kill in major air campaigns throughout the 1990s.

In contrast, because of their different historical experiences, Germany and Sweden have been hampered in employing military power in real operations and in moving away from a threat-based paradigm to a more proactive defence policy, taking on security challenges at a distance. Consequently, the GAF's performance has been constrained considerably in real operations. German out-of-area contributions, apart from air mobility, have been relegated to SEAD and reconnaissance. Though readied to do so, the GAF has not so far dispatched ground-attack aircraft armed with LGBs. In the case of the SwAF, only the transport unit has been dispatched on out-of-area missions up to this writing. Both Germany and Sweden prefer the employment of their offensive capabilities against purely military targets to avoid collateral damage. Accordingly, the SwAF's rapid reaction units were first prepared solely for reconnaissance and air-to-air tasks, and the

NBG Gripen detachment is geared up for CAS. Nevertheless, both the GAF and the SwAF have exposed crews to appreciable risks entailing potential fatalities.

Real operations in the post–Cold War era have highlighted significant European air power shortfalls. Some of these deficits have not been due to a lack of capabilities but to national caveats. The GAF, for instance, has a potent SEAD component that was used to great effect in Allied Force, where it released nearly a third of all anti-radiation missiles. In light of the small size of the German detachment, this was a momentous achievement. While this capability was already available during Deliberate Force in 1995, it could not be used to its full effect because of political constraints. Participation in deployed operations that might entail combat are still highly sensitive in Germany, as demonstrated by the recent debate about German crews manning NATO AWACS aircraft patrolling Turkish airspace against the backdrop of Operation Iraqi Freedom.

Real operations have been a catalyst for cooperative endeavours, doctrine development, and procurement in the post–Cold War era. The Franco-British Air Group, the predecessor of the European Air Group, for instance, developed from Franco-British experiences in the air operations over Bosnia. The operational dimension itself has been shaped by the requirements of combined operations, which have been dominated by US air power and which meant the adoption of a doctrine geared up for medium-altitude operations. Thus, while regular low-level training in Canada was abandoned by both the GAF and the RNLAF, integrated exercises such as Red Flag in the United States have become of pivotal importance. These exercises allowed German and Dutch pilots to plug seamlessly into combined air campaigns over Bosnia and Kosovo, despite their lack of combat experience throughout the Cold War. Even the SwAF, which is not a NATO alliance air force, came to recognise the importance of integrated air exercises for deployed operations. Real operations and integrated exercises led to a convergence of European air power. This thrust has been further facilitated through ventures such as the NATO Response Force in a NATO context or, more recently, the EU Air Rapid Response Concept in an EU context, underlining the importance of European air power cooperation. In the setting of these ventures, major European air forces, such as the RAF or the FAF, have gained leadership experience by taking on responsibility for combined air component commands. Yet, despite the adoption of a common operational doctrine, certain distinct features have been retained by the air forces examined. The RNLAF has employed its swing-role concept to great effect, and the FAF has excelled in deploying small contingents for integrated operations in Africa.

In Kosovo in 1999, air power was the weapon of choice for not only the United States but also for European participants. Yet unlike USAF doctrine, a fragile political environment has resulted in a gradualist Continental European approach to the use of air power—demonstrated in Allied Force. Continental Europeans are reluctant to get involved in "shock and awe" air campaigns, given the casualty sensitivity in the context of wars of choice. Germany and Sweden—with their particular but different legacies—prefer stabilisation operations, where air power is used in an integrated mode and plays a rather ancillary role. In fact, a transatlantic division of labour was apparent. While European nations provided the bulk of ground forces in Bosnia and Kosovo, Americans provided most of the assets for air operations. For instance, the Clinton administration steadfastly refused to commit ground forces to UNPROFOR in Bosnia, but the United States was the key player in Deny Flight and Deliberate Force.[8] France has often embedded air power into integrated operations. In 1995 the French military promoted the presence of a heavy ground component in Bosnia that could be used in conjunction with the air strikes. Sweden also puts a premium upon integrated operations. The NBG Gripen detachment is primarily geared up for the joint battle. This emphasis upon jointness partly explains why European air forces do not receive the same priority in the overall defence structure as American air power does. In contrast, the USAF considers the independent long-range strike role as one of its raisons d'être.

Yet, growing involvement in out-of-area operation can cause increased expenditure on air power. In the case of Germany, the A400M programme together with the Eurofighter is about to consume a significant share of the defence equipment budget. Deployed operations in general have had far-reaching implications for procurement, serving as a major catalyst in force protection, armament, and C4ISTAR. The issue of casualty aversion was addressed by a new emphasis on force protection, while highly complex combined air campaigns demanded an advanced and interoperable C2 infrastructure—as the FAF painfully learned in Desert Storm.

Apart from deployed air campaigns at the upper end of the spectrum of military force, the post–Cold War era has seen so-called non-kinetic air power used to great effect in humanitarian operations. Sweden and Germany have particularly felt more at ease with these kinds of operations in light of their particular legacies.

New Intellectualism in Air Power Thinking and Doctrine

Developments at the grand-strategic level have influenced air power thinking and doctrine. Sweden's and Germany's major defence reforms triggered, or at least supported, a more sophisticated approach to the doctrinal dimension of air power. The German Air Power Centre was set up against the backdrop of the 2002–3 defence reform, and Germany also acted as the lead nation in setting up NATO's Joint Air Power Competence Centre, which became operational in early 2005. Similarly, the Swedish doctrinal debate experienced a major impetus after the country's defence political process began broadening, starting in 1999, and with the SwAF readying units for rapid deployment. From 2002 onwards, the SwAF organised a series of major air power conferences annually. It became involved in a doctrine development process that in 2004 resulted in the SwAF's first published air power doctrine. In addition, the air faculty at the Swedish National Defence College started to produce its own publications on air power theory. Moreover, doctrine and proactive air power thinking provided the SwAF with a means to foster NATO interoperability. Despite Sweden's non-membership in NATO, the Swedish armed forces in general and the SwAF in particular have deliberately transferred NATO concepts into their military doctrine manuals.

Besides developments at the grand-strategic level, real operations have triggered doctrine development and a broader air power debate, particularly in the case of the RNLAF and the FAF. The launching of the air power journal *Penser les ailes françaises* can be understood as a response to the FAF's increased involvement in combined air operations. In line with its goal of being a lead European air force, the FAF has felt it necessary to also shape the air power debate. The RNLAF published its first air power doctrine in 1996 as a reaction to its involvement in the air operations over Bosnia and other deployed operations. Two major motivations can be discerned for doctrinal innovation within the RNLAF—staying compatible with its US counterparts as well as codifying and disseminating lessons learned from recent deployed operations. The RNLAF has viewed a doctrinal debate as contributing to its professionalism and enhancing its performance in combined air operations. As previously mentioned, General Manderfeld, chief of the Air Staff from 1992 to 1995, attributed the RNLAF's outstanding performance over Bosnia not only to training but also to education.

Accordingly, the RNLAF's efforts were well received, as General Horner, the Desert Storm air component commander, indicated. He particularly underlined the RNLAF's outstanding qualities with regard to professionalism, equipment, training, and attitude.

Courtesy Netherlands Institute for Military History; photo by Frank Visser

RNLAF F-16 at Kandahar Airfield in November 2008

RNLAF leadership envisions that a sound grasp of doctrine and air power theory will empower Dutch airmen to be on a par with the Netherlands' most advanced allies and maximize their operational effectiveness.

As revealed by Germany's response to Desert Storm, the GAF was uneasy with offensive air operations in the wake of the Cold War. This stance also reflected on air power doctrine. US operational dominance naturally led to US doctrinal dominance, which put a premium upon offensive air operations. Germany's difficulty in reconciling offensive air power doctrine with its historical legacy resulted in doctrinal stagnation throughout the 1990s. Right after Desert Storm in March 1991, the GAF published its first air power doctrine. The document was doctrinally anchored in the later stages of the Cold War. Unlike the RAF, however, which also published its first edition of AP 3000 in 1991 and saw a need to revise the doctrine to codify the lessons of Desert Storm, the GAF was reluctant to incorporate those lessons. Whereas in the later stages of the Cold War, GAF officers regularly expressed their views on issues such as the GAF's role within NATO's defence doctrine in various journals, in the 1990s fewer GAF officers wrote about doctrinal issues. Only in the environment of the 2002–3 defence reform did the GAF again approach doctrinal and conceptual issues more proactively. The establishment of the Air Power Centre and the GAF's lead role in the formation of NATO's Joint Air Power Competence Centre are part of this reorientation, perceived as necessary in light of recent German commitments to cooperative endeavours such as the NATO Response Force or the EU Air Rapid Response Concept.

Doctrine has not only been a response to the challenges of deployed operations but has also been a carrier of institutional interests. The SwAF and FAF responded to joint doctrine manuals by developing single-service doctrines. The GAF started to revise its air power doctrine when faced with a new armed forces reform that put a premium upon the joint level. Its leadership perceived it necessary to express explicitly air power's capabilities so as not to lose out relative to the other services.

The senior command echelon plays a crucial role in the doctrine development process, as particularly exemplified by the Royal Australian Air Force. Recognising the importance of doctrine, the chief of the Air Staff ordered the formation of the RAAF Air Power Studies Centre in 1989, providing Australia with an edge in doctrine development in the post–Cold War era. In France, the opposite occurred throughout the 1990s. Attempts at formalising doctrine were primarily driven by exogenous factors, and the chief of the Air Staff considered formal doctrine dogmatic, inhibiting innovative thinking. Only recently has doctrine development in France received top-level cover. Germany's uneasiness with deployed offensive operations also might have led to a lack of top-down support for doctrine development. Only the defence reform starting in 2002 put an end to doctrinal stagnation.

In contrast, the Netherlands entered the post–Cold War era with the least historical and political baggage. Partly as a consequence of that and due to natural inclination, the RNLAF senior command was receptive to the importance of sound education and doctrine. As argued above, General Manderfeld regarded education as vital for operational effectiveness. His successor, General Droste, directly linked deployed operations to the need for adequate air power doctrine in his foreword to the RNLAF's first published version of its air power doctrine in 1996.[9] In general, the RNLAF's senior command echelon has been receptive to doctrine development as well as to education.

The senior command's attitude to doctrine also proved to be crucial in the case of the SwAF. Whereas the first and second attempts at developing a national air power doctrine were bottom-up approaches, only with sufficient top-down backing and resource allocations did the third attempt result in a published doctrine.

The Continental European air forces—particularly the senior command levels—have gradually recognised the value of air power doctrine in the post–Cold War era and have started to emulate their Anglo-Saxon counterparts. This is reflected not only in the air forces' doctrine development but also in the dissemination and teaching of air power thought. However, they have not yet come up with truly innovative thinking. As the air power conferences in Sweden and the Netherlands reveal, the Western air power debate is still dominated by Anglo-Saxon thinkers. Moreover, though the defence colleges in the Netherlands and Sweden developed rigorous air power reading lists for their students, none of the four air forces has yet produced an intellectual air power study course comparable to that of SAASS in the United States. While ventures such as the EAG or NATO's JAPCC are commendable and help European air forces to become more interoperable, they do not really address challenges on an intellectual level.

Genuine European air power also requires a conceptual grasp of air power theory and history. No European air power school of thought has yet emerged and encapsulated modern Europe's gradualist approach to air war. This mindset is a result of Europe's rather tenuous political environment and preference for integrated and stabilisation operations. If European air forces intend to go beyond mere pooling of forces and to be an effective tool for common military and foreign political goals, a conceptual grasp of air power is essential.

Procurement

Continental European states generally have pursued three approaches to procurement—national programmes, cooperative programmes, and buying off the shelf. The low impact of politico-industrial concerns in the Netherlands has facilitated its ability to buy off the shelf, allowing the procurement of advanced American aircraft at relatively low unit costs. As a result, the RNLAF has been able to concentrate almost exclusively upon operational requirements and thereby to generate—in the short term at least—relevant air power. Buying off the shelf freed resources which could be invested in swing-role training.

In contrast, the path of strategic independence, as pursued by France, has its price in terms of operational effectiveness. Ambitious national development programmes in times of constrained defence budgets have led to significant delays in acquiring new capabilities. Particularly, the French air transport fleet has been suffering from chronic shortcomings and overstretch. This leads to a further conclusion which has affected all of the four air forces to a larger or lesser degree. Essentially, they ignored the impact of cost escalation on procurement and adopted overambitious plans. Plans had to be subsequently cancelled, as was, for instance, the case with a planned Franco-German SEAD missile

Despite the consequent loss of capabilities, such as dedicated SEAD, the FAF has been able to retain a balanced force structure, allowing France to stage smaller operations independently in deployed scenarios. For France, an independent defence industrial base is pivotal to further French ambitions. In the post–Cold War era, France has been the only European state to independently develop and produce not only combat aircraft but also a coherent range of subsystems from laser-designator pods to air-to-air and air-to-ground weapons. Only in cases where autonomous approaches have become excessively expensive, such as in the case of airborne early warning, has this principle been abandoned.

Courtesy French Air Force, Sirpa Air

Rafale flying over Afghanistan armed with AASM precision-guided munitions. The French AASM is an all-weather precision-guided bomb propelled by a rocket booster. Depending on its release altitude, it can engage targets at close or medium ranges, exceeding 50 km, with various options of terminal impact angles. In April 2008, Rafales engaged Taliban positions with AASMs for the first time, with the target coordinates provided by a Canadian forward air controller.

Germany, participating in the multilateral Eurofighter programme, has been successful in evolving towards a more balanced air force by striving for an improved ratio between strike assets and force enablers. The GAF's force transformation was further facilitated through multilateral development and procurement programmes. This approach has been deliberately pursued to strengthen the interrelationship within both the NATO and EU alliances. It has enabled Germany to be less concerned about economy of scale—particularly in the area of combat aircraft—with the burden being shared by several partners. It can hence be argued that the advantages of cooperation outweigh the handicaps. Yet, the GAF's evolvement into a more balanced air force is unlikely to be fully translated into effective operational output as a dichotomy between operational potential and political acceptance still exists.

Note that the Netherlands—which did not have any major politico-industrial obligations in European cooperative aircraft programmes—has not placed a larger emphasis upon the force-enabling component of air power. Instead, it has put a premium upon acquiring state-of-the-art strike assets such as the AH-64 Apache. Kinetic assets are still seen as providing for a country's defence, whereas an overemphasis on force enablers is seen as role specialisation—supposedly only providing a narrow bandwidth of political options. The nation-state still dominates the shaping of air power; this results in a significant imbalance between the shaft and the spear, not only on a national level but particularly on a European level. Despite the fact that, for instance, the procurement of a limited number of C-17 Globemasters in the late 1990s by a medium-sized European state would have generated an international leverage out of proportion to the costs involved, this option has not been pursued.

Different approaches to procurement lead to a further conclusion concerning the importance of European air power cooperation. Buying off the shelf, while promising cost-effective, short-term bargains, has not contributed to the buildup of European strategic autonomy. In the case of France and to a lesser degree of Germany, politico-industrial concerns were paramount over operational requirements. In contrast, the Netherlands—with its strong transatlantic orientation—has primarily equipped its Air Force with sophisticated American equipment at relatively low unit costs. Yet, in the absence of a European aerospace industry, American suppliers might have had no incentive to offer their products at a competitive price. Hence, it can be argued that the Netherlands has indirectly benefited from a European aerospace industry. Moreover, if Meilinger's tenth air power proposition that "air power includes not only military assets but an aerospace industry and commercial aviation" holds true, an aerospace industry is a critical requirement for at least the larger European nations.[10]

While the Netherlands' approach has generated effective air power at low costs, such an approach might paradoxically widen the transatlantic air power capability gap in the long term if pursued by the major European countries. On the one hand, a number of European constituencies might become reluctant to spend large sums on air power assets not produced domestically and on equipment that can only be used effectively with US forces. On the other, US suppliers might potentially become less inclined to sell their products at competitive prices. To strengthen the transatlantic partnership, Europe must be able to cover a significant spectrum of the air power spectrum on its own. In this regard, European cooperation is essential, as relevant European air power is no longer a primarily national matter.

From a grand-strategic point of view, only approaches taking into account the importance of a credible European aerospace industry can serve a common European defence effort in the long term. One British commentator argued that "it would be difficult to envisage any credible European security policy in the second century of air power without a credible aerospace industrial base to sustain it."[11] This does not imply that all specialised assets, such as HALE RPAs, must be developed and manufactured in Europe.

In the post–Cold War era, technological advancements have compensated for reductions and limited defence resources, particularly through the increased attention upon multirole-capable air power assets. A 1992 RAND study's assumption that the number of multirole squadrons would decline substantially did not materialise at all—quite the opposite has been the case.[12] All major Continental European air forces have primarily opted for single-seat, multirole fighter-bombers. Technological trends in the man-machine interface have made the single-seat fighter-bomber a viable option. Despite this trend towards multirole aircraft, technological niche capabilities in the areas of SEAD or TBMD have provided single European air forces with leverage in deployed operations.

Technological trends have particularly played into the hands of the RNLAF and have allowed it to overcome its limited defence resources. In the 1980s highly sophisticated and specialised aircraft such as the Tornado were required for the all-weather attack role. But the shift towards medium altitude combined with bolt-on solutions, such as laser-designator pods or GPS devices, transformed lighter fighters such as the F-16 into potent multirole aircraft. The RNLAF F-16s are expected to remain in the front line up to 2020. Through this extended life span, significant cost savings can be achieved. Given the relatively permissive air environment of the post–Cold War era, bolt-on solutions have also turned less advanced aircraft into relevant attack aircraft. In Allied Force, where no major air superiority battles had to be fought, French Jaguars equipped with laser-designator pods made significant contributions. The asymmetry between Western air power and potential opponents in general led to a new emphasis upon the air-to-ground role as opposed to the air-to-air role.

Technological advancements have also appreciably enhanced precision and firepower. The Rafale, for instance, can release a standard air-to-ground weapons load of four to six PGMs. In Allied Force, advanced strike aircraft such as the Tornado, the Mirage 2000D, or the F-16 carried a standard weapons load of only two LGBs. Such an increase in firepower puts a British commentator's argument, made in the wake of Kosovo—that Europe would have to increase its 500 all-weather bombers by 50 per cent—into a different perspective.[13] Moreover, the FAF and GAF have made quantum leaps in the acquisition of cruise missiles. During Operation Allied Force in 1999, when American and British forces released a total of 329 cruise missiles, the Continental Europeans were not poised to contribute to the cruise missile campaign. Since then, the GAF and the FAF have together acquired over one thousand missiles.

Several technological benchmarks have been established. Minimum requirements have to be met to plug and play, particularly in the areas of interoperable C2 and precision strike. These common criteria no longer allow for autonomous approaches. In this regard, the way that air power has been generated has become more convergent. Advances in the technological dimension also helped to computerise C2 functions and hence facilitated leaner overall command structures. Only through these technological advances could a long-held air power tenet—unity of command over all air assets—be put into practice.

France's leading role in the development of the Link-16 terminal MIDS is a good example of how grand-strategic ambitions and operational requirements have influenced each other. Its ambition to be a lead nation in European defence and its shortcomings in Desert Storm in interoperability led to France's decision to become a key player in this crucial technological endeavour. The demands of interoperability also influenced the SwAF, which had to supplant its advanced national data links with NATO interoperable Link-16 terminals.

In general, the emphasis upon airframes has shifted to a more balanced approach that gives more consideration to air power software. In fact, air power software gained unprecedented importance at the outset of the post–Cold War era. In the wake of Desert Storm, Martin van Creveld argued that "no other country possesses the hardware, much less the 'software,' needed for mounting an air campaign that will even remotely compare with US capabilities in this field."[14] Since then France and Germany have been building up computerised C2 systems for conducting combined air campaigns.

While the technological dimension is without doubt of utmost importance for the generation of relevant air power, it has been less important for Continental European air forces as compared to the USAF. During Operations Deny Flight and Deliberate Force, for instance, the RNLAF alongside other European allies would have preferred to employ air power more flexibly instead of relying upon rigid force packages with standoff jammers and dedicated SEAD assets. This more human-centric approach was also revealed in logistics when the FAF deployed to Manas in Central Asia in early 2002. In contrast to the American detachment, the FAF made up for limitations in power projection by drawing as much as possible upon regional resupply. Accordingly, Continental Europe has not produced technological quantum leaps in air power but has pursued a path of gradual steps.

Assessment

Given air power's dependence upon technology and the almost exponential rise in the costs of air power assets, one might expect materiel and defence budget resources to be the primary shapers

The ASC 890 Erieye is the SwAF's NATO interoperable airborne early warning and control aircraft, shown escorted by Gripen C and D fighter aircraft.

of air power. Contrarily, the evidence suggests that adaptations to the altered strategic environment and a nation's willingness to share risks across the spectrum of military operations have largely influenced Continental European air power from 1990 to 2005. These characteristics effectively offset economic and material limitations.

Political adaptations to the new post–Cold War realities could potentially be realized faster than major development programmes, which could not be so easily adjusted in this transition period. Consequently, force size was not the predominant factor in producing relevant European air power—as exemplified by the RNLAF. Size will be more relevant once new capabilities have been strengthened, particularly in the force-enabling areas.

It is in technology and operations where Continental European air power has become most convergent. Combined operations have required common doctrinal and technological criteria. Only by meeting these criteria can an air force plug and play.

At the outset of the post–Cold War era, Desert Storm gave a vivid demonstration of modern air power. According to their national circumstances, the four air forces examined reacted differently to the air war in the Gulf. Yet in the medium and long term, each of the air forces began to implement the lessons of modern air warfare. Their approaches to force structuring, operations, procurement, and doctrine reflected these lessons.

Buying American—combined with a willingness to actually employ air power across the spectrum of military force—has facilitated effective air power at a relatively low price. In contrast, European autarchy in developing new proficiencies has led to shortcomings in the operational dimension, particularly in inter-theatre airlift. Nevertheless, this approach is pivotal for fortifying European defence. As discussed above, however, meaningful European air power hinges not only upon physical but also upon conceptual components. A doctrinal grasp of air power that reflects Europe's political environment and distinct approaches to air power is therefore a prerequisite.

Each air force has responded to the air power challenges of the post–Cold War era according to its context. Hence, the single air forces are not representative and are to a large degree sui generis. Yet, the four air forces combined set the benchmark for Continental European air power in its various dimensions, including a transatlantic, national, and European emphasis upon air power. While European shortfalls have led to various cooperative ventures, leading to converging European air power at the technological and operational levels, these developments are not likely to end up in a wider European air force. The nation-state still dominates defence policy at the strategic level.

Overall, Continental Europe places less merit upon air power than does the United States. Nevertheless, Continental Europe has been strengthening air power capabilities that have the potential of taking on a more prominent role on the international stage. Provided that planned development and procurement programmes are implemented, Continental Europe together with the UK will be capable of autonomously conducting a potential operation similar to Deliberate Force in Europe's vicinity. Deliberate Force saw the employment of approximately 300 combat and support aircraft, mainly from well-prepared air bases in Italy. This number was slightly below the requirements stated in the EU Helsinki Headline Goal of 1999, which conceived the ability to deploy up to 400 combat aircraft for EU operations but proved too ambitious at the time. Yet a campaign similar to Allied Force, involving about

1,000 aircraft in the final stages—including American long-range bombers—is still beyond Europe's capability alone despite technological advances and requires significant American involvement.[15] Nevertheless, in an environment of decreasing American air assets, European contributions to deployed operations might become more relevant—not only from a political point of view but also from an operational one. Continental European air forces remain natural allies for their Anglo-Saxon counterparts, and understanding their different approaches is indispensable for combined Western air operations.

Notes

1. Mathieu, *La Défense nationale*, 54; and Portz, "The Future of Air Power," 14.

2. JAPCC, *NATO Flightplan for Unmanned Aircraft Systems*, "Annex B: NATO Unmanned Aircraft Systems—Operational," accessed 4 August 2008, http://www.japcc.de/nato_flightplan_uas.html.

3. Gething and Sweetman, "Air-to-Air Refuelling," 45; and International Institute for Strategic Studies, *The Military Balance 1995-96* (Oxford, England: Oxford University Press for the IISS, 1995), 66.

4. See, for instance, Dörfer, *Arms Deal*, 108.

5. Freedman, "Can the EU Develop an Effective Military Doctrine?" in Everts et al., *European Way of War*, 23.

6. Gnesotto et al., *European Defence*, 49.

7. Lt Col Christophe Taesch, FAF, "Air Power and Rapid Reaction Interventions in Africa" (address, "Europe, NATO, and Air Power," conference, Allied Air Component Command Headquarters, Ramstein, Germany, 7 October 2008).

8. Daalder and O'Hanlon, *Winning Ugly*, 217; and Steven R. Bowman, *Bosnia: U.S. Military Operations* (Washington, D.C.: Congressional Research Service, updated 8 July 2003), 1.

9. *Royal Netherlands Air Force Air Power Doctrine.*

10. Phillip S. Meilinger, *10 Propositions regarding Air Power* (Washington, D.C.: Air Force History and Museums Program, 1995), 61.

11. Mason, *Air Power*, 249.

12. Lorell, *Future of Allied Tactical Fighter Forces*, xi.

13. Garden, "European Air Power," in Gray, *Air Power 21*, 114.

14. Martin van Creveld, Steven L. Canby, and Kenneth S. Brower, *Air Power and Maneuver Warfare* (Honolulu, HI: University Press of the Pacific, 2002), xiv.

15. See Lambeth, *NATO's Air War for Kosovo*, 35.

Abbreviations

AAM	air-to-air missile
AAP	Australian air publication
AAR	air-to-air refuelling
AASM	*armement air-sol modulaire* (modular air-to-surface armament)
ACCS	air command and control system
ACE	Allied Command Europe
ACLANT	Allied Command Atlantic
ACO	Allied Command Operations
ACT	Allied Command Transformation
AD	air division
AFDC	Air Force Doctrine Center
AFDD	Air Force doctrine document
AFM	Air Force manual
AFSOUTH	Allied Forces Southern Europe
AGS	Alliance Ground Surveillance
AIM	air-intercept missile
AJP	allied joint publication
ALARM	air-launched anti-radiation missile
AMRAAM	advanced medium-range air-to-air missile
AP	air publication
ASMP	*air-sol moyenne portée* (medium range air-to-ground weapon)
ATARES	Air Transport and Air-to-Air Refuelling Exchange of Services
ATP	allied tactical publication
AU	Air University
AWACS	Airborne Warning and Control System
BMVg	Bundesministerium der Verteidigung (Federal Ministry of Defence)
C2	command and control
C4ISTAR	command, control, communications, computers, information/intelligence, surveillance, target acquisition, and reconnaissance
CAOC	combined air operations centre
CAP	combat air patrol
CAS	close air support
CESA	Centre d'études stratégiques aérospatiales (Centre of Strategic Aerospace Studies)

CESAM	Cellule études et stratégie aérienne militaire (air power strategy and studies cell)
CFSP	Common Foreign and Security Policy
CICDE	Centre interarmées de concepts, de doctrines et d'expérimentations (Joint Forces Centre for Concept Development, Doctrine, and Experimentation)
CID	Collège interarmées de défense (Joint Forces Defence College)
CJEX	Combined Joint European Exercise
CJFAC	combined joint force air component
CJFACC	combined joint force air component commander
CJTF	combined joint task force
CSAR	combat search and rescue
DAG	doctrine advisory group
DATF	Deployable Air Task Force
DCI	Defence Capabilities Initiative
DMAW	deployable multinational air wing
EAC	European Airlift Centre
EACC	European Airlift Coordination Cell
EAD	extended air defence
EAG	European Air Group
EATC	European Air Transport Command
EBO	effects-based operations
ECR	electronic combat/reconnaissance
EDA	European Defence Agency
EPAF	European Participating Air Forces
EPAF EAW	European Participating Air Forces' Expeditionary Air Wing
ESDI	European Security and Defence Identity
ESDP	European Security and Defence Policy
ESGA	École supérieure de guerre aérienne (Advanced School of Air Warfare)
EU	European Union
EU Air RRC	EU Air Rapid Response Concept
EUFOR	European Union Force
EURAC	European Air Chiefs Conference
EU RRAI	EU Rapid Response Air Initiative
FAF	French Air Force
FBEAG	Franco-British European Air Group

FBW	fighter-bomber wing
FLA	Future Large Aircraft
FLIR	forward-looking infrared
FMV	Swedish Defence Materiel Administration
FüAk	Führungsakademie (German Armed Forces Command and Staff College)
FW	fighter wing
GAF	German Air Force
GBAD	ground based air defence
GDP	gross domestic product
GNP	gross national product
GPS	Global Positioning System
GS	General Staff (German)
GSO	general staff officer
HALE	high-altitude long-endurance
HARM	high-speed anti-radiation missile
HG	Headline Goal
HHG	Helsinki Headline Goal
IDS	interdiction-strike
IFF	identification, friend or foe
IFOR	Implementation Force
IISS	International Institute for Strategic Studies
IRIS-T	Infrared Imaging System—Tail/Thrust Vector-Controlled
ISAF	International Security Assistance Force
ISTAR	intelligence, surveillance, target acquisition, and reconnaissance
JAPCC	Joint Air Power Competence Centre
JDAM	Joint Direct Attack Munition
JFACC	joint force air component commander
JSCSC	Joint Services Command and Staff College
JSF	Joint Strike Fighter
JSO	junior staff officer
JSTARS	Joint Surveillance Target Attack Radar System
km	kilometer
LANTIRN	low-altitude navigation and targeting infrared for night
LGB	laser-guided bomb
LMZ	Luftmachtzentrum (Air Power Centre)

LPM	*Loi de Programmation Militaire* (French military programme bill of law)
MALE	medium-altitude long-endurance
MANPADS	man-portable air defence system
MCCE	Movement Coordination Centre Europe
MEADS	Medium Extended Air Defence System
MIDS	multifunctional information distributions system
MLU	midlife update
MOD	Ministry of Defence
MOU	memorandum of understanding
MRT	multirole transport
MRTT	multirole transport tanker
NAEW&CF	NATO Airborne Early Warning and Control Force
NASAMS	Norwegian Advanced Surface-to-Air Missile System
NATD	Netherlands Apache Training Detachment
NATO	North Atlantic Treaty Organization
NBG	Nordic Battle Group
NDC	Netherlands Defence College
NDD	Netherlands Defence Doctrine
NDSO	Netherlands Detachment Springfield Ohio
NFZ	no-fly zone
NG	*nouvelle génération* (new generation)
NRF	NATO Response Force
PfP	Partnership for Peace
PGM	precision-guided munition
QRA	quick reaction alert
RAAF	Royal Australian Air Force
RAF	Royal Air Force
RFAS	Reaction Forces Air Staff
RNLAF	Royal Netherlands Air Force
ROE	rule of engagement
RPA	remotely piloted aircraft
RRC	Rapid Response Concept
RRF	rapid reaction force
RW	reconnaissance wing
SAASS	School of Advanced Air and Space Studies
SACEUR	supreme allied commander, Europe

SACT	Supreme Allied Commander Transformation
SALIS	Strategic Airlift Interim Solution
SAM	surface-to-air missile
SAMP-T	*sol-air moyenne portée terrestre* (land-based medium range surface-to-air missile)
SAR	search and rescue
SCCOA	*système de commandement et de conduite des opérations aériennes* (air operations command and control system)
SE01	Combat Aircraft Unit 01
SEAD	suppression of enemy air defences
SHAPE	Supreme Headquarters Allied Powers, Europe
SIDM	*système intérimaire de drone MALE* (interim MALE RPA system)
SIGINT	signals intelligence
Sirpa Air	Service d'information et de relations publiques de l'Armée de l'Air (FAF Information and Public Relations Service)
SNDC	Swedish National Defence College
SwAF	Swedish Air Force
SWAFRAP	Swedish Air Force Rapid Reaction Unit
TACTESS	tactical training, evaluation, and standardisation squadron
TBMD	theatre ballistic missile defence
THG	Tactical Helicopter Group
UK	United Kingdom
UN	United Nations
UNMEE	United Nations Mission in Ethiopia and Eritrea
UNPROFOR	United Nations Protection Force
UNSCOM	United Nations Special Command
USAF	United States Air Force
USMC	United States Marine Corps
WEU	Western European Union
WMD	weapon of mass destruction

Selected Bibliography

This selected bibliography is adapted from the author's PhD dissertation that contained a more extensive listing. The following references are those that the author consulted most often in conducting the research for this book.

Interviews, E-Mails, and Letters

Andersson, Bengt (senior analyst, Swedish Defence Research Agency [FOI], Stockholm). Interview by the author, 11 February 2005.

Annerfalk, Anders, communications manager, Gripen International. To the author. E-mail, 7 July 2005.

Björs, Lt Col Christer, Swedish Air Force (SwAF) (chief, Air Operations [A3], Swedish Air Component Command). Telephone interview by the author, 3 August 2005.

Brolén, Lt Col Pepe, SwAF, retired (commander, Air Transport Unit, 1990–98). Telephone interview, 2 November 2005.

Brose, Lt Col Manfred (German Air Force [GAF], Führungsakademie [German Armed Forces Command and Staff College], Hamburg, Germany). Interview by the author, 3 May 2006.

De Winter, Rolf, and Erwin van Loo (Royal Netherlands Air Force [RNLAF] History Unit, The Hague). Interview by the author, 24 June 2004.

Duance, Wg Cdr Richard, and Wg Cdr Pete York. "An Interview with Air Chief Marshal Sir Glenn Torpy, Chief of the Air Staff, Royal Air Force [RAF]." *Transforming Joint Air Power: The Journal of the JAPCC* 3 (2006): 26–29.

Facon, Patrick, director, AIR, historical service of the armed forces, Château de Vincennes, Paris. To the author. E-mail, 19 April 2006.

Godderij, Lt Gen P. J. M., RNLAF, military representative of the Netherlands to NATO, Brussels. To the author. E-mail, 4 January 2006.

Harryson, Maj Tobias, SwAF, 2002 air power doctrine project team member. To the author. E-mails, 12 October 2005, 29 March 2006, and 24 April 2006.

Havenith, Lt Col (General Staff [GS]) Armin, GAF (Federal Ministry of Defence [MOD], Bonn, Germany). Interview by the author, 26 April 2006.

Höglund, Lt Col Bertil, SwAF, commander, Air Transport Unit. To the author. Letter, 2 August 2005.

Janssen Lok, Joris, international editor, *Jane's International Defence Review*, and special correspondent, *Jane's Defence Weekly*. To the author. E-mail, 10 December 2004.

Jarosch, Lt Gen Hans-Werner, GAF, retired. Telephone interview by the author, 16 July 2008.

Jertz, Lt Gen Walter, GAF, retired. To the author. E-mails, 23 June 2007 and 26 June 2008. (In 1995 he was the GAF contingent commander for air operations over the Balkans. He left his final posting as commandant, German Air Force Command, in June 2006.)

Kent, John R., F-35 Lightning II Communications, Lockheed Martin Aeronautics Co. To the author. E-mail, 3 August 2009.

Kleppien, Lt Gen Axel, GAF, retired (commander, 5th Air Division, 1 October 1991–mid-1993). Telephone interview by the author, 16 July 2008.

Leiwig, Lt Col (GS) Guido, GAF (MOD, Bonn, Germany). Interview by the author, 26 April 2006.

Mercier, Col Denis, French Air Force (FAF), Air Staff, Paris. To the author. E-mail, 1 September 2006.

Persson, Lt Col Anders P., SwAF, Armed Forces Headquarters, Stockholm. To the author. E-mails, 10 May 2005, 28 October 2005, and 27 November 2005.

———, chief of staff, Air Combat Training School, Uppsala, Sweden. To the author. E-mail, 25 August 2009.

Radke, Lt Col (GS) Nicolas, GAF, MOD, Bonn, Germany. To the author. E-mail, 28 November 2006.

Reuterdahl, Lt Col Jan, SwAF, head, air faculty, Department of War Studies, Swedish National Defence College (SNDC), Stockholm. To the author. E-mails, 17 February 2005, 9 and 10 June 2005, and 11 November 2005.

Rietsch, Lt Col (GS) Carsten, GAF, assistant air attaché, German Embassy in London. To the author. E-mails, 5 August 2009 and 28 August 2009.

Schmidt, Col (GS) Lothar, GAF (MOD, Bonn, Germany). Interview by the author, 26 April 2006.

Schön, Col (GS) Hans-Dieter, GAF (MOD, Bonn, Germany). Interview by the author, 27 April 2006.

Silver, Brig Gen Anders, SwAF, commander, Air Component Command. To the author. E-mail, 5 February 2006. Forwarded by Maj Tobias Harryson, 8 March 2006.

Soeters, Joseph, professor for organization studies and social sciences, Royal Netherlands Military Academy and War College. To the author. E-mail, 13 December 2004.

"Sustaining Modern Armed Forces." Interview with Dr. Jan Fledderus, Netherlands national armaments director. *Military Technology*, no. 8 (1997): 9–19.

Svedén, Brig Gen Åke, SwAF, retired, commander, Air Transport Unit, 1983–90. To the author. E-mail, 22 January 2006.

Tankink, Lt Col Peter, RNLAF, commander, 323 Tactical Training, Evaluation, and Standardization Squadron. To the author. E-mails, 11 July 2004, 21 January 2005, and 8 January 2006.

Tode, Brig Gen Göran, SwAF, retired. To the author. E-mails, 7 June and 14 June 2005.

Trautermann, Lt Col (GS) Michael, GAF, MOD, Bonn, Germany. To the author. E-mail, 30 May 2006.

Vallance, Air Vice-Marshal Andrew, RAF (MOD, London). Interview by the author, 24 January 2007.

Van Ermel, Col Lex Kraft, RNLAF, former director of studies, Netherlands Defence College, The Hague. To the author. E-mails, 13 June and 22 June 2005.

Van Loo, Erwin. RNLAF History Unit, The Hague. To the author. E-mails, 9 September 2004 and 21 October 2004.

Vogt, Col (GS) Reinhard, GAF, Air Power Centre (*Luftmachtzentrum* or LMZ), Cologne Wahn, Germany. To the author. E-mail, 26 October 2006.

"We Want the Best Materiel for the Best Price." Interview with Dr. Jan Fledderus, Netherlands national armaments director. *Military Technology* 24, no. 12 (December 2000): 17–30.

Official Documents and Speeches

Air Force Doctrine Document (AFDD) 1. *Air Force Basic Doctrine*, September 1997.

———. *Air Force Basic Doctrine*, November 2003.

Air Force Manual 1-1. *Basic Aerospace Doctrine of the United States Air Force*. Vols. 1 and 2, March 1992.

Air Publication 3000. *British Air and Space Power Doctrine*, 4th ed., 2009.

———. *British Air Power Doctrine*, 3d ed., 1999.

———. *Royal Air Force Air Power Doctrine*, 1991.

———. *Royal Air Force Air Power Doctrine*, 2d ed., 1993.

Allied Joint Publication 3.3. *Joint Air and Space Operations Doctrine*, July 2000.

Armée de l'Air. "Audition du général Wolsztynski, Commission de la Défense nationale et des forces armées," 22 October 2004. Accessed 19 May 2007. http://www.defense.gouv.fr/air/archives /22_10_04_audition_du_general_wolsztynski.

Bank, Col Henk. "Development of the EPAF [European Participating Air Forces] Expeditionary Air Wing." Presentation. European Air Chiefs Conference, Noordwijk Aanzee, Netherlands, 20 November 2003.

Bergé-Lavigne, Maryse, and Philippe Nogrix. *Rapport d'information fait au nom de la commission des Affaires étrangères, de la défense et des forces armées (1) à la suite d'une mission sur le rôle des drones dans les armées*. No. 215. Paris: Sénat, 22 February 2006. http://www.senat.fr/rap/r05-215/r05-2151.pdf.

Bundesministerium der Verteidigung (MOD, Germany). *Luftwaffendienstvorschrift: Führung und Einsatz von Luftstreitkräften*, LDv 100/1 VS-NfD. Bonn, Germany: MOD, March 1991.

———. *Outline of the Bundeswehr Concept*. Berlin: MOD, 10 August 2004.

———. *Verteidigungspolitische Richtlinien für den Geschäftsbereich des Bundesministers der Verteidigung*. Bonn, Germany: MOD, 26 November 1992.

———. *Verteidigungspolitische Richtlinien für den Geschäftsbereich des Bundesministers für Verteidigung: Erläuternder Begleittext*. Berlin: MOD, 21 May 2003.

———. *Weißbuch 1985 zur Lage und Entwicklung der Bundeswehr* [White book 1985: the state and development of the Bundeswehr]. Bonn, Germany: Federal Ministry of Defence, 1985.

———. *Weißbuch 1994 zur Sicherheit der Bundesrepublik Deutschland und zur Lage und Zukunft der Bundeswehr* [White book 1994: the security of the Federal Republic of Germany and the state and future of the Bundeswehr]. Bonn, Germany: MOD, 1994.

———. *White Paper 2006: German Security Policy and the Future of the Bundeswehr.* Berlin: Federal Ministry of Defence, August 2006.

Clark, Gen Wesley K., supreme allied commander, Europe, and Lt Gen Michael Short. Testimony before the Senate Subcommittee on Armed Services. "Lessons Learned from Military Operations and Relief Efforts in Kosovo," 21 October 1999.

Concept de l'Armée de l'Air. Paris: Sirpa air, September 2008.

Defence Commission. *Defence for a New Time: Introduction and Summary.* Stockholm: MOD, 1 June 2004.

———. *Summary: Secure Neighbourhood—Insecure World.* Stockholm: MOD, 27 February 2003.

De Villepin, Sénateur Xavier. *Rapport d'information fait au nom de la commission des Affaires étrangères, de la défense et des forces Armées (1) sur les premiers enseignements de l'opération "force alliée" en Yougoslavie: quels enjeux diplomatiques et militaires?* [Information report on behalf of the commission of foreign and defence affairs and the armed forces on the first lessons learned of Operation Allied Force in Yugoslavia: what kind of diplomatic and military challenges?]. No. 464. Paris: Sénat, 30 June 1999. http://www.senat.fr/rap/r98-464/r98-464_mono.html.

Doktrin för luftoperationer 2004. Stockholm: Swedish Armed Forces, 2004.

European Union. "Declaration on European Military Capabilities." Military Capability Commitment Conference, Brussels. Approved by the General Affairs and External Relations Council, 22 November 2004. http://www.consilium.europa.eu/uedocs/cmsUpload/MILITARY%20CAPABILITY%20COMMITMENT%20CONFERENCE%2022.11.04.pdf.

————. *Headline Goal 2010.* Approved by General Affairs and External Relations Council on 17 May 2004. Endorsed by the European Council of 17 and 18 June 2004. http://www .consilium.europa.eu/uedocs/cmsUpload/2010%20Headline %20Goal.pdf.

Försvarsmaktens doktrin för luftoperationer. Stockholm: Swedish Armed Forces, 2005.

Godderij, Air Commodore P. J. M. "The Evolution of Air Power Doctrine in the Netherlands." Paper presented at the RNLAF Air Power Symposium, 15–19 April 1996.

Kamp, Henk, minister of defence. Address. Royal Netherlands Association of Military Science, Nieuwspoort Press Centre, The Hague, 1 March 2004.

Loi de Programmation Militaire (LPM) [French military programme bill of law] 1997–2002. *Rapport fait au nom de la commission des Affaires étrangères, de la défense et des forces armées (1) sur le projet de loi, adopté par l'Assemblée Nationale après déclaration d'urgence, relatif à la programmation militaire pour les années 1997 à 2002* [Report on behalf of the commission of foreign and defence affairs and the armed forces on the military programme bill of law 1997–2002, approved by the national assembly after a declaration of urgency]. No. 427. Paris: Sénat, 1996.

LPM 2003–2008. *La politique de défense* [Defence policy]. Paris: Journaux officiels [Official journals of the French Republic], May 2003.

LPM 2009–2014. *Projet de loi relatif à la programmation militaire pour les années 2009 à 2014 et portant sur diverses dispositions concernant la défense* [Law project pertaining to military programme bill of law 2009–2014 and to different directives regarding defence]. Assemblée Nationale [National Assembly], 29 October 2008.

Ministère de la Défense [Ministry of Defence], France. *Defence against Terrorism: A Top-Priority of the Ministry of Defence.* Paris: Délégation à l'information et à la communication de la défense, April 2006.

————. *Défense et Sécurité nationale: le Livre blanc.* Paris: Odile Jacob/La Documentation française, June 2008.

———. *Les enseignements du Kosovo: analyses et références.* Paris: MOD, November 1999.

———. *Livre blanc sur la défense* [White book on defence]. Paris: MOD, 1994. http://lesrapports.ladocumentationfrancaise.fr/BRP/944048700/0000.pdf.

Ministry of Defence, Netherlands. *Defence Priorities Review.* Abridged version. The Hague: MOD, 1993.

———. *Defence White Paper 1991: The Netherlands Armed Forces in a Changing World.* Abridged version. The Hague: MOD, 1991.

———. *Defence White Paper 2000.* English summary. The Hague: MOD, November 1999.

———. *The Prinsjesdag Letter—Towards a New Equilibrium: The Armed Forces in the Coming Years.* English translation. The Hague: MOD, 2003.

Ministry of Defence, Sweden. "Continued Renewal of the Total Defence." Fact sheet, October 2001.

———. "A Functional Defence: Government Bill on the Future Focus of Defence." Fact sheet no. 2009.06, March 2009. http://www.sweden.gov.se/content/1/c6/12/30/22/3ed2684c.pdf.

———. "A Functional Defence—with a Substantially Strengthened Defence Capability." Press release, 19 March 2009. http://www.regeringen.se/sb/d/10448/a/123010.

———. "The New Defence." Summary of Government Bill 1999/2000:30. Presented to Parliament on 25 November 1999. http://www.sweden.gov.se/sb/d/574/a/25639.

———. *Our Future Defence: The Focus of Swedish Defence Policy 2005–2007.* Summary of the Swedish Government Bill 2004/05:5. Stockholm: MOD, October 2004.

———. *Regeringens proposition 1996/97:4: Totalförsvar i förnyelse—etapp 2* [Government defence bill 1996/97:4: renewal of total defence—stage 2]. Stockholm: MOD, 12 September 1996.

Olsen, John. "Effects-Based Targeting through Pre-Attack Analysis." Paper presented at the RNLAF Air Power Symposium, 19 November 2003.

Organisation for Security and Cooperation in Europe. *Annual Exchange of Information on Defence Planning: The Kingdom of Sweden* [*According to the*] *Vienna Document 1999.* Stockholm: The Kingdom of Sweden, March 2005.

Parliamentary Defence Committee. *Försvarsutskottets betänkande 1991/92 FöU12: Totalförsvarets fortsatta utveckling 1992/93–1996/97*. Stockholm: Parliamentary Defence Committee, 1992.

Royal Netherlands Air Force. *Airpower Doctrine*. The Hague: RNLAF, April 2002.

Royal Netherlands Air Force Air Power Colloquium. "Air Power: Theory and Application." The Hague, 24–28 November 1997.

Royal Netherlands Air Force Air Power Doctrine. English translation of *KLu Air Power doctrine voor het basis- en operationele niveau* [RNLAF basic and operational level doctrine], The Hague: RNLAF, 1996. The Hague: RNLAF, 1999.

Royal Netherlands Air Force Air Power Symposium [15–19 April 1996]. The Hague: RNLAF, April 1996.

Royal Netherlands Air Force Air Power Symposium [19 November 2003]. The Hague: RNLAF, November 2003.

Sarkozy, Nicolas, president of the French Republic. "Foreword." In Ministère de la Défense, *Défense et Sécurité nationale: le Livre blanc*.

Scharping, Rudolf. *Die Bundeswehr sicher ins 21. Jahrhundert: Eckpfeiler für eine Erneuerung von Grund auf*. Berlin: MOD, 2000.

Starink, Lt Gen Dirk, commander in chief, RNLAF. Briefing. Shephard Air Power Conference 2005. London, 25–27 January 2005.

Struck, Peter. *Defence Policy Guidelines for the Area of Responsibility of the Federal Minister of Defence*. Berlin: MOD, 21 May 2003.

Syren, Gen Hakan, supreme commander of the Swedish armed forces. "The Swedish Armed Forces Today and towards the Next Defence Decision." Presentation. National Conference on Folk och Försvar [people and defence], Sälen, Sweden, 21 January 2004.

"Une Défense Nouvelle 1997–2015." Hors série [special edition]. *Armées d'aujourd'hui*, no. 208 (March 1996).

Von Weizsäcker, Richard, director. *Gemeinsame Sicherheit und Zukunft der Bundeswehr: Bericht der Kommission an die Bundesregierung*. Berlin: Kommission Gemeinsame Sicherheit und Zukunft der Bundeswehr, May 2000.

Books and Articles

Agüera, Martin. "Zwischen Hoffnung und Verzweiflung: Strategischer Lufttransport für Europa." *Europäische Sicherheit*, no. 6 (June 2002): 47–50.

Ahlgren, Jan, Anders Linnér, and Lars Wigert. *Gripen: The First Fourth Generation Fighter*. Sweden: Swedish Air Force, FMV and Saab Aerospace, 2002.

Airbus Military. "A400M." Accessed 4 February 2008. http://www.airbusmilitary.com.

Andersson, Jan Joel. *Armed and Ready? The EU Battlegroup Concept and the Nordic Battlegroup*. Report no. 2. Stockholm: Swedish Institute for European Policy Studies, March 2006.

Andersson, Raymond, Kurt Karlsson, and Anders Linnér, eds. *Flygvapnet: The Swedish Air Force*. Stockholm: Swedish Air Force, 2001.

Armée de l'Air 2007: enjeux et perspectives. Paris: Sirpa air, 2007.

Åselius, Gunnar. "Swedish Strategic Culture after 1945." *Cooperation and Conflict: Journal of the Nordic International Studies Association* 40, no. 1 (2005): 25–44.

Asenstorfer, John, Thomas Cox, and Darren Wilksch. *Tactical Data Link Systems and the Australian Defence Force (ADF): Technology Developments and Interoperability Issues*. Rev. ed. Edinburgh, South Australia: Defence Science & Technology, February 2004. http://www.dsto.defence.gov.au/publications/2615.

Bailes, Alison J. K. "European Security from a Nordic Perspective: The Roles for Finland and Sweden." In Huldt et al., *Strategic Yearbook 2004*, 59–81.

Baldes, Eugénie. "Les Mirages français passent le relais." *Air actualités*, no. 556 (November 2002): 34–37.

Bitzinger, Richard A. *Facing the Future: The Swedish Air Force, 1990–2005*. Santa Monica, CA: RAND, 1991.

Bombeau, Bernard. "L'Armée de l'Air étudie l'achat de C-130J." *Air et Cosmos*, no. 2166 (April 2009): 32–33.

Bourdilleau, Gen François. "Evolution de l'Armée de l'Air vers le modèle Air 2015." In Pascallon, *L'Armée de l'Air*, 241–59.

Boyne, Walter J. *Operation Iraqi Freedom: What Went Right, What Went Wrong, and Why*. New York: Tom Doherty Associates, 2003.

Bundesministerium der Verteidigung (BMVg) (Federal Ministry of Defence). "Verlässlicher Zugriff auf 'fliegende Güterzüge,'" 23 March 2006. Accessed 9 August 2006. http://www.bmvg.de.

Carlier, Claude. "L'aéronautique et l'espace, 1945–1993." In *Histoire militaire de la France*, Tome 4, *De 1940 à nos jours*, edited by André Martel, director, 449–80. Paris: Presses Universitaires de France, 1994.

Chamagne, Col Régis, FAF, retired. *L'art de la guerre aérienne.* Fontenay-aux-Roses, France: L'esprit du livre éditions, 2004.

Clark, Wesley K. *Waging Modern War: Bosnia, Kosovo, and the Future of Combat*. Oxford, England: PublicAffairs, 2001.

Collège interarmées de défense [CID] (Joint Forces Defence College). "Enseignement opérationnel: Euro exercice CJEX." Accessed 28 March 2006. http://www.college.interarmees .defense.gouv.fr.

Cordesman, Anthony H. *The Lessons and Non-Lessons of the Air and Missile War in Kosovo*. Rev. ed. Washington, D.C.: Center for Strategic and International Studies, 20 July 1999.

Corum, James S. "Airpower Thought in Continental Europe between the Wars." In *The Paths of Heaven: The Evolution of Airpower Theory*, edited by Philip S. Meilinger, 151–81. Maxwell AFB, AL: Air University Press, 1997.

Daalder, Ivo H., and Michael E. O'Hanlon. *Winning Ugly: NATO's War to Save Kosovo*. Washington, D.C.: Brookings Institution Press, 2000.

D'Abramo, Michael P. *Military Trends in the Netherlands: Strengths and Weaknesses*. Washington, D.C.: Center for Strategic and International Studies, 31 August 2004.

"Das Kommando Operative Führung Luftstreitkräfte." In *CPM Forum: Luftwaffe 2004*, managing directors/editors Harald Flex and Wolfgang Flume, in cooperation with the GAF staff, MOD, Bonn, Germany, and editorial coordinator Lt Col (GS) Rainer Zaude, 56–58. Sankt Augustin: CPM Communication Presse Marketing GmbH, 2004.

De Durand, Etienne, and Bastien Irondelle. *Stratégie aérienne comparée: France, Etats-Unis, Royaume-Uni*. Paris: Centre d'études en sciences sociales de la défense, 2006.

Defense Industry Daily. "nEUROn UCAV Project Rolling Down the Runway," 14 June 2007. Accessed 27 December 2007. http://www.defenseindustrydaily.com.

De Rousiers, Gen Patrick. "Contribution de l'Armée de l'Air à la construction de la défense européenne: nature, perspectives, importance." In Pascallon, *L'Armée de l'Air*, 537–45.

Deptula, Maj Gen David A. "Effects-Based Operations: Change in the Nature of Warfare." In Olsen, *Second Aerospace Century*, 135–73.

Dörfer, Ingemar. *Arms Deal: The Selling of the F-16*. New York: Praeger Publishers, 1983.

———. *The Nordic Nations in the New Western Security Regime*. Washington, D.C.: Woodrow Wilson Center Press, 1997.

Dreger, Paul. "JSF Partnership Takes Shape: A Review of the JSF Participation by Australia, Canada, Denmark, Israel, Italy, the Netherlands, Norway, Singapore, Turkey and the UK." *Military Technology* 27, no. 4 (April 2003): 28–34.

Droste, Lt Gen B. A. C., commander in chief, RNLAF. "Decisive Airpower Private: The Role of the Royal Netherlands Air Force in the Kosovo Conflict." *NATO's Nations and Partners for Peace*, no. 2 (1999): 126–30.

———. "Shaping Allied TBM Defence." Special issue. *NATO's Sixteen Nations and Partners for Peace* 42 (1997): 47–52.

EADS. "A400M: Erfolgreicher Erstflug des weltweit modernsten Transportflugzeugs," 11 December 2009. Accessed 4 January 2010. http://www.eads.net.

Elliot, James. "From Fighters to Fighter-Bombers." *Military Technology* 23, no. 8 (August 1999): 69–76.

Eriksson, Arita. "Sweden and the Europeanisation of Security and Defence Policy." In Huldt et al., *Strategic Yearbook 2004*, 119–37.

European Air Group (EAG). "History of the EAG." Accessed 15 May 2007. http://www.euroairgroup.org/index.php?s=history.

"European Military Aircraft Programmes Revisited." *Military Technology*, no. 6 (1997): 6–18.

Evers, Roger. "Transportflieger in humanitärem Auftrag." In Goebel, *Von Kambodscha bis Kosovo*, 86–100.

Everts, Steven, Lawrence Freedman, Charles Grant, François Heisbourg, Daniel Keohane, and Michael O'Hanlon, eds. *A*

European Way of War. London: Centre for European Reform, 2004.

Finn, Gp Capt Chris. "British Thinking on Air Power—The Evolution of AP 3000." *Royal Air Force Air Power Review* 12, no. 1 (Spring 2009): 56–67.

"Frankreich attackiert Deutschland wegen A400M." *Handelsblatt,* 31 January 2011. http://www.handelsblatt.com/unternehmen /industrie/airbus-frankreich-attackiert-deutschland-wegen -a400m;2743458.

Freedman, Lawrence. "Can the EU Develop an Effective Military Doctrine?" In Everts et al., *European Way of War,* 13–26.

"Further Development of the Luftwaffe—Luftwaffe Structure 6." In *CPM Forum: Luftwaffe 2005—The German Air Force Today and Tomorrow,* managing director Wolfgang Flume and project director Jürgen Hensel, 44–47. Sankt Augustin, Germany: CPM Communication Presse Marketing GmbH, 2005.

Garden, Timothy. "European Air Power." In Gray, *Air Power 21,* 99–122.

Gates, David. "Air Power: The Instrument of Choice?" In Gray, *Air Power 21,* 23–39.

Gautier, Louis. *Mitterrand et son armée 1990–1995.* Paris: Grasset, 1999.

"The German Air Force Structure 5: Principles, Structures, Capabilities." In *CPM Forum: German Air Force—Structure and Organisation, Equipment and Logistics, Programmes and Perspectives,* edited by Wolfgang Flume in cooperation with the GAF staff, MOD, Bonn, Germany, and editorial coordinator Lt Col (GS) Rainer Zaude, 19–26. Sankt Augustin, Germany: CPM Communication Presse Marketing GmbH, 2003.

"German Eurofighters Arrive for Baltic Air Policing Debut." *Flight International* 176, no. 5205 (8–14 September 2009): 18.

Gething, Michael J., and Bill Sweetman. "Air-to-Air Refuelling Provides a Force Multiplier for Expeditionary Warfare." *Jane's International Defence Review* 39 (February 2006): 43–51.

Gnesotto, Nicole (chair), Jean-Yves Haine, André Dumoulin, Jan Foghelin, François Heisbourg, William Hopkinson, Marc Otte, Tomas Ries, Lothar Rühl, Stefano Silvestri, Hans-Bernhard Weisserth, and Rob de Wijk. *European Defence: A Proposal for a White Paper*. Report of an independent task force. Paris: European Union Institute for Security Studies, May 2004.

Goebel, Peter. "Von der Betroffenheit zur Selbstverständlichkeit." In *Von Kambodscha bis Kosovo*, 11–20.

Goebel, Peter, ed. *Von Kambodscha bis Kosovo: Auslandeinsätze der Bundeswehr*. Frankfurt am Main, Germany: Report Verlag, 2000.

Gräfe, Lt Col (GS) Frank. "Die Mehrrollenfähigkeit des Waffensystems EUROFIGHTER." *Europäische Sicherheit*, no. 4 (April 2010): 50–55.

Grant, Charles. "Conclusion: The Significance of European Defence." In Everts et al., *European Way of War*, 55–74.

Gray, Peter W., ed. *Air Power 21: Challenges for the New Century*. London: The Stationery Office, 2000.

Gregory, Shaun. *French Defence Policy into the Twenty-First Century*. London: MacMillan Press, 2000.

Groß, Jürgen. "Revision der Reform: Weiterentwicklung des Bundeswehrmodells '200F.'" In *Europäische Sicherheit und Zukunft der Bundeswehr: Analysen und Empfehlungen der Kommission am IFSH*, edited by Jürgen Groß, 117–23. Baden-Baden, Germany: Nomos Verlagsgesellschaft, 2004.

Hagena, Brig Gen (Dr.) Hermann, GAF, retired. "Charter oder Leasing? Zwischenlösung für neue Transportflugzeuge." *Europäische Sicherheit*, no. 4 (April 2002): 35.

Hagman, Hans-Christian. *European Crisis Management and Defence: The Search for Capabilities*. Adelphi Paper no. 353. Oxford, England: Oxford University Press for IISS, 2002.

Haine, Jean-Yves. "ESDP: An Overview." European Union Institute for Security Studies. Accessed 14 May 2005. http://www.iss-eu.org/esdp/01-jyh.pdf.

Helsing, Stefan, Swedish Air Force Rapid Reaction Unit (SWAFRAP) AJS-37 information officer. "Pilots for Peace." In Andersson et al., *Flygvapnet*, 62–65.

Hewish, Mark, and Joris Janssen Lok. "Connecting Flights: Datalinks Essential for Air Operations." *Jane's International Defence Review* 31, no. 12 (28 November 1998): 41–47.

Honig, Jan Willem. *Defense Policy in the North Atlantic Alliance: The Case of the Netherlands*. Westport, CT: Praeger, 1993.

Hoppe, Reinhart. "First In and Last Out: Lufttransport im weltweiten Einsatz." In Jarosch, *Immer im Einsatz*, 211–16.

Hoyle, Craig. "Baltic Exchange." *Flight International* 175, no. 5191 (2–8 June 2009): 30–33.

Huldt, Bo, Teija Tiilikainen, Tapani Vaahtoranta, and Anna Helkama-Rågård, eds. *Finnish and Swedish Security: Comparing National Policies.* Stockholm: SNDC and the Programme on the Northern Dimension of the CFSP conducted by the Finnish Institute of International Affairs and the Institut für Europäische Politik, 2001.

Huldt, Bo, Tomas Ries, Jan Mörtberg, and Elisabeth Davidson, eds. *Strategic Yearbook 2004: The New Northern Security Agenda—Perspectives from Finland and Sweden.* Stockholm: SNDC, 2003.

Janssen Lok, Joris. "Budget Crunch May Hit Swedish Air Operations." *Jane's Defence Weekly,* 5 September 2001. Accessed 25 May 2007. http://search.janes.com.

———. "RNLAF Concerns Increase over Targeting Pod Acquisition Being Slow to Hit the Mark." *Jane's International Defence Review,* April 2006, 6.

Janssen Lok, Joris, and J. A. C. Lewis. "New French Air Power." *Jane's International Defence Review* 36 (June 2003): 41–45.

Jarosch, Hans-Werner, ed. *Immer im Einsatz: 50 Jahre Luftwaffe.* Hamburg, Germany: Verlag E. S. Mittler & Sohn, 2005.

Jertz, Lt Gen Walter. "Einsätze der Luftwaffe über Bosnien." In Goebel, *Von Kambodscha bis Kosovo,* 136–53.

———. *Im Dienste des Friedens: Tornados über dem Balkan.* 2d rev. ed. Bonn, Germany: Bernard & Graefe Verlag, 2000.

———. "Unser Schwerpunkt ist der Einsatz: Das Luftwaffenführungskommando auf dem Weg in die Zukunft." *Strategie & Technik,* March 2006, 18–25.

Joint Air Power Competence Centre (JAPCC). "History." Accessed 23 March 2007. http://www.japcc.de/fileadmin/user_upload /History/ JAPCC_History.pdf.

Jonsson, Maj Gen Jan, chief of staff, SwAF. "The Future of Air Power (II): Sweden." *Military Technology,* no. 7 (1999): 15–20.

Jorgensen, Jan. "Battle Group Gripens." *Air Forces Monthly,* August 2007, 42–46.

Karlsson, Kurt. "The History of Flygvapnet." In Andersson et al., *Flygvapnet,* 84–144.

Keaney, Thomas A., and Eliot A. Cohen. *Gulf War Air Power Survey: Summary Report*. Washington, D.C.: Dept. of the Air Force, 1993.

Kreuzinger-Janik, Lt Gen Aarne, chief of the Air Staff. "Die Luftwaffe im Gesamtsystem Bundeswehr." *Europäische Sicherheit*, no. 6 (June 2010): 23–31.

Kromhout, Gert. "The New Armée de l'Air." *Air Forces Monthly*, no. 118 (January 1998): 40–49.

Kuebart, Jan. "Air Policing Baltikum." *Europäische Sicherheit*, no. 12 (December 2005): 32–36.

Lambeth, Benjamin S. *NATO's Air War for Kosovo: A Strategic and Operational Assessment*. Santa Monica, CA: RAND, 2001.

———. "Operation Allied Force: A Strategic Appraisal." In Olsen, *Second Aerospace Century*, 101–34.

———. *The Transformation of American Air Power*. Ithaca, NY: Cornell University Press, 2000.

Lanata, Gen Vincent. "Faire face: l'ère des nouveaux défis." *Défense nationale*, no. 8/9 (August/Septembre, 1993): 9–18.

Lange, Sascha. *Neue Bundeswehr auf altem Sockel: Wege aus dem Dilemma: SWP-Studie*. Berlin: Stiftung Wissenschaft und Politik, 2005.

Lange, Wolfgang. "Lufttransport—Ansätze zur Kompensation von Defiziten." *Europäische Sicherheit*, no. 10 (October 2002): 36–41.

Lebert, Lt Col (GS) Jörg. "Einrichtung eines Europäischen Lufttransportkommandos." *Europäische Sicherheit*, no. 7 (July 2001): 20–24.

Lehmann, Mathias. "The Further Development of Luftwaffe's Equipment." *Military Technology*, special issue, no. 2 (1999): 50–55.

Lewis, J. A. C. "French Connection: A New President and Preparations for a New Defence White Paper Promise Changes for the French Military." *Jane's Defence Weekly* 44, no. 27 (4 July 2007): 22–30.

Liander, Peter. "On Alert: Continued Focus on the New Gripen and New Helicopters." *The Swedish Armed Forces Forum: Insats & Försvar*, English ed., no. 1 (2004): 22–25.

Lorell, Mark A. *Airpower in Peripheral Conflict: The French Experience in Africa*. Santa Monica, CA: RAND, 1989.

———. *The Future of Allied Tactical Fighter Forces in NATO's Central Region*. Santa Monica, CA: RAND, 1992.

"Luftwaffenstruktur 6." *Wehrtechnischer Report: Fähigkeiten und Ausrüstung der Luftwaffe*. No. 3. Bonn, Germany: Report Verlag GmbH, 2006, 12–15.

Lundmark, Martin. "Nordic Defence Materiel Cooperation." In Huldt et al., *Strategic Yearbook 2004*, 207–30.

Lutgert, Wim H., and Rolf de Winter. *Check the Horizon: De Koninklijke Luchtmacht en het conflict in voormalig Joegoslavië 1991–1995*. The Hague: Sectie Luchtmachthistorie van de Staf van de Bevelhebber der Luchtstrijdkrachten, 2001.

Mason, Tony. *Air Power: A Centennial Appraisal*. Rev. ed. London: Brassey's, 2002.

Mathieu, Jean-Luc. *La Défense nationale*. Paris: Presses Universitaires de France, 2003.

Meiers, Franz-Josef. *Zu neuen Ufern? Die deutsche Sicherheits- und Verteidigungspolitik in einer Welt des Wandels 1990–2000*. Paderborn, Germany: Ferdinand Schöningh, 2006.

Moulard, Col Jean-Pierre. "L'organisation de la campagne aérienne dans le cadre de la NRF 5?" *Penser les ailes françaises*, no. 8 (January 2006): 39–45.

Näsström, Maj Gen Staffan, chief, Air Force Materiel Command. "The Swedish Air Force 2000—Materiel Aspects." *Military Technology: Defence Procurement in Sweden*, special issue, 1996, 34–42.

NATO. "NATO Airborne Early Warning and Control Force: E-3A Component." Fact sheet, 2010. Accessed 30 September 2010. http://www.e3a.nato.int/html/media.htm.

———. "NATO's Allied Ground Surveillance Programme Signature Finalised," 25 September 2009. Accessed 4 January 2010. http://www.nato.int/cps/en.

———. "Strategic Airlift Capability: A Key Capability for the Alliance," 27 November 2008. Accessed 2 January 2009. http://www.nato.int/issues/strategic-lift-air-sac/index.html.

NATO Headquarters. *NATO Handbook*. Brussels: NATO Office of Information and Press, 2001.

Newdick, Thomas. "A Change in the Air." *Jane's Defence Weekly* 47, no. 22 (2 June 2010): 29–32.

Nijean, Jean-Laurent. "L'Armée de l'Air et l'Europe: gros plan—au coeur de l'Europe de la défense." *Air actualités*, no. 605 (October 2007): 36–39.

———. "L'Armée de l'Air et l'Europe: opération EUFOR RDC." *Air actualités*, no. 605 (October 2007): 34–35.

Olsen, John Andreas, ed. *A Second Aerospace Century: Choices for the Smaller Nations*. Trondheim, Norway: Royal Norwegian Air Force Academy, 2001.

Ouisse, François. "Présence française en Centrafrique: l'Armée de l'Air sans frontières." *Air actualités*, no. 493 (June 1996): 50–51.

Owen, Col Robert C., ed. *Deliberate Force—A Case Study in Effective Air Campaigning: Final Report of the Air University Balkans Air Campaign Study*. Maxwell AFB, AL: Air University Press, January 2000.

Pape, Robert A. "The True Worth of Air Power." *Foreign Affairs* 83, no. 2 (March/April 2004): 116–30.

Pascallon, Pierre, ed. *L'Armée de l'Air: les armées françaises à l'aube du XXIe siècle*. Tome II. Paris: L'Harmattan, 2003.

Patoz, Jacques, and Jean-Michel Saint-Ouen. *L'Armée de l'Air: survol illustré dans les turbulences du siècle*. Paris: Editions Méréal, 1999.

Peters, John E., Stuart E. Johnson, Nora Bensahel, Timothy Liston, and Traci Williams. *European Contributions to Operation Allied Force: Implications for Transatlantic Cooperation*. Santa Monica, CA: RAND, 2001.

Portz, Lt Gen Rolf, Inspector, GAF. "The Future of Air Power (IV): Germany." *Military Technology*, no. 10 (1999): 11–14.

Ripley, Tim. *Air War Bosnia: UN and NATO Airpower*. Osceola, WI: Motorbooks International, 1996.

———. *Air War Iraq*. Barnsley, South Yorkshire, England: Pen and Sword Aviation, 2004.

———. *Operation Deliberate Force: The UN and NATO Campaign in Bosnia 1995*. Lancaster, England: Centre for Defence and International Security Studies, 1999.

"Rote Liste für den Baron." *sueddeutsche.de*. Accessed 7 July 2010. http://www.sueddeutsche.de/politik/sparkurs-im-verteidig ungsministerium-rote-liste-fuer-den-baron-1.971018.

Sargent, Lt Col Richard L. "Weapons Used in Deliberate Force." In Owen, *Deliberate Force*, 257–77.

Schetilin, Maj (GS) Markus. "Material- und Ausrüstungsplanung der Luftwaffe." *Europäische Sicherheit*, no. 10 (October 2005): 23–34.

Schmidt, Wolfgang A. *Die schwedische Sicherheitspolitik: Konzeption—Praxis—Entwicklung, Beiträge und Berichte.* Monograph no. 88. St. Gallen, Switzerland: Hochschule St. Gallen, December 1983.

Schmitt, Burkard. *European Capabilities Action Plan (ECAP).* Paris: European Union Institute for Security Studies. Accessed 5 March 2007. http://www.iss-eu.org/esdp/06-bsecap.pdf.

Schön, Col (GS) Hans-Dieter, and Maj (GS) Markus Schetilin. "Material- und Ausrüstungsplanung der Luftwaffe." *Europäische Sicherheit*, no. 10 (October 2006): 38–50.

Soubirou, Lt Gen André, French Army, retired. "The Account of Lieutenant General (Ret) André Soubirou, Former Commanding General of the RRF [Rapid Reaction Force] Multinational Brigade in Bosnia from July to October 1995." *Doctrine Special Issue*, February 2007, 23–29. http://www.cdef.terre.defense.gouv.fr/publications/doctrine/no_spe_chefs_francais/us/art07.pdf.

Souvignet, Cdt José, and Lt Col Stéphane Virem. "L'Armée de l'Air dans la tourmente: la campagne aérienne du Golfe." *Penser les ailes françaises*, no. 9 (February 2006): 92–96.

Thompson, Michael E. *Political and Military Components of Air Force Doctrine in the Federal Republic of Germany and Their Implications for NATO Defense Policy Analysis.* Santa Monica, CA: RAND, September 1987.

United Nations, Peacekeeping Best Practices Unit, Military Division. *Operation Artemis: The Lessons of the Interim Emergency Multinational Force.* UN: October 2004.

United States European Command. "Operation Northern Watch," 20 February 2004. Accessed 3 November 2004. http://www.eucom.mil/Directorates/ECPA/Operations/onw/onw.htm (site discontinued).

Utley, Rachel. *The Case for Coalition: Motivation and Prospects: French Military Intervention in the 1990s.* The Occasional, no. 41. New Baskerville, England: Strategic and Combat Studies Institute, 2001.

Van der Mark, Kees. "Firing Patriots in Greece." *Air Forces Monthly*, February 2010, 40–45.

———. "Joint Helicopter Ops Dutch Style." *Air Forces Monthly*, August 2008, 64–68.

———. "Tall Ambitions for the Lowlands." *Air Forces Monthly*, July 2009, 40–48.

Van Loo, Erwin. *Crossing the Border: De Koninklijke Luchtmacht na de val van de Berlijnse Muur*. The Hague: Sectie Luchtmachthistorie van de Staf van de Bevelhebber der Luchtstrijdkrachten, 2003.

Van Staden, Alfred. "The Netherlands." In *The European Union and National Defence Policy*, eds. Jolyon Howorth and Anand Menon, 87–104. London: Routledge, 1997.

Von Hoyer-Boot, Bernd Baron. "Erweiterte Luftverteidigung." *Europäische Sicherheit*, no. 9 (September 1992): 506–11.

Von Wintzingerode-Knorr, Lt Col (GS) Eberhard Freiherr. "Weiterentwicklung des 'Jägers' Eurofighter für den Luft-/Boden-Einsatz." *Europäische Sicherheit*, no. 11 (November 2006): 38–43.

Wagevoort, Maj Gen Marcel. "Materiel Projects in the RNLAF." Defence Procurement in the Netherlands. *Military Technology*, no. 8 (August 1997): 44–50.

Wedin, Lars. "Sweden in European Security." In Huldt et al., *Strategic Yearbook 2004*, 319–35.

Wennerholm, Bertil, and Stig Schyldt. *1990-talets omvälvningar för luftstridskrafterna: Erfarenheter inför framtiden* [The air power revolution of the 1990s: experiences for the future]. *Krigsvetenskapliga Forskningsrapporter* [War scientific research report] no. 3. Stockholm: SNDC, 2004.

Willett, Susan, Philip Gummet, and Michael Clarke. *Eurofighter 2000*. London Defence Studies. London: Brassey's, 1994.

Winquist, Claes. "We See with Our Ears." In Andersson et al., *Flygvapnet*, 32–35.

"Wirken gegen Ziele in der Luft." *Wehrtechnischer Report: Fähigkeiten und Ausrüstung der Luftwaffe*, no. 3 (2006): 41–46.

Wolsztynski, Gen Richard, chief of the Air Staff. "La contribution de l'Armée de l'Air à la construction de l'Europe de la défense." *Défense nationale*, no. 7 (July 2004): 5–12.

Yost, David S. "The U.S.-European Capabilities Gap and the Prospects for ESDP." In *Defending Europe: The EU, NATO and the Quest for European Autonomy*, edited by Jolyon Howorth and John T. S. Keeler, 81–106. Basingstoke, England: Palgrave Macmillan, 2003.

Zaude, Lt Col (GS) Rainer, editorial coordinator, in cooperation with the GAF staff, MOD, Bonn, Germany. *CPM Forum: Luftwaffe*. Sankt Augustin, Germany: CPM Communication Presse Marketing GmbH, 2002.

Index

air-to-ground surveillance, 31, 54–55, 61
Air Transport and Air-to-Air Refuelling Exchange of Services, 25
Air University (USAF), 46, 89, 251–53, 311, 341
Air University Balkans Air Campaign Study, 253
Air War College, 252, 311
AJ 37 Viggen. *See under* aircraft by designation
AJP. *See* allied joint publication
AJP-01, *Allied Joint Doctrine*, 188, 308
AJP-3.3, *Joint Air and Space Operations Doctrine*, 188, 308, 310
AJP-5, *Allied Joint Doctrine for Operational Planning*, 188
AJS 37 Viggen. *See under* aircraft by designation
AJSF 37 Viggen. *See under* aircraft by designation
AJSH 37 Viggen. *See under* aircraft by designation
ALARM. *See* air-launched anti-radiation missile
Algerian War, 100
Al Khafji, 29, 61
Alliance Ground Surveillance Core project, 54, 61, 200, 263
Allied Air Forces Central Europe, 168
Allied Command Atlantic, 20–21, 94
Allied Command Channel, 20
Allied Command Europe, 20–21, 94
Allied Command Operations, 21
Allied Command Transformation, 21
Allied Forces Southern Europe, 94, 170
allied joint publications, 188, 308, 310
Alpha Jet. *See under* aircraft by designation
al-Qaeda, 37
Amendola Air Base, Italy, 242
American-led, 62, 102, 109, 121, 176, 218, 343
AMRAAM. *See* advanced medium-range air-to-air missile
Air Publication (AP) 3000. See *British Air and Space Power Doctrine*, 4th ed., 2009, and *British Air Power Doctrine*, 3d ed., 1999
Anatolian Eagle (multinational integrated air exercise), 100
An-124. *See under* aircraft by designation
Apache combat helicopter. *See under* aircraft by designation
Apache cruise missile, 125
APG-65 radar, 161, 195
Arab Emirates, 100, 106, 110
Arizona Air National Guard, 237
Armed Forces 2000 project (France), 87–88
Armed Forces Helicopter Wing (Sweden), 287, 291, 298, 319
Armed Forces Model 2015, 83–84, 87, 89–91, 131

armement air-sol modulaire (modular air-to-surface armament), 57, 124–26, 135, 355
Army air wing (*Aviation légère de l'armée de terre*), 91
Aconit (Operation), 103
Allied Force (Operation), 4–5, 8, 10, 16, 19, 27, 35–36, 38, 59, 82, 99, 105–6, 121, 124, 126, 128, 130, 133, 177, 181, 192, 195, 204–5, 221, 235, 238, 242–44, 246, 253, 256–57, 312, 339, 347–49, 358, 361
Anaconda (Operation), 107
Artemis (Operation), 109–10, 183–84, 302, 345
Article 5 (NATO), 21, 46, 225, 345
AS-30-L laser-guided air-to-ground missile, 102, 124
ASC 890. *See under* aircraft by designation
ASMP. *See air-sol moyenne portée* (medium range air-to-ground weapon)
Atalanta (Operation), 291
ATARES. *See* Air Transport and Air-to-Air Refuelling Exchange of Services
Ateliers de l'Armée de l'Air (Air Force conference), 117–18
Ateliers du CESA (CESA conference), 117–18, 120
AU. *See* Air University
AWACS. *See* Airborne Warning and Control System

BAE Systems, 315
Baghdad, 28, 30, 55, 119
Bahrain, 182
Balkan, 3, 157
Baltic region, 111, 185, 247, 281–83, 287, 294, 296, 300–302, 321, 324
Baltic Sea, 282, 284, 287, 293, 296, 300–2
Bam, Iran, 246
Belgian Air Force, 234
Belgian Congo, 285, 301. *See also* Democratic Republic of Congo
Belgian Staff College, 252
Belgium, 21, 23–24, 28, 34, 36, 99, 109, 173, 236, 243, 252, 254–55, 341
Belgrade, Serbia, 106
Berlijn, D. L., 236, 250
Berlin Plus agreement, 18
Björklund, Leni, 283
Blair, Tony, 106
Bosnia, 3–4, 31–32, 34–35, 50, 61, 92, 105, 110–11, 124, 177, 182, 194, 206, 231, 240–42, 253, 266–67, 306, 348–49, 351
Bosnian civil war, 178
Bosnian Muslims, 33
Bosnian Serb army, 42
Bosnian Serbs, 32, 35, 105, 179, 182, 241

The Quest for Relevant Air Power

Continental European Responses to the Air Power Challenges of the Post–Cold War Era

Air University Press Team

Chief Editor
Jeanne K. Shamburger

Copy Editor
Carolyn Burns

Cover Art and Book Design
L. Susan Fair

Illustrations
Daniel Armstrong

*Composition and
Prepress Production*
Ann Bailey

Print Preparation and Distribution
Diane Clark